Energy Policies, Politics and Prices

snova
New York

Energy Policies, Politics and Prices

Pipeline Safety: Background, Practices and Regulatory Changes
Michael E. Massaro (Editor)
ISBN: 979-8-89530-058-9
eBook ISBN: 979-8-89530-066-4

Sustainable and Green Energy Policies
Kürşad Emrah Yıldırım, PhD
and Eser Gemici, PhD (Editors)
ISBN: 979-8-89113-773-8
eBook ISBN: 979-8-89113-852-0

The Development of Crude Oil and Natural Gas Prices and the Energy Security of the European Union
Matej Korček, PhD
and Saleh Mothana Obadi, PhD
ISBN: 979-8-89113-357-0
eBook ISBN: 979-8-89113-357-0

The Future of Energy: Challenges, Perspectives, and Solutions
Thomas Valone, PhD (Editor)
ISBN: 978-1-53618-186-9
eBook ISBN: 978-1-53618-348-1

A Strategic Evaluation of Energy Security in the Eastern Mediterranean
Michail Chalaris, PhD
ISBN: 978-1-53619-273-5
eBook ISBN: 978-1-53619-346-6

Caspian: Status, Challenges, and Prospects. An Analysis into the Legal Classification, Security and Environmental Concerns, Geopolitics and Energy Flow Impact of the Caspian Plateau
Anis H. Bajrektarevic,
Mak Bajrekatarevic
and Tanvi Kamlesh Chauhan
ISBN: 978-1-53618-956-8
eBook ISBN: 978-1-53619-001-4

More information about this series can be found at
https://novapublishers.com/product-category/series/energy-policies-politics-and-prices/

Victoria Whitehead
Editor

The Struggle for Energy Independence in America

Regulations, Overreach, and Security

snova
New York

Library of Congress Cataloging-in-Publication Data

ISBN: 979-8-89530-558-4

Published by Nova Science Publishers, Inc. † New York

Contents

Preface

For nearly 50 years, U.S. energy policy has been dominated by concerns over the scarcity and depletion of oil and natural gas. Dependence on foreign oil was often framed as a national security crisis, prompting the creation of a wide array of policies, regulations, and subsidies designed to address this issue. Fuel efficiency mandates, ethanol subsidies, renewable energy incentives, export controls, and domestic production subsidies were all introduced in the name of achieving energy independence and security. Yet, despite these efforts, the desired outcomes have largely remained elusive.

Today, however, we stand on the cusp of achieving true energy security. Ironically, it is not the decades of government intervention that have brought us to this point, but the innovation and efficiency of the private sector.

Chapter 1

Energy Independence: How Burdensome Regulations are Crushing Offshore Small Energy Producers*

Committee on Small Business

Thursday, September 28, 2023

House of Representatives, Committee on Small Business,
Subcommittee on Rural Development,
Energy, and Supply Chains,

Washington, DC.

The Subcommittee met, pursuant to call, at 10:04 a.m., in Room 2360, Rayburn House Office Building, Hon. Wesley Hunt [chairman of the Subcommittee] presiding.

Present: Representatives Hunt, Luetkemeyer, Meuser, Stauber, Mann, Gluesenkamp Perez, and Golden.

Also Present: Representative Williams.

Chairman HUNT. Good morning, everyone. If you do not mind will you please stand for the Pledge of Allegiance?

I pledge allegiance to the flag of the United States of America. And to the Republic for which it stands, one nation, under God, indivisible and justice for all.

---

* This is an edited, reformatted and augmented version of Hearing Before The Subcommittee on Rural Development, Energy, And Supply Chains of the Committee on Small Business United States House of Representatives, Publication No. 118–028, dated September 28, 2023.

In: The Struggle for Energy Independence in America
Editor: Victoria Whitehead
ISBN: 979-8-89530-558-4

Good morning, everyone. I now call the Subcommittee on Rural Development, Energy, and Supply Chains to order.

Without objection, the Chair is authorized to declare a recess of the Committee at any time.

The Committee is here today to hear testimony about the role of small businesses in our domestic energy production and the regulatory hurdles drillers and service providers are facing today.

I want to thank you all for being with us here today and I am looking forward to today's conversation.

I now recognize myself for my opening statement.

First, I want to thank all of our witnesses for being here today, this morning. I understand you traveled pretty far to be here and we appreciate your time. This is a very important issue. It is near and dear to my heart being in Houston, the energy capital of the world. Energy corridors in my district makes me the energy congressman of the entire world. I tell myself that every morning when I wake up. It makes myself feel better.

We have had an issue with the oil and gas industry I think for the last couple of years. We are seeing gas prices soaring and I think it is time for us as a country to unleash American energy. I have spoken to many of you in person and no one has an issue with renewable energy. No one has an issue with the next abun dant, affordable source for the future but I think we have an understanding that we cannot get there without us.

Domestic oil and gas is the lifeblood of our economy. And when American energy production is strong our country is strong. Our allies are strong. American families are strong. The world is a stronger and a better place.

Independent oil and gas producers are feeling the negative impacts of regulatory hurdles every single day. It is not just regulatory issues; it is also the higher tax rates that we are paying and I am sure you will discuss that here in a little bit.

A recent example evolves around BOEM's institution of a new rule in June regarding the capital reserve requirements for Outer Continental Shelf oil and gas leases and grantees. Requirements such as having an investment grade credit rating, which are large, upfront expenses for any small business place additional burdens exclusively on smaller, offshore oil and gas companies. BOEM's rule further tips the scale in favor of large companies and small businesses lack the same luxury of access to capital and credit.

Another rule we will focus on today is NOAA's Rice's whale rule—I am not going to have Nessie today. Do not worry about that—which places undue burdens on offshore operators. The National Oceanic and Atmospheric Administration's expansive rule about the Rice's whale limits oil exploration and dramatically increases cost for producers in the Gulf. This rule continues numerous anti-oil and gas provision which the sole aim of dismantling our country's offshore production. All over a possible sighting of a Rice's whale almost a decade ago.

As a candidate for president, Joe Biden made a campaign promise to end fossil fuels. As president, his administration has worked every day to do just that. As I said earlier, it is not about ending our industry; it is about finding ways to unleash American energy. We need more electrons in the world, not fewer.

I represent Congressional District 38 here in Texas. As I told you, and I will probably say it three more times, Energy Congressman of the World. That is something that is very important to me, and if you are looking at the international and the safety of the world, it is literally relying in the hands of energy production.

Look at what is happening in Russia and the Ukraine. Look at what is happening all over the world. We have the opportunity to lift nations and countries out of poverty by just giving them American oil and gas.

But these oppressive rules and regulations are being forced upon our domestic oil and gas producers and are crushing production, future investment, and the benefits that American consumers could enjoy if we were energy dominant once again. And I am confident that we will get there.

It is my opinion that the word "transition" is a myth. There is no transition. It is only energy addition. Over the course of the next decade we are going to add 2 to 3 billion more people to this earth. Last year we had the highest demand on earth for oil and gas, and being from Houston and that Houston heat, I am not going back. I will continue to turn on my air conditioning; otherwise, I will not be very comfortable in Houston at all.

We should be talking about energy addition, not transition. And again, this is an 'all hands on deck" approach.

I want to thank you all for being here this morning. Greatly appreciate it. Thank you for your time. I look forward to having a very spirited debate and discussion.

I now recognize Ranking Member Ms. Gluesenkamp Perez for her opening statement. And thank you, ma'am, for being here.

Ms. GLUESENKAMP PEREZ. Thank you, Mr. Chairman. The success of our small businesses and the economic security of working families relies on abundant sources of affordable energy. As we all know, the high cost of oil and gas over the past 18 months has caused significant financial distress to many families, particularly those in rural communities like mine where we have to drive significant distances to access simple services. That is why I am supportive of efforts to make our nation more energy independent and take advantage of the wealth of natural resources available to power our economy long into the future.

Unfortunately, many of the sources of energy that we rely on are vulnerable to foreign manipulation. Just 2 weeks ago we saw Russia and Saudi Arabia collude to continue production cuts through the end of the year. As a result, oil prices spiked. And despite the record amount of oil being produced in the U.S. right now, gas prices remain elevated for most of the consumers. This is because the price of oil is set on a global market as a global commodity. And no matter how much is produced at home, it will still be vulnerable to manipulation by dictators and despites abroad.

But what is clear is that while we transition to more sustainable and cost-effective renewable energy, we still need oil and gas in the meantime.

One potential source of oil and gas is found off our shores. Unfortunately, the process of extracting it comes with significant tradeoffs, particularly for coastal communities. We know that as offshore oil and gas infrastructure ages it can corrode and release harmful chemicals into the water. Idle and unplugged wells can leak oil that is toxic to fish, shellfish, and marine life, harming local fishermen and rural economies. In the event of large spills, oil can reach the shore, polluting beaches and threatening coastal tourism.

One study found the Deep Water Horizon spill cost tourism and fishing industries in Alabama, Louisiana, Mississippi, and Florida between $15 and $30 billion. This is why I opposed offshore drilling off the coast of Washington, and my first amendment as a Member of Congress was to prohibit drilling off the shore of Oregon and Washington that would adversely affect our fisheries.

We are here today to discuss new actions by the Bureau of Ocean Energy Management (BOEM) to better understand these tradeoffs to both protect coastal communities and shield taxpayers from incurring additional costs when oil companies that drill offshore go bust.

New proposed financial insurance regulations would require oil and gas companies conducting business offshore to cover the costs of decommissioning upfront in the event their companies go bankrupt. While decommissioning wells is already required by law, companies often seek to shirk their duties to do so. As a result, it is taxpayers who are often left with the expense.

According to the GAO, the Department of Interior held less than $3 billion in bonds to cover the estimated cost of $38.2 billion for decommissioning these wells in 2015. While I sincerely sympathize with the concerns of my colleagues that this could negatively impact independent oil producers, it is essential that our regulations address the real cost of offshore drilling and ensure those costs are borne by the companies that profit from the practice, not taxpayers and coastal communities. I hope this regulation will adequately protect coastal communities from any downstream consequences of offshore drilling.

With that, I thank all of the witnesses for joining us and look forward to a productive conversation.

Thank you. I yield back.

Chairman HUNT. Thank you, ma'am.

With that I would like to recognize our Full Committee Chair, Mr. Roger Williams from Texas for his opening statement.

Mr. WILLIAMS. Thank you. And good morning to all our witnesses. We appreciate you being here today.

I want to thank Chairman Hunt for holding today's hearing and we greatly appreciate all of your time.

You know, it is no secret that the Biden administration has been at war with the oil and gas industry since day one. Countless regulations have created a burdensome environment for our nation's small oil and gas producers that are a critical component of our domestic energy supply. Some of the regulations that we will be discussing today are a continuation of this trend that is doing nothing but crippling and harming our independent oil and gas producers. Our small businesses across every industry are the engines of our economy and without them America will struggle and has struggled. Requirements like obtaining an investment grade credit rating are expensive and put small businesses at a competitive disadvantage.

So as Americans are facing extremely high inflation and energy prices we need to be making it easier. I repeat, easier, not more difficult for domestic energy companies to succeed.

I look forward to hearing from our witnesses today. And with that, Mr. Chairman Hunt, I want to thank you again, and I want to thank our witnesses for being with us. And I yield my time back.

Chairman HUNT. Thank you, Chairman Williams.

At this time I would like to introduce our witnesses that are with us today.

Our first witness is Mr. Mike Minarovic. He is the CEO of Arena Energy, cofounded by Mr. Minarovic in 1999. Arena Energy focuses on pursuing the lower risk expiration and production opportunities that remain in the Gulf after 50 years of drilling by larger oil companies. Since the company startup in 1999, it has invested over $7 billion in the Gulf of Mexico and has paid production royalties of over $1.4 billion to the U.S. Treasury. Mr. Minarovic is a graduate of the University of Texas in Austin. They are having a pretty good football season this year, sir. And his Bachelor of Science in Petroleum Engineering. Mr. Minarovic, thank you so much for joining us today. We greatly appreciate it. We will look forward to your testimony, sir.

Our second witness is Mr. Paul Danos, who serves as Chairman of the Board for NOIA, the National Ocean Industries Association. Founded in 1972, NOIA represents all facets of the domestic energy related industries. Mr. Danos was reelected as Chairman of the Board this past May. Mr. Danos is also the owner and CEO of Danos LLC, a family-owned and managed energy services provider. He joined the company in 2005 and has been the co-CEO through 2020. Mr. Danos graduated from Louisiana State University with a bachelor of science in finance and then went to Stanford University Graduate School of Business where he earned his MBA. Mr. Danos, thank you so much for being here today. Greatly appreciate it. Thank you so much for your time.

Our third witness is Mr. Alex Epstein. He is a philosopher, journalist, and author that has covered the energy industry for over 2 decades and is the founder and president of the Center for Industrial Progress based on San Diego, California. The Center for Industrial Progress was founded by Mr. Epstein in 2011 to offer a positive, pro-human alternative to the Green Movement. He is the author of the Moral Case for Fossil Fuels and New York Times Best Seller arguing that human flourishing and progress requires the use of more fossil fuels, not less. Mr. Epstein has spoken to employees and leaders and dozens of Fortune 500 energy companies, including Exxon, Mobil, Chevron, and Phillips 66. He has also made his moral case for fossil fuels at dozens of campuses including Harvard, Yale, Stanford, and Duke. Mr. Epstein graduated from Duke University with a bachelor of arts in philosophy. Mr. Epstein, thank you, sir, for being here. Greatly appreciate your time.

I now recognize the Ranking Member to introduce the minority witness for today's hearing.

Ms. GLUESENKAMP PEREZ. I am pleased to introduce our final witness today, Ms. Kathleen Nisbet Moncy. Ms. Nisbet Moncy is a native of Willapa Bay in my district where she is a second generation oyster farmer. Working as chief organizing officer at her family's business Goose Point

Oysters, she grew up in a family business farming oysters and working in the processing plant before attending college at the University of Hawaii at Hilo where she studied marine biology and Spanish. After college, she returned to her family business and has worked there for the past 15 years. Throughout her time working for Goose Point Oysters, she has worked with a variety of organizations such as serving on the board of the Pacific Coast Shellfish Growers Association and establishing the Willapa Bay Grays Harbor Ecosystem Collaborative. She is currently the president of the Willapa-Grays Harbor Growers Association with the mission to uphold the legacy of environmental stewardship and water quality through sustainable shellfish farming. Thank you, Ms. Nisbet Moncy for taking the time out of your schedule and traveling all the way to D.C. to testify today. We are truly honored to have somebody that really works for a living and represents their community in the best ways. Thank you so much.

Chairman HUNT. Thank you for being here. Thank you, ma'am.

Before recognizing the witnesses I would like to remind them that their oral testimony is restricted to 5 minutes in length. If you see the light turn red in front of you it means your 5 minutes have concluded and you should wrap up your testimony.

I now recognize Mr. Minarovic for his 5 minute opening remarks.

Statements of Mike Minarovic, Chief Executive OffiCer, Arena Energy, LLC; Paul Danos, Chairman of the Board, Owner & Chief Executive Officer, Danos, LLC; Alex Epstein, President and Founder, Ceter for Industrial Progress; Kathleen Nisbet Moncy, Chief Operating Officer, Goose Point Oyster

Statement Of Mike Minarovic, Chief Executive Officer, Arena Energy, LLC

Mr. MINAROVIC. Thank you, Chairman Hunt, and Ranking Member Gluesenkamp Perez, for the opportunity to testify this morning.

My name is Mike Minarovic, CEO of Arena Energy. We are an offshore exploration production company focused on the Gulf of Mexico, one of the most active drillers in the Gulf of Mexico. Over the past 24 years, as Chairman Hunt mentioned, we have invested $7 billion in the Gulf and paid the federal government over $1.4 billion in royalties. Last year alone we paid the federal government $230 million in combined royalties and corporate taxes.

I also speak to you today on behalf of the Gulf Energy Alliance. It is a coalition of independent Gulf producers. Although we are not household names, independents are the backbone of the offshore industry, producing 35 percent of the oil and natural gas from this basin.

I am here to discuss a new regulation recently proposed by the Department of Interior. This hearing is the appropriate place to discuss this issue as the Bureau of Ocean Energy Management, also known as BOEM, stated that 76 percent of the businesses operating offshore are defined as small businesses. This totals 407 businesses according to BOEM, and this rule poses an existential threat to those companies.

Also, the SBA Office of Advocacy submitted a very strong letter denouncing BOEM's proposed rule. BOEM repeatedly states that the goal of this new rule is the protection of the American taxpayers from exposure to financial loss associated with offshore decommissioning. We firmly believe that it is the primary responsibility of the current operators to abandon their platforms and wells and that taxpayers should never be responsible for abandonment obligations.

Arena and GEA are willing to work with BOEM to find a solution to achieve that goal but this rule as currently drafted is a massive overreach to a limited problem. It is devastating for small businesses and it is completely unworkable in its current form.

BOEM states that the cases where taxpayers have actually paid costs for decommissioning are rare and that since 2009, more than 30 corporate bankruptcies have occurred involving offshore oil and gas leases with unbonded decommissioning liability is $7.5 billion. Those are big numbers.

So how much has been paid by the American taxpayer for their abandonment obligations in the Gulf? Well, that was not addressed in the rule. But in July of 2023, BOEM communicated to the House Subcommittee on Energy and Mineral Resources that U.S. taxpayers have incurred obligations totaling $58 million net of bonds. Fifty-eight million dollar total after more than 30 recent bankruptcies.

Again, the American taxpayers should never pay for offshore abandonment obligations.

But to put this in perspective, in the last 20 years, offshore oil and gas production has generated $124 billion in royalty and leased bonuses for the U.S. Treasury, $6 billion a year, $500 million a month on average. And how has the taxpayer been protected? By law, by regulation, and by practice. All current and previous owners stand in front of the taxpayer and remain collectively responsible for decommissioning obligations. Selling a property in no way relieves the previous owner of its continuing decommissioning obligations.

To address the $58 million in taxpayer exposure BOEM now seeks an additional $9.2 billion in new bonding which would cost small businesses in the Gulf $5.7 billion over the next 20 years. Those are BOEM's own numbers. We feel they are much higher than that. That money could be used to invest in projects and actually decommission these platforms and wells. Overkill barely begins to describe BOEM's solution.

But you do not need to believe an oil and gas guy. The comment letters have been submitted. This is what the State of Louisiana said. "The proposed rule falls within the unpleasant category of a solution in search of a problem."

In reality, there is no problem within taxpayer funded decommissioning. There is certainly no problem that would justify eviscerating the oil and gas industry. The SBA Office of Advocacy said that BOEM's analysis intentionally ignores joint and several liability as the way that the taxpayers are protected from these unfunded liabilities. It presumes that small businesses impose a significant risk to the taxpayer despite the full backing of companies that BOEM has already exempted. As a result, small businesses are not only disproportionately harmed by the proposal, but only small businesses are harmed by the proposal. And the Surety and Fidelity Association of America, which is a trade organization of more than 400 companies that writes 98 percent of the surety and fidelity bonds in the United States and are the group that BOEM is looking to to solve this problem, what do they say about it? They fear that the proposed regulations might have unforeseen and possible negative consequences leading to an uptick in bankruptcies among small independent operators ultimately worsening unfunded decommissioning liabilities. Such financial failures will have cascading negative impacts that undermine BOEM's stated purpose for increasing access to these bonds and alleviating the burden on the taxpayer.

Given the increased risk in the market created by recent bankruptcy rulings, the proposed regulations could have an even stronger, chilling effect on the surety industry's appetite to participate.

As is, players are already exiting the marketplace, further exacerbating capacity concerns.

This rule is not workable. The group they are looking to for a solution says it will not work. Additionally, CAC Specialty, which is a surety bonding brokerage firm said, "Like a Greek tragedy, the BOEM's actions could expediate the outcomes that it wished to avoid." And once again the State of Louisiana said, it is Louisiana's view that supplemental bonding is unnecessary. As long as a credit-worthy party exists in the chain of title there exists no meaningful decommissioning burden on the federal government.

The solution resides in the BOEM's own analysis. The proposed rule states that if BOEM were to take into account the financial capacity of the predecessor lessees, the financial burden on small companies would be substantially reduced compared to that result- ing from the proposed rule. And this would mitigate the impact of the proposed rule on small businesses.

We remain optimistic that we can work together to create a sensible solution that truly protects the taxpayer and maintains a strong and viable small business community along the Gulf Coast. Thank you.

Chairman HUNT. Thank you, Mr. Minarovic.

I now recognize Mr. Danos for his 5 minutes. Thank you, sir.

Statement of Paul Danos, Chairman of The Board, Owner and Chief Executive Officer, Danos LLC, Testifying on Behalf of National Ocean Industries Association

Mr. DANOS. Thank you, Chairman Hunt, Ranking Member Gluesenkamp Perez, and Members of the Subcommittee. Thank you for having me here today to testify.

My name is Paul Danos. I am president and CEO of Danos, LLC. We are a 76-year-old family business headquartered in south Louisiana. I am also the Chair of the National Ocean Industries Association, which for more than 50 years has represented the all segments of the offshore industry, including oil and gas, offshore wind, offshore minerals, and offshore carbon sequestration.

So in 1947, my grandfather started our company. He had one partner. So there were two of them. That is definitely a small business. And most of the work happened at my grandmother's kitchen table and a boat offshore. They sort of took turns between the table and the boat offshore. And we have done a pretty good job of preserving history. We put a book out kind of capturing some of our history last year on the 75th year anniversary. And there are some great stories about why the founders of our company started this business. And it was certainly like many entrepreneurs they wanted to better themselves. They had an opportunity to do that. But also for their extended family and their community. And that is the fabric of our economy and so many small businesses. People who take a risk. My grandfather borrowed $2,000 from his sister, bought a boat, and got started. And took that risk and provide jobs to people in an industry that provides just an amazingly valuable service, the innovation and quality of life that we enjoy because of the affordable and reliable energy that comes from oil and gas woven into that fabric of our small business and so many others.

So today I am here, our footprint has grown in our company. I am representing my family and certainly our employees for our business but the employees for our industry which total about 370,000 direct and indirect employees for the offshore oil and gas. So I bring a message from the employees. I spend time at crew changes and offshore locations, some of our land locations talking to our folks.

And kind of two things that I would like to convey on behalf of our people and our industry. First, I hope you realize the massive impact that the regulation that comes out of Washington has on the people on the frontlines; the frontline heroes who are providing that reliable and affordable energy for our country. And I know, sometimes we can get disconnected and it is words on paper and it is legislation and it is regulatory acts, but at the end of the day we are talking about real people on the front lines who are making this work happen. So I represent them and just want to just again, you know, make sure that we understand the impact that happens on families, on people.

And then the second thing is just when we talk about any policy that limits the production of oil and gas here in the U.S., I would say that is bad for our country. We have some of the lowest carbon intensity barrels in the world produced in the Gulf of Mexico. An ICF study that came out recently said that the average carbon intensity of barrels produced in the Gulf of Mexico are 46 percent lower than the average of the rest of the world in terms of carbon intensity. So when we restrict supply here in the Gulf—and we have to understand something about oil and gas demand and the way it works—it is a fungible resource. It is abundant around the world. So if we restrict supply, sort of like squeezing a balloon, that supply will show up somewhere else and it will be produced somewhere else. And most often it is going to be places like Venezuela, Iran, Russia.

So first what we are going to do is we are going to send those barrels there. And those are higher carbon barrels. So it is a negative impact on the world's environment. We also send those jobs somewhere else; right? Back to those guys, those people, men and women on the front lines. When we restrict supply here in the U.S., we are sending those jobs, we are sending the royalties that Mr. Minarovic talked about, $6 billion in royalties every year. We are sending those to other counties. And then we are giving up our energy security at the same time. So the idea of moving barrels outside of our country, producing them elsewhere is bad policy. Limiting production here in the U.S. is bad policy.

So what do we need? We need good regulations. Part of why the barrels that we produce in the U.S. are low carbon barrels is because of the way we produce. The regulations that we have worked together within industry and regulators over the decades that are common sense, that protect our environment, protect our people, protect our costs.

I am from a place called Cut Off, Louisiana. It is a coastal south Louisiana community. I grew up fishing and enjoying the coast. Some of the best places to fish are around platforms and old platforms. I had my daughter and my wife in the Gulf of Mexico fishing on an old platform early this summer. We had a fantastic experience. So we can co-exist environmentally but we need common sense policy. We need a regulatory environment that promotes good, common sense regulations and policy to protect our environment but encourage domestic production.

So I look forward to the conversation that we will have today and any questions that we have going forward. Thanks again.

Chairman HUNT. Thank you, Mr. Danos.

I now recognize Mr. Epstein for his 5 minutes for your opening remarks. Thank you, sir.

Statement of Alex Epstein, President, Center for Industrial Progress

Mr. EPSTEIN. Thank you.

So my name is Alex Epstein. I am a philosopher and energy expert representing only myself, and everything I say today is my own independent conclusion, not influenced by anyone from any industry including the fossil fuel industry but which I am a very big admirer of that industry for reasons you will see.

Because energy is a major input in every business, the lower cost and more reliable energy is, the lower cost and more reliable everything is. The higher cost and less reliable energy is, the higher cost and less reliable everything is, and the less efficient and less competitive U.S. businesses are, including small businesses.

Today we are hearing how irrational policies are crippling one area of the energy business: offshore oil production, which provides almost one-sixth of our domestic oil production, production without which gasoline would be far more expensive. And I am from California where we are around $7 right now, and it would be far, far less secure. So it is really crucial to have this production.

But today I would like to put the irrational attack on offshore drilling in its broader context, which is that it is just a tiny piece of today's rabidly irrational global anti-fossil fuel agenda, an agenda that poses as rational and scientific but is actually the exact opposite.

So the dozens of discrete attacks of the Biden administration on the fossil fuel industry, they are not isolated incidents. They are all attempts to impose an agenda that says fossil fuel use must be essentially eliminated by the year 2050, which is now very, very close.

At the beginning of his administration, President Biden announced a U.S. "whole of government" approach to climate change, which in practice means an overarching commitment to rapidly eliminate fossil fuels. Every agency was told figure out new ways to attack fossil fuels. And that is what they have done.

The Biden administration is systematically committed to attacking fossil fuels on four fronts.

So the first front of attack is fossil fuel investment. Fossil fuel investments are long-term commitments that require confidence in a payoff. Our government, often colluding with activists, incessantly discourages fossil fuel investments, directly and by threatening the industry's future, like running on "I guarantee you we are going to end fossil fuel" and then being elected president. The federal government has been a major driver of ESG anti-fossil fuel climate policies with regulators pressuring financial institutions to declare their commitment to getting off fossil fuels. The SEC's recent "climate disclosure rules" so called are taking the damage to the next level.

Now, front two of the attack is on fossil fuel production are common. So punishing restrictions and onerous taxes specifically on fossil fuel production are common across developed countries, unfortunately. And these suppress supply from the freest nations on the globe. And unfortunately, we are leading this right now.

One of the many punishments President Biden has inflicted on the supposedly preferred fossil fuel industry, which the opposite is true, was his early moratorium on issuing oil and gas leases on federal lands. And the Biden administration continues to delay and decline holding lease sales.

Now, the third front of the attack on fossil fuels is fossil fuel transport. One of the most effective ways of punishing the fossil fuel industry has been attacking critical transport infrastructure, including pipelines and export/import facilities with delays, regulations, and lawsuits.

Now, we all know about Keystone XL, which has been hugely destructive, but perhaps the most destructive way of singling out fossil fuel transport for punishment has been killing natural gas pipelines. We have a virtually limitless supply of natural gas in this country, particularly in Appalachia, and we have an incredible ability to ramp up production quickly. But we are blocking pipeline after pipeline.

Now, the final front of the attack on fossil fuels is punishing fossil fuel use. So our government's "whole of government attack" on fossil fuels seeks out every way it can to punish fossil fuel use.

Perhaps the most destructive and underdiscussed form of this is punishing reliable fossil fueled electricity generation and thereby destroying our grid.

So as a Californian, I am well aware of the costs of punishing fossil fuel electricity generation. Gavin Newsom and others prematurely shutting down vital fossil fuel plants led to devastating statewide blackouts in 2020. And 5 days after pledging to all EVs by 2035, Newsom told us there was not enough power to charge or EVs. From his administration, "avoid using major appliances and charging electric vehicles."

Nationally, we are in an electricity crisis with reliable power plants shutting down far faster than they are being built. And the Biden EPA insanely plans to make things much worse with seven policies that gravely threaten 10 to 20 percent of our reliable capacity in the next 7 years.

So every one of these hundreds of actions the Biden administration is taking in terms of its "whole of government" approach is harmful to American businesses, including small businesses who have to struggle with higher energy costs and lower reliability.

Now, the final thing I want to address quickly is just the lie that solar and wind can rapidly replace fossil fuels. If this were true they would not need massive government preferences of their own and they would not need massive government punishments of fossil fuels. And if solar and wind could actually replace fossil fuels, China, which dominates the production of solar and wind components, would not be using coal to produce these components and they would not have 300 planned new coal plants—new coal plants—more than all we have in the United States right now to last over 40 years.

So if you are thinking we are going to have more security with this stuff, we are not going to have much energy and it is all going to be dependent on China.

So what I want to encourage everyone here to do, particularly small businesses and politicians, pro-freedom politicians, is use your powerful voices to advocate for liberating all forms of energy, including fossil fuels, not limited to fossil fuels, so that America and the world can be a better place. Thank you very much.

Chairman HUNT. Thank you, Mr. Epstein.

I now recognize Ms. Nisbet Moncy for her 5 minute opening remarks. Thank you for being here.

Statement of Kathleen Nisbet Moncy, Chief Operating Officer, Goose Point Oysters

Ms. NISBET MONCY. Thank you, Chairman Hunt, Ranking Member Gluesenkamp Perez, and Members of the Subcommittee for this opportunity to testify today. My name is Kathleen Nisbet Moncy. I am the COO of Goose Point Oysters. I am a second-generation shellfish farmer and a mother to five children.

Our company is located in the heart of Willapa Bay, Washington, where we own 2,000 acres of tidelands. Willapa Bay is one of two coastal estuaries in Washington that produce over 25 percent of the nation's oysters.

The shellfish industry in Willapa Bay and the surrounding areas generates a tremendous economic impact. As the largest private employer in Pacific County, local shellfish farmers in Pacific County and Grays Harbor are responsible for nearly 2,000 family-wage jobs, and contribute to over $102 million in economic impact.

Shellfish farmers, while they are the largest private employer in our county, our tidelands are the backbone of our area. We are locally owned and our tidelands are passed down from generation to generation. We have a family farm that has been farming there for over 150 years.

Shellfish farming in Washington State predates statehood. We are the backbone of our community in supporting local schools, providing employment and, participating in community programs. Shellfish farming is a way of life in our community.

The ecosystem in which we farm is of upmost importance to us, as shellfish farmers we depend on clean water and a healthy habitat. A single submerged oyster will filter up to 7 gallons of water an hour, our responsibility is to protect that clean water. We are dedicated to managing the health of our ecosystem, and as a result are regarded as one of the healthiest working estuaries in the U.S. I would promote stricter regulations on offshore energy no matter what the company size is. Our oceans are extremely fragile environments, and shellfish farmers feel even the slightest changes in our ocean conditions. The most recent one we have been struggling with is ocean acidification. It came up in 2008, and according to NOAA, we are going to continue to see shifts in ocean acidification into the future.

Some of the negative impacts of ocean acidification on farmers have been pH shifts, lack of calcium in the water, and harmful algal blooms. We must continue to fight for our waters and learn how to evolve with the changes instead of adding more complications with offshore energy.

Along with more stringent regulations for offshore energy, there also must be adequate funds set aside for any disasters or even small leaks that would arise. The Gulf was not prepared for the oil spill it endured years ago and the shellfish industry is still impaired due to it. We should learn from the mistakes and ensure that they do not happen again.

As an example we have been working with the State of Mississippi to assist them in creating hatchery systems so they can rebuild their reefs because the industry down there is still struggling today. We have sustainable farming mechanism in Willapa Bay and they are trying to mimic that down in the Gulf States. Knowing how to sustainably farm shellfish will provide a consistent food source for the future and water quality is a key part of that.

An oil spill off of our coast would destroy our estuaries and the communities that are supported by them. Based on the way the waters flow, we get two full tidal fluctuations of water a day, anything that happens off of our coastline would enter into our estuaries and have a significant impact on shellfish, crab, eel grass, and other species that live there.

Besides some of the local issues that would arise with offshore energy, our location does not have adequate services to respond to a spill. The Willapa Bay entrance and channels have not been properly managed for decades by the Department of Army Corps and would pose a danger for any type of response team. There is no positive scenario for our fragile ecosystems if there were to be any sort of spill off of our coastline.

Along with the impacts to shellfish in our coastal estuaries, we must also consider the Dungeness crab and salmon fishing communities. Shellfish tidelands are the rearing habitat for Dungeness crab. Dungeness crab fishing grounds are directly located off the Washington and Oregon Coast. Access to these fishing grounds is currently limited and further limitations would risk the livelihood of the fleet and the communities that process and sell the seafood. The cumulative impacts to the fishing grounds would make it near impossible for future generations to be successful in the fishing industry. We should be focused on increasing seafood production in the U.S. instead of polluting the very waters that provide it.

In conclusion, I would not support reduced regulations to offshore energy at the risk of the livelihoods of our rural communities. Shellfish is a sustainable seafood that has a huge benefit by providing many ecosystem services, including purifying the water, reducing carbon and many other essential services in the areas we farm. Our fishing communities must be able to retain the available fishing grounds for the success of their industry so we must stand up and protect the waters from contamination and be responsible stewards of the ecosystems in which we work in.

Thank you for your time.

Chairman HUNT. Thank you, so much for being here. And again, I want to thank the witnesses.

We now move to the Member questions for the under 5 minute rule.

I recognize myself for 5 minutes.

The first question is for you, Mr. Minarovic. The Financial Assurance Rule is said to be put in place to protect the taxpayers. Is there currently a real risk to the taxpayer to be held accountable for these costs? And if so, what would you propose as a solution?

Mr. MINAROVIC. Thank you, Chairman Hunt.

You know, I think this is really important. And I have got a graphic that I think is in front of each of you that helps explain this complicated issue. As Ranking Member Gluesenkamp Perez mentioned, the BOEM estimates that current abandonment obligations are from the Rule $42.9 billion. I think her estimate was in the high 30s from a 2015 study.

What you need to understand in order to understand the exposure to the taxpayer is it is broken up into these three buckets here. The dark green is investment grade companies. The government is exempting that. Their numbers suggest that is $22.6 billion of the abandonment. The bottom number in red is noninvestment grade companies, both current and prior owners. No investment grade company in the chain of title totals $750 million. That is where we really need to focus our efforts here, and the GEA and Arena Energy support additional bonding associated with properties with no investment grade companies in the chain of title.

The thing that is really critical is the box in between, $19.5 billion. This is where a previous owner has an investment grade rating. And thing is that the government has stated that under this proposed rule the government retains the authority to pursue predecessor lessees for the performance of decommissioning obligation. When a company sells an asset they remain in the chain of title. That is what protects the taxpayer. Any company that owns or previously owned leases is responsible. If Arena sold its assets to another company, we would stay in the chain of title and be responsible for that abandonment. That is what protects the taxpayer. So none of these dark green boxes or light green boxes should be considered for bonding. The only risk to the taxpayer is this red box and that is what has really resulted in such a small loss relative to the exposure in the Gulf.

Chairman HUNT. Thank you, sir.

This question is for you, Mr. Danos. America energy companies are leaders in innovation and produce the cleanest oil and gas in the entire world. And I know. I have been on these rigs and I have seen them actually personally. How would the rule on the Rice's whale and others like it affect the industry's ability to operate at its full capacity and reach its full potential?

Mr. DANOS. Thank you, Chairman. It is a great question.

Any new regulation needs to make sense. It needs to be balanced across all stakeholders and in terms of what it is trying to do. And it needs to be effective. And I think the Rice's whale, for instance, a couple of notes about that. There is very little data that exists that we have the Rice's whale in the area that was originally pulled out of the last lease sale. So the balance, the effective components of that regulation just really did not add up.

And one thing related to marine operations that I think is really important to point out when we think about balancing the different stakeholders, oftentimes in the Gulf of Mexico what we do is we operate vessels at night so that we can do the higher risk work that we do on the platforms during the day. So with the Rice's whale regulation prohibiting vessels from moving at night what you are doing is you are flipping that on its head. And the unintended consequence of that is that the people who work for my company and other small businesses and large businesses in the Gulf of Mexico, we are increasing the risk to those people with the work that they are doing because they would have to move more of that work at night as opposed to doing it during the daylight hours. And there is plenty of work that goes on at night but again, we are increasing that exposure. So that is something that needs to be thoroughly examined and thought out as we proceed.

Chairman HUNT. Thank you so much.

Mr. Epstein, my last question is for you. What do you think of Biden's energy policy, and is it good for the country? You are allowed to take a little bit more time here if you would like. I am not going to need 45 seconds.

Mr. EPSTEIN. I addressed a lot of it. I mean, the thing people need to get is the danger of this whole of government approach because it really has a terrorizing effect on industry because industry knows that it is not just any given individual restriction that is a problem; it is that the government is literally trying to shut them down. You see this happening with coal right now where the government is trying to shut down all of our coal plants which are preventing us from the abyss right now. We have catastrophic reliability problems and now we have Mike Bloomberg who I offered to publicly debate so far has not stood up yet. Jordan Peterson offered to host. So Mike, let's do it.

But he is spending $500 million on top of billions of other dollars to shut down our coal plants and he is not willing to publicly defend it.

One other thing. Since the UAW is in the news right now, how are you going to get higher wages if you have Joe Biden pressuring companies to sell vehicles where they are losing billions of dollars? Companies do not give higher wages when they are losing billions of dollars due to forced policies by the government. So those are just two specific examples but the "whole of government" approach is the overall disaster.

Chairman HUNT. Thank you very much.

I now recognize the Ranking Member for 5 minutes for questions. Thank you, ma'am.

Ms. GLUESENKAMP PEREZ. Ms. Nisbet Moncy, you highlighted in your testimony the vital role oysters play in filtering water in Willapa Bay. Can you explain the fragility in the system and the outcome to even minor changes in water quality?

Ms. NISBET MONCY. Yes, I would love to do so.

Shellfish are filter feeders and as a result are a product of their environment which make them very susceptible to environmental changes. Most recently, due to climate change and ocean acidification, we are seeing an increase in harmful algal blooms or HABs in our estuaries which are having a direct impact on the productivity of our shellfish beds. We are noticing that after a harmful algal bloom event the oysters and other shellfish will perish within 2 to 6 weeks.

Therefore, if we have pollution coming into our water ways, our shellfish, eelgrass, Dungeness crab, salmon and other species will be negatively impacted. For many decades we have been the continued stewards of the estuaries collaborating with upland timber companies, private home owners and mills to mitigate impacts and protect our clean water.

Ms. GLUESENKAMP PEREZ. So any minor spill would have a negative impact?

Ms. NISBET MONCY. Any minor spill would have a significant negative impact.

Ms. GLUESENKAMP PEREZ. So fishing off the coast is an important implication of our economy alongside oyster fishing in the bay. Could you explain any form of offshore energy production that Washington endangers commercial fishing industry?

Ms. NISBET MONCY. Currently, there is a limited amount of area that our commercial fishermen can fish, with the introduction of offshore energy you reduce the areas even further. So whether it is a platform, wind energy, or even cables underground, you are disrupting the Dungeness fishing grounds. Any infrastructure impacts fishing off our coastline. The fisheries are very strictly managed and cannot withstand any further reduction in the available area to fish.

Ms. GLUESENKAMP PEREZ. Yeah. We were talking in the elevator on the way here and we have talked out in the bay. People from more urban communities or other parts, they assume nothing is going on. There is this arrogance of like there is nothing going on out there. Like, we will just like move it somewhere where it is not in my view shed. Or it is somebody else's problem. And I think this loss of a sense of agency and determination in our rural communities, what happens here, how we prefer to make our livelihoods. So I am deeply grateful for your advocacy of a way of life that I think is critical to our culture as Americans and as Washingtonians.

So farming oysters is one of the most sustainable sources of seafood. Not only purifying the water but also absorbing and sequestering carbon emissions. Can you talk about the importance of sustainability broadly? Because this is not just about dollars and cents but to your business and to your philosophy as a business owner?

Ms. NISBET MONCY. Yes, for us it is our way of life. We have always had the mentality that if you take care of the estuaries in which you work, that they will take care of you. We sustainably farm our products. We have hatcheries that produce our own seed which we plant and cultivate to be harvested. We are considered a sustainable seafood not a wild fishery.

Our perspective is not just that of a shellfish farmer but as a community member with everyone else who works in and around the estuary, if we all collaborate our ecosystems will flourish. We want to retain the rights to be able to continue what we are doing for generations to come. While it has been very difficult with climate change and OA (ocean acidification), we continue to figure out how to evolve. We are huge proponents of finding solutions and must have a voice at the table when rules are being made, especially here in DC. I recognize, I am a small fish in a big pond but the pond has ripple effects and those can cripple our industry and the communities we support. Making sure that we are being smart with our regulations and understanding the impacts they will have down the road is key.

Ms. GLUESENKAMP PEREZ. It feels like a lot of industries you can sort of take the money and run. But it feels like, what I really value about our shellfish growers is when you talk about the generational wealth, you are talking about the health of the estuary. That is where the money is. That is where the culture and the community is is in the actual rootedness and health of your ecosystem.

So I cannot say how grateful I am to all the smart, hardworking people who have chosen to make and continue to make their family livelihoods in these rural communities. And thank you. Sincerely, thank you for taking time away from your family and your business to be here today.

Ms. NISBET MONCY. Thank you.

Chairman HUNT. Thank you, Ranking Member.

I now recognize Representative Meuser from the great state of Pennsylvania for 5 minutes.

Mr. MEUSER. Thanks. Thank you, Chairman, very much. Thank you very much to our witnesses for your work, your business, and for taking the trip being here with us.

So Mr. Minarovic—how do we pronounce that?

Mr. MINAROVIC. Minarovic.

Mr. MEUSER. Minarovic, thank you.

You have got a business. You have got perhaps family members and you have a small business. And that business is important to you?

Mr. MINAROVIC. Yes, sir.

Mr. MEUSER. Okay. So we have small businesses here. Some in the energy sector. Some in other industries. I do not think it is up to this Committee to decide who is more important. You know, that is just me.

So this BOEM rule, when it perhaps goes into effect, you called it existential to you and to, you know, I am just reading here, the billions of dollars and other small businesses that will be wiped out. Existential means wiped out.

Mr. MINAROVIC. Absolutely.

Mr. MEUSER. Okay. You are a highly regulated business obviously; right?

Mr. MINAROVIC. Absolutely.

Mr. MEUSER. Yeah. And have there been any problems lately? I mean, I come from the Marcellus shale region, a new part of my district. You can go to a rig site, the level of regulation which is great. That is how it should be in the United States. And even outdoor rig sites, it is like walking into a clean room.

Mr. MINAROVIC. Right.

Mr. MEUSER. It is perfect; right? There have not been issues. So, and that is a good thing. There should not be. And so that is the way it should be.

So in your business, I expect, not abiding by some of the pollutant control requirements which I am all for because this is the United States. We want to do things better than every other country. But if you were to cross those regulations that would also be existential to your business, would it not?

Mr. MINAROVIC. Absolutely. Yes, sir.

Mr. MEUSER. Okay. so this BOEM, we will work on this because once again these sort of government actions hurt those that supposedly it is trying to help. By increasing the level of bonds that you need to secure, you expressed it in your opening testimony, but is it completely impossible under the current rules?

Mr. MINAROVIC. Yes, sir. Let me respond to your first statement.

It is absolutely the case that we are a heavily regulated industry. The Gulf of Mexico has been heavily regulated for 75 years. The federal waters of the Gulf of Mexico. And there is no better place to get our oil than the Gulf of Mexico.

Mr. MEUSER. Not Venezuela? Not Russia?

Mr. MINAROVIC. It is going to be a lot cleaner coming from the Gulf of Mexico.

Mr. MEUSER. Do you think the Middle East, do you think they have got the same pollutant controls as the United States of America?

Mr. MINAROVIC. Lets get our barrels from the place where the states that surround it—Mississippi, Alabama, Texas, and Louisiana support it. Lets get our barrels from the place that the federal government gets a royalty check.

Mr. MEUSER. How about Venezuela? Do you think Venezuela's standards——

Mr. MINAROVIC. Absolutely not.

Mr. MEUSER. We live in a round world; right?

Mr. MINAROVIC. Absolutely.

Mr. MEUSER. So those pollutants cause problems everywhere?

Mr. MINAROVIC. Ranking Member Gluesenkamp Perez said that, you know, people do not realize if you do not do it here it is going to be done somewhere else. Things move around. If we are going to get our oil from somewhere—Iran, Venezuela, Russia, Iraq——

Mr. MEUSER. And what about the costs associated with all this? By losing our energy independence, forget about energy dominance.

Mr. MINAROVIC. Yes, sir.

Mr. MEUSER. Okay, with gasoline 5 bucks a gallon, $4 a gallon,

$7 a gallon out in Pennsylvania making us highly vulnerable to world events. I mean, man, we should have learned that when you and I were kids in the '70s. Right? World events. I mean, Ukraine, you name is. OPEC decides to cut back, leaves us, completely wrecks our economy, causes inflation.

Mr. MINAROVIC. Absolutely.

Mr. MEUSER. Or partially causes inflation. That and wildly excessive spending. Hurts small business. Hurts the oyster business. Right? As your costs go way up you just cannot pass all that along. So, yes. So I would say we are in a devolving initiative. What we should be focused on is energy dominance with the highest levels of pollutant controls. You could probably handle that. Right?

Mr. MINAROVIC. Absolutely.

Mr. MEUSER. I mean, as opposed to having to follow this BOEM rule?

Mr. MINAROVIC. Absolutely. We need to do that. I mean, oil prices are being driven by supply and demand imbalances. Anything we do to restrict domestic production is going to have an impact on the world oil price ultimately. And that is going to be a regressive tax on the lower income people of this country. We need to move to alternatives. We need to move to all the above. The best of the above is the right answer. And we need to pursue everything. But we cannot limit fossil fuels. We are going to impair our country and our economy.

Mr. MEUSER. We are going to go to work for you.

Mr. Danos and then Mr. Epstein and then, I am sorry, Ms.—I should know—Ms. Moncy. I am running out of time but I was just going to ask what came out of Washington that harmed you the most and what has come out of Washington that has helped you the most? And maybe we could follow up with that afterwards if we have a second round.

I yield back, Mr. Chairman.

Chairman HUNT. Thank you so much.

I now recognize Representative Golden from Maine for 5 minutes.

Mr. GOLDEN. Thank you.

Ms. Nisbet Moncy, I know most of the hearing is focused on oil and gas and I wanted to talk a little bit about BOEM and offshore wind development. As an oyster farmer specifically, do you have any concerns about offshore wind development and how could it impact local aquaculture operations?

Ms. NISBET MONCY. Yes, I do. The proposals that they are working on in Oregon and off the Washington coast, we are not proponents of those. Our current cost of power that we have from hydro in the northwest is about 4 cents per kilowatt-hour that goes into our PUDs. The cost for wind would be about a dollar. You are looking at a significant increase across the board for power to those of us who live in the Northwest, and it has been shown that there is zero net gain of wind energy off of our coastline. By the time you install it, run it, and decommission it, there is really no added benefit for our region to have offshore power based on the cost of all those components. Our PUDs provide power to other areas such as California and profit significantly. As for the fishing grounds, you are looking at limiting the area available to fish in our region. The infrastructure required for the windmills such as cables, diesel/oil tanks and increased boat traffic all elevate the risk to our fishing grounds and estuaries.

Mr. GOLDEN. Do you have any concerns that it could also be potentially harmful to the ecosystem?

Ms. NISBET MONCY. Any sort of offshore spill without a proper response would have a negative impact on our communities.

Mr. GOLDEN. Offshore wind that is.

Ms. NISBET MONCY. Offshore windmills still have diesel on them and must be serviced so you will increase the boat traffic in our area. There have been several ships that have run aground on the Columbia River bar so if you increase boat traffic to be more than what it currently is, you increase the risk of a spill. We currently do not have an adequate response team to the Willapa Bay entrance if there were to be any sort of spill.

Mr. GOLDEN. For aquaculture communities and also fisheries in your region, do you feel like BOEM has been a good partner in hearing your concerns? Is there any evidence that they are working to address those concerns?

Ms. NISBET MONCY. In our region specifically there has not been a direct nexus. No. Down in southern Oregon with the Crab Fisherman Association there has been a nexus with BOEM.

Mr. GOLDEN. Okay. Thank you. That is helpful.

I guess I will try, Mr. Epstein, but anyone else is welcome to try and answer this. And this comes from a genuine place. I am not contesting that some of the policies you see coming out of the White House are concerning on oil and gas production here domestically. You see some good things like approval of Willow up in Alaska and then at the same time you see the cancellation of permits for leases up in Alaska. Sometimes going in the right direction; other times, not. So at the same time I have seen no shortage of reports that oil and gas production in the United States is up rather than down currently compared to let's say 2020. Is that true or is that not true?

Mr. EPSTEIN. I mean, 2020 is very misleading selected time.

Mr. GOLDEN. Sure. I said general.

Mr. EPSTEIN. So in general——

Mr. GOLDEN. You pick your years. Is oil/gas production in the United States up or down?

Mr. EPSTEIN. It is lower. Well, it just depends on your starting date but it is up lower than it should be if we were free. So all of these things are, you have to look at it, what is the opportunity cost.

Mr. GOLDEN. Sure. I agree.

Mr. EPSTEIN. So in this case we have a global anti-fossil fuel movement that is trying to bring us to net zero, which essentially means the elimination of fossil fuel.

Mr. GOLDEN. Have we ever produced more oil and gas in the United States than we are today?

Mr. EPSTEIN. I am not exactly sure year to year but the point is if you look at what has happened, by slowing the growth of oil and gas in a world that needs far more energy, we have caused a global energy crisis. And yet this administration and this global movement is trying to eliminate it. So the net zero movement has not even succeeded at reducing fossil fuels. They have only succeeded at slowing the growth. And already, we have this huge vulnerability to the Russia-Ukraine situation. Already, we have these cost increases. So that should be a wakeup call that this whole agenda is ill conceived and if you want to address emissions you should instead do so with energy freedom.

Mr. GOLDEN. We are running out of time. I just want to be clear just that—I am on your side, buddy. Like I want all of the above, I want America to be an energy——

Mr. EPSTEIN. So will you condemn the Biden energy agenda?

Mr. GOLDEN. I just told you—wow, you are asking me questions now. It is like you came here with an agenda yourself, my friend.

Mr. EPSTEIN. I am just a curious guy.

Mr. GOLDEN. What I would say is that I think that this administration is governing over an increase in oil and gas production while simultaneously seemingly unwilling to celebrate that which they should be celebrating. And I would agree with you that some of their policies as I stated earlier as an example up in Alaska are the wrong calls. And I would condemn that.

Mr. EPSTEIN. Thank you.

Mr. GOLDEN. Thank you.

Chairman HUNT. Thank you so much.

I now recognize Representative Stauber from Minnesota for 5 minutes. Thank you, sir.

Mr. STAUBER. Thank you very much for holding this hearing. You know, since taking office, President Biden has sought to shut down the oil and gas industry at every turn putting our economy and energy security at risk. The week President Biden took office, gas on average was $1.87 a gallon in my home district in Minnesota. Today, it is nearing $4 a gallon and President Biden did that. Rather than develop our abundant resources here safely where we have the greatest environmental labor standards, this administration would rather beg OPEC, Saudi Arabia, and others for more oil. This administration's model on energy and mining is clear. Anywhere but America, any worker but American.

The Biden administration has completely neglected its duty to support oil and gas development and has failed to properly carry out an offshore leasing plan. Even to the point that my democratic colleagues had to force the Biden administration to do so in the Inflation Reduction Act last year. And still, this administration is failing to follow along. Rather than offering a robust lease sale, this administration tried to subvert the law and pull millions of acres from the proposal for lease sale 261 due to unfounded concerns over the Rice's whale. Luckily, the Western District of Louisiana has intervened in which Judge James Cain stated was "a weaponization of the Endangered Species Act." After litigation and delay form this administration and the environmental activists, they teamed up with the courts and have directed the lease sale to move forward without the administration's ridiculous conditions by early November. I was proud to join Chairman Hunt and several of my republican colleagues in filing an Amicus brief in this case.

Mr. Chair, I ask unanimous consent to enter into the record a copy of this brief, as well as a copy of the District Judge Cain's September 21 order.

Chairman HUNT. Without objection.

Mr. STAUBER. Mr. Minarovic, can you share with the Committee the history of transactions for offshore properties and how smaller producers like Arena Energy came to own properties they are developing in the Gulf of Mexico today?

Mr. MINAROVIC. Thank you, sir. I appreciate the question. It is very important to this rule.

As major oil companies develop this basin in the '60s, '70s, and '80s, they installed thousands of platforms, 7,000 in total, drilled 50,000 wells in the Gulf of Mexico. At some point those assets deplete to a point that they are not important to those big companies.

And they have a choice to produce them and develop them or to abandon themselves. Most of them chose to sell them in the marketplace. When a company decides to sell, it knows that it stays in the chain of title. There is no question about that, so it is well-established in regs and law and in practice. No one is fighting that by the companies that sold. And so when they do that, they sell to a private company like ours or a small public company, they put in a certain amount of protection, some bonding, a total of about $3 billion of additional private bonds are in place. For instance, Arena has about 600 million of abandonment obligations and we have 400 million of private bonds. The rule that BOEM is proposing does not even account for that 400 million of bonds.

And then additionally, the reason those bonds are put in place is because the seller, the bigger company, recognizes that they have an ongoing chain of title liability for abandoning those assets. So there is a negotiation that takes place. The selling company decides do they want more insurance? Do they want more purchase price? Because the more insurance they want, the lower the purchase price. It goes back and forth. Ultimately, they decide what risk they want to take. The first thing they need to do is make sure they have a good buyer that is going to carry through on their obligations but that is how it happens.

Mr. STAUBER. So what is a workable solution in your eyes? One that protects the American taxpayers from decommissioning liabilities without putting an undue burden on offshore operators? What is your——

Mr. MINAROVIC. Well, it really falls out of what I just said on these transactions. And it goes back to that red box that I presented earlier. These are the properties that are really the taxpayer risk. We need to address those. We need to bond those. We need to make sure that the government is protected. We should not have taxpayer losses associated with those assets.

The SBA recognized this. And again, this is a Small Business Subcommittee meeting and SBA said the BOEM should narrowly tailor this rule to cover only those liabilities for which there is no predecessor leaseholders that BOEM considers creditworthy. That is it. That fixes the problem right there.

Mr. STAUBER. Thank you.

Mr. Danos, real quick. If lease sale 261 moved forward as the Biden administration and the environmental activists had hoped, what kind of impact would the conditions around the Rice's whale have on oil and gas companies' ability to operate in the Gulf of Mexico?

Mr. DANOS. It would be significant. There is a significant portion of the lease sale that was carved out. And again, the question is we are making policy related to one sighting of one whale. So let me say this about the Rice's whale. The oil and gas industry is in favor of protecting the Rice's whale. There is a known habitat area for the Rice's whale that has protection and we should maintain that. But the area that we are carving our or it is being carved out by the administration is an area where there is not good science that the whale is there and it is a huge impact.

Mr. STAUBER. Thank you very much, Mr. Chairman. My time is up and I yield.

Chairman HUNT. Thank you so much.

I now recognize Chairman Williams from Texas for 5 minutes.

Mr. WILLIAMS. Thank you, Mr. Chair. And I want to thank again the witnesses for being here today.

Full disclosure, I am a car dealer in Texas so I guess I can say I deal with oil and gas quite a bit.

The Gulf of Mexico produces the cleanest barrels of oil on the planet. Instead of acknowledging that, this administration, the Biden administration is more interested in virtual signaling and has placed burdensome regulations and attempted to restrict the sale of oil and gas leases in the Gulf.

Mr. Minarovic, you highlight the irony that the compliance costs, the administration rules, and the BOEM rule in particular has required your company's capital to be used for compliance as opposed to decommissioning efforts.

The question is, if regulation and compliance costs were eliminated could you elaborate on how your company would use this additional capital? Because we see this government regulating, regulating, regulating, taking capital away from what we can do to create jobs. So could you talk about that?

Mr. MINAROVIC. Yes, sir. Thank you, Chairman Williams. It is an important issue.

Obviously, every extra dollar that we spend on regulations—and there needs to be a level of regulations—but every extra dollar we spend on that is a dollar that we cannot invest in these properties, continue their life, and also, investment that we can make in decommissioning of these assets over time. And some of the issues that we got into over the last decade with 30 bankruptcies related to overregulation of those companies and their inability to perform in that environment.

Mr. WILLIAMS. One of the things that my colleague, Mr. Stauber likes to talk about is the Biden administration has come in and brought $75 billion worth of regulations. This is a real problem. This administration's attempts to rapidly do away with and punish the use of fossil fuel is harming Americans.

Mr. Epstein, you highlight the anti-fossil fuel agenda—reliable fossil fuels in states like California which is often under threat of rolling blackouts. So should the federal government adopt these policies at large? And what do you think will happen to the U.S. energy supply and our electric grid?

Mr. EPSTEIN. So should they adopt California energy policies? Oh, man. It is rough. I mean, you saw part of what happened in 2020 was we saw that other neighboring states were starting to import our policies but they did not realize we have a parasitical policy. So we shut down reliable power plants and then we depend on our neighbors. But the neighbors said, hey, we want to look good and we are going to do that, too. But then we had a heat wave, the wind died down, of course, and then we did not have enough power from our neighbors. So, no, it is a parasitical thing. And in general, anyone who claims to be net zero or anything like that is lying. And these are not things that can scale at all. So we need to recognize that fossil fuels are essential for decades to come, and if you want alternatives you need to do things like liberate nuclear. Liberate mining. I think Representative Stauber know, I mean, look at the Biden administration saying, hey, we have got an amazing mining property in Minnesota. Let's ban mining on that for 20 years and become even more dependent on China for crucial minerals. So we need energy freedom, not net zero.

Mr. WILLIAMS. All right. The Biden administration has prioritized regulations that hurt the United States energy security. That is what we are talking about.

Mr. Danos, in your testimony you highlight that the environmental stewardship and energy process are not mutually exclusive. So can you quickly elaborate on how you believe the Biden administration has missed the nark on this balancing act?

Mr. DANOS. Yes. Thank you, Mr. Williams.

Again, we have a long history in the Gulf of Mexico of regulations that are common sense and support good production of affordable, low cost energy, reliable energy. And we have worked that way together for a long time. And today what we are seeing is the president getting up and saying that he wants to end production on federal lands; right? And this maybe goes back to why production is higher today, and Mr. Golden, it was a great question. Much of the increase in production is coming from state lands. A lot of that is coming from the Permian Basis in Texas. So those are not federal lands. And a lot of these investment decisions were years back. Right? So what we need going forward is predictability and reliability so that those multimillion, even billion dollar investments that will create a higher production in the future and we have that certainty going forward and the regulatory environment is a big part of that.

Mr. WILLIAMS. In the time I have got left, Ms. Moncy, I am from Texas so I listened to your testimony. You have got quite an industry. Do you also in your industry run into the regulations, this massive amount of regulations the Biden administration is putting on Main Street America? I am in the car business. We see it in the oil industry. Do you see it, too, in your industry, the regulations?

Ms. NISBET MONCY. We have seen regulations change in our industry as well. Yes.

Mr. WILLIAMS. And it probably affects the cost of goods sold for you, too?

Ms. NISBET MONCY. We do have impacts associated with that.

Mr. WILLIAMS. Okay.

Mr. Chairman, I yield my time back. Thank you.

Chairman HUNT. Thank you so much. I think we are good here. We are not going to do another round if I am not mistaken. And I want to thank you all so much for being here.

I want to say this, Ms. Moncy, thank you so much for being here. I have learned a lot about your industry. I actually had no idea. This is why I love this job. And I have been on Mr. Minarovic's rigs. I have spoken to you, sir. We have done interviews together in the past, and I would love to come see your operation as well to learn more about it and to see exactly how we can all coincide and live together.

What is really important to us, and what is really important to me is, as you have heard from these gentlemen, it is an "all hands on deck" approach is what is necessary. We cannot restrict any electrons in the discussion. This is not up to the government to pick winners and losers. This is about supplying, not just the United States, not just Texas where I am from, not just California, but supplying the entire world with affordable, clean, and abundant energy from our country that does it better than anywhere else in the entire world. From somebody that spent 2 years in Saudi Arabia when I was deployed there, and they have an abundance of resources there, let me tell you. But I was also told that the Marcellus shale is actually known as the Saudi Arabia of natural gas. To put that in perspective, there is no such thing as an endless amount of fossil fuels but Saudi Arabia is pretty doggone close to it. And the fact that we are sitting on the Marcellus shale, part of it runs into New York, and we cannot drill for any of it is actually ridiculous. We as a country have got to get out of our own echo chamber and we forget sometimes and in this country we have champagne problems. I can tell you that China and India, Taylor Swift is pretty popular right now. Oh, Tay Tay. I feel sorry for Travis right now. But we do. And if you look at what is happening in China, India, and Russia, a lot of these countries of the Middle East, OPEC countries, Iran, they really do not care about the world's climate at all. I am a military guy. I have seen it. This country cares immensely. And while we should continue to lead from the front in this, we also have got to realize that we are the bridge. Oil and gas, natural gas, energy, we are the bridge and it is everything.

I visited almost every single energy company in my district, and there are a lot of them. And I can tell you I have never heard a single executive say anything disparaging about renewable energy. I have heard nothing but actually I want to reinvest in it. Let's talk more about nuclear. Let's talk more about hydrogen. I have heard nothing but those things. And I can attest to that. It is genuine. It is real. It is not just corporate, big oil trying to spin off their ideas for world dominance because they really want to keep this going forward. These companies are actually literally investing in these renewable energy sources because I have seen their books. And I have seen what they are trying to actively do. I just want to make sure that we are smart about it, that we are savvy about it, and we do it in a way where I can continue to eat oysters because I love oysters. I had some last night and I am not going to stop eating them. And no, I do not want to ruin your business and your ecosystem for your small business. And also, sir, I want you to continue to strive to provide reliable and clean energy for our country.

So with that said, thank you all so much for being here. Greatly appreciate it. I think at some point we will have you back. To me this is the most important issue that we are facing in our world today.

And without objection, Members have 5 days to submit additional materials and written questions for the witnesses to the Chair which will be forwarded to the witness. I ask the witness to please respond promptly.

If there is no further business, without objection the Committee is adjourned. And thank you so much.

[Whereupon, at 11:18 a.m., the subcommittee was adjourned.]

Appendix

Written Testimony of Mike Minarovic Chief Executive Officer Arena Energy, LLC
Before the U.S. House of Representatives Committee on Small Business Rural Development, Energy & Supply Chains Subcommittee
Energy Independence: How Burdensome Regulations Are Crushing Offshore Small Energy Producers

September 28, 2023

Chairman Hunt, Ranking Member Perez, and Members and staff of the Committee and Subcommittee, thank you for the opportunity to testify this morning. My name is Mike Minarovic, and I am the Co-Founder and CEO of Arena Energy, an employee-owned independent exploration and production company exclusively focused on offshore oil and natural gas development in the shallow waters of the U.S. Gulf of Mexico (the "GOM").

Arena Energy

Arena is one of the largest independent operators in the GOM. Every year, we invest drilling and facilities capital of $250-300 million in the federal waters of the Gulf of Mexico. Arena currently owns 130 platforms on federal leases and is the second most active oil and gas company in the GOM based on wells drilled over the last ten years. Since inception, Arena has safely drilled more than 350 wells in federal waters with a commercial success rate exceeding ninety-four percent, contributing to over $1.4 billion in royalties paid to the U.S. government. Equally important, Arena has decommissioned over three hundred wells and forty-five platforms and has funded its decommissioning costs with not a single dollar falling to the U.S. taxpayer.

Despite the challenging regulatory environment over the past several years, we have expanded our commitment to the U.S. Gulf of Mexico by capitalizing new companies to provide drilling rigs, pipeline transportation, and environmental remediation services to the offshore industry. Arena made these investments to ensure safer operations, mitigate pollution risk, and to provide needed capacity to decommission wells, pipelines, and platforms in the GOM.

The Gulf Energy Alliance

I also speak today on behalf of the Gulf Energy Alliance, a coalition of leading independent offshore producers formed in 2016 to work with regulators, elected officials, and other stakeholders to develop a reasonable framework for financial assurance requirements that protect the U.S. taxpayer in a manner that does not threaten the viability of independent offshore producers. All GEA members, like Arena, are small businesses almost exclusively focused in the GOM. Arena and other independent offshore producers are not household names and have distinctly different business models than the major oil and gas companies. But collectively, independent producers were responsible for approximately 35% of Outer Continental Shelf oil and natural gas production in 2022.[1]

[1] Office of Natural Resources Revenue; Bureau of Safety and Environmental Enforcement (2023).

I. The Biden Administration Aims to Shut Down Offshore Oil and Natural Gas Production

There are many examples I could cite today to illustrate the burdensome and unnecessary regulations crushing small businesses operating in the GOM. And every single one of those examples must viewed through the lens of a President intent on shutting down domestic oil and natural gas production when and wherever possible.

In a recent interview, President Joe Biden reiterated his campaign pledge to do so:

> I wanted to stop all the drilling on the east coast, and the west coast, and in the Gulf, but I lost in court. *But, we're still pushing, we're still pushing really, very hard.*[2]

On his first day in office, President Biden issued an Executive Order pausing permits for new oil and gas drilling on federal lands and waters, and the cascade of overly burdensome executive actions and new regulations targeting domestic production has accelerated over the past year. The administration's hostility to domestic production, particularly in federal waters, is difficult to comprehend, particularly when you consider the administration's stated goal of reducing global greenhouse gas emissions. As global oil demand continues to grow, the cleanest barrels on the planet should be produced first, and those barrels are produced in the GOM.[3]

I am here today to discuss a new regulation recently proposed by the Bureau of Ocean Energy Management (BOEM) that poses an existential threat to small offshore businesses. Specifically, BOEM recently proposed new regulations (the "Proposed Rule") related to bonding requirements for offshore decommissioning liabilities that, very simply, cannot be met.[4] BOEM's rule relies entirely on the surety industry as a "solution" to a massively overstated problem— i.e., taxpayer exposure to these liabilities— and the surety market has responded that BOEM's "solution" will not work.

---

[2] President Joe Biden. Television interview. *The Weather Channel*, August 8, 2023 (https://weather.com/news/climate/news/2023-08-09-president-biden-the-weather-channel-climate-comments).

[3] GHG Emission Intensity of Crude Oil and Condensate Production, ICF (May 8, 2023).

[4] *Risk Management and Financial Assurance for OCS Lease and Grant Obligations* (June 29, 2023)(RIN 1010-AE14) (to be enacted at 30 CFR Parts 550, 556, and 590) ("BOEM Proposed Rule").

Notably, the U.S. Small Business Administration's Office of Advocacy (the "SBA") has also weighed in on this rulemaking, concluding that the proposal explicitly favors major oil and gas companies over small businesses without any corresponding benefit for taxpayers.[5]

Proposing a rule with a solution that is irrational and unworkable—and unjustifiably harms small business—is not in the best interests of the government, the industry, or taxpayers. As the State of Louisiana concluded, "the Proposed Rule is fundamentally flawed in multiple ways, but what makes it uniquely troubling is its obvious insincerity. While the Proposed Rule reflects a valiant attempt at facial neutrality, it cannot be read without considering the Biden Administration's pervasive animosity toward domestic oil production."[6]

II. BOEM's Proposed Rule is a Massive Overreach to a Limited Problem

Arena and the GEA support policies and regulations that provide workable solutions for real problems. BOEM's Proposed Rule focuses on a potential problem—protecting taxpayers from financial liability for the costs of decommissioning offshore oil and natural gas wells, platforms, and related infrastructure in federal waters of the GOM, which BOEM acknowledges as a "rare" occurrence."[7] Indeed, BOEM recently disclosed that potential taxpayer liability for decommissioning after seventy-five years of offshore production amounts to $58 million.[8]

To address the $58 million in potential taxpayer exposure, BOEM now seeks an additional $9.2 billion in new bonding, which will cost small

---

[5] U.S. Small Business Administration Office of Advocacy Comment Letter ("SBA Comment Letter"), https://www.regulations.gov/comment/BOEM-2023-0027-1699, at p. 3.

[6] State of Louisiana Comment Letter, https://www.regulations.gov/comment/BOEM-2023-0027-1985, at p. 4.

[7] 88 Fed. Reg. at 42141.

[8] On July 26, 2023, BOEM communicated to the House Subcommittee on Energy & Mineral Resources that the estimated decommissioning costs for current orphan well and infrastructureinventory on the OCS is approximately $73 Million with an estimated $25 Million available for decommissioning activities, which includes approximately $15 Million BOEM has collected in forfeited bonds and BSEE's $6 Million in appropriated funds and $4 Million in BipartisanInfrastructure Law funding.

businesses in the GOM $5.7 billion.[9] "Overkill" barely begins to describe BOEM's "solution"; this is cracking a nut with a sledgehammer.

Since 2004, there have been more than thirty-two bankruptcies of offshore oil and gas lessees involving decommissioning liabilities of approximately $17 billion—and less than 0.5% of that total might fall to U.S. taxpayers. Contrast that with the financial benefits to U.S. taxpayers resulting from GOM oil and gas development over the same period. Since 2004 offshore oil and natural gas production has generated over $124 billion in royalties, lease bonuses, and rentals for the U.S. Treasury.

A. The Proposed Rule Explicitly Favors Major Oil and Gas Companies Over Small Businesses

Offshore oil and gas operations began in the Gulf of Mexico in 1947.[10] Offshore drilling at that time was exclusively conducted by major oil and gas companies. From the late 1940s into the 1990s, major oil and gas companies installed almost all the platforms, pipelines, and facilities in the GOM, creating the decommissioning obligations at issue in BOEM's Proposed Rule.

Beginning in the late 1980s, after recovering most of the oil and natural gas and taking considerable profits along the way, major oil and gas companies sold their shallow water properties to smaller, independent producers to chase bigger returns in more lucrative fields in the deepwater GOM and elsewhere.[11] And in every single one of those transactions, the major oil and gas companies—among the most sophisticated buyers and sellers of assets on the planet— had absolute clarity as to their continuing liability for their accrued decommissioning obligations as defined by statute, regulation, common law, and the lease forms.[12] In other words, when selling

---

[9] This is according to BOEM's own analysis, which is an undiscounted number and includes erroneous assumptions about the cost of capital and collateral requirements for the new bonds.Correcting for BOEM's flawed assumptions, the actual costs to small businesses is closer to $870million annually.

[10] Kerr-McGee installed the first offshore platform that was out of sight of land in 1947.

[11] In many of these sales, major oil and gas companies required bonding, trust accounts, or other forms of security in case the successor owner(s) defaulted on its decommissioning obligations. In many other sales, the majors made the commercial decision to maximize the sales price and did not require the purchasers to provide security to cover their potential default on decommissioning liabilities.

[12] *See* 30 CFR § 250.1703 (joint and several liability for all lessees and grant holders); 30 CFR § 556.710 (joint and several liability for all predecessors) ("If you assign your record title interest,as an assignor you remain liable for all obligations, monetary and non-monetary, that accrued in connection with your lease during the period in which you owned the record title..."). See also,30 CFR § 250.1702 (when decommissioning obligations accrue). Indeed, in the largest offshorebankruptcy, Fieldwood Energy LLC (Case No. 20-33948) (S.D.

these properties, major oil and gas companies made risk-adjusted commercial decisions to balance cash proceeds versus protection against a potential default by immediate and subsequent purchasers of these assets.

As the State of Louisiana and the SBA point out in their respective comment letters, the Proposed Rule completely ignores a bedrock principle of offshore production: By law, regulation and practice, all current and previous owners stand in front of the taxpayer and remain collectively responsible for decommissioning obligations. Selling a property in no way relieves an owner of its continuing decommissioning obligations under the law, the terms of its lease, and its contractual obligations to buyers. But solely for the purposes of this rulemaking, BOEM chooses to ignore this foundational principle.[13]

As the SBA notes:

> BOEM's proposed rule distorts the marketplace by ignoring a key element of the leases, joint and several liability. All participants in the market know and understand this liability in their business dealings. … This proposed rule explicitly favors large businesses over small businesses without providing the taxpayer theintended protection from unfunded liabilities.[14]

Even though the existing regulatory regime has consistently protected the American taxpayer, major oil and gas companies have seized upon BOEM's rulemaking as an opportunity to retroactively re-trade past transactions. That would get things exactly backwards. The singularfocus of this rulemaking should be protecting the American taxpayer, and not major oil and gas companies, who reaped the substantial profits of exploiting offshore properties before selling these properties to independents, often at substantial additional profits.

B. Compliance Costs Under the Proposed Rule Will be Borne Almost Entirely by Small Businesses

BOEM concedes that small businesses will bear the staggering costs imposed by the new bonding requirements:

Texas), involving $9 billion (government's estimate) of decommissioning liability, predecessors assumed 100% of the decommissioning liability, resulting in zero exposure to the taxpayer. Instead, abandonment liability devolved to predecessors in the chain of title who were aware of that risk when they sold the assets to Fieldwood.

[13] 88 Fed. Reg. at 42141.

[14] SBA Comment Letter at p. 3.

> Based on these criteria, approximately 407 (76 percent) of the businesses operating on the OCS subject to this proposed rule are considered small; the remaining businesses are considered large entities. All of the operating businesses meeting the SBA 'small business' classification is potentially impacted; therefore, BOEM expects that the proposed rule would affect a substantial numberof small entities.[15]

Moreover, the SBA explained

> ...BOEM's analysis intentionally ignores joint and several liability as the way that the taxpayers are protected from these unfunded liabilities. It presumes that thesesmall businesses impose a significant risk to the taxpayer despite the full backingof companies that BOEM has exempted. *As a result, small businesses are not only disproportionately harmed by the proposal, but only small businesses are harmed by the proposal.*[16]

The SBA raised several additional concerns in its comments to the Proposed Rule, including how the rule will relieve sellers of the need to perform due diligence on the buyer, encouraging—not discouraging— "moral hazard" in transactions:

> This proposed rule imposes an unreasonable burden for financial assurances on small businesses. BOEM should include a waiver from this requirement for leaseswith credit worthy predecessor leaseholders. Small businesses should not be required to stand alone and assume the full responsibility for making the taxpayer whole, especially where large businesses are equally liable under DOI regulationsand the terms of the lease.[17]

Consequently, the SBA urged BOEM to withdraw the Proposed Rule in favor of a solution that would require supplemental bonding only by the first leaseholder and allow the market to allocate risk through private security required in transactions, which would acknowledge that large companies remain contingently liable for making the taxpayer whole under the regulations and the terms of the lease.

---

[15] 88 Fed. Reg. at 42157.
[16] SBA Comment Letter at p. 4.
[17] *Id.* at p. 5.

III. The Proposed Rule is Devastating for Small Businesses on the Gulf Coast

A. The Cost of the Proposed Rule Vastly Exceeds Any Potential Benefit

The Proposed Rule claims that the total decommissioning liabilities are approximately $42.8 billion.[18] This exposure is overstated.[19] According to BOEM, the total decommissioning liabilities associated with properties in which majors or large independents (representing $1.8 trillion in combined market capitalization) are not part of the current ownership or previous chain-of-title is only $750 million, which represents a small fraction of the total decommissioning liability in the GOM.

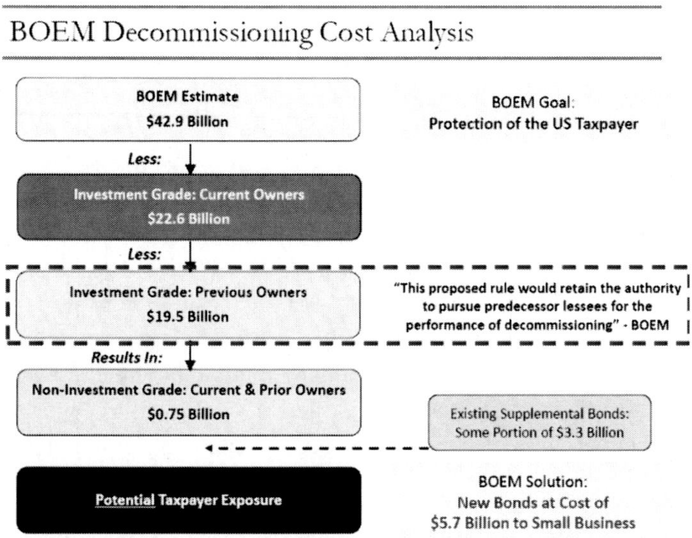

And the total decommissioning liability is rapidly decreasing. As the below diagram illustrates, of the seven thousand platforms installed in the GOM over the past seventy-five years, just over 1,500 remain. Over the past fifteen years, industry has removed structures at a rate of 168 per year and has plugged and abandoned approximately 735 wells per year over the same period. Independent producers have performed most of the decommissioning.

---

[18] 88 Fed. Reg. at 42137.

[19] *A Cost-Benefit Analysis of Increased OCS Bonding*, Opportune LLP (July 2023) (the "OpportuneStudy"), Stated differently, 93% of the total decommissioning liability has a major oil and gas orlarge independent in the chain of title and who jointly and severally liable with the current owners.

Gulf of Mexico Infrastructure: 1947 - 2023

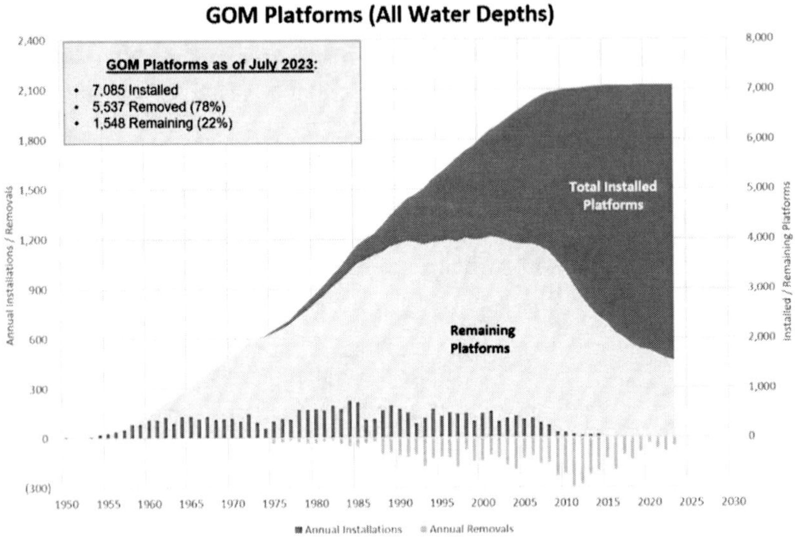

Faced with the enormous costs of complying with the Proposed Rule, independent oil and gas companies will have no choice but to direct capital away from ongoing decommissioning efforts, increasing the potential risk to the government and taxpayer. As for independents who are financially unable to provide the costly supplemental bonding required under the Proposed Rule, the Surety and Fidelity Association of America, a national trade association whose members provide most bonds that secure regulatory obligations, stated

> It is unknown whether the oil and gas operators will retain sufficient ability to provide such collateral to the sureties in support of the new and existing obligations. Given that "less" (relatively speaking) creditworthy lessees will necessarily have restricted access to capital and would be the ones subject to theproposed financial assurance requirements, having their capital tied up in collateral obligations will present a substantial burden on these operators and likely lead to more financial failures. Such financial failures will have cascading negative impacts that undermine BOEM's stated purpose of increasing access to these bonds and alleviating the burden on the taxpayer.[20]

---

[20] Surety and Fidelity Association of America Comment Letter ("SFAA Comment Letter"), https://www.regulations.gov/comment/BOEM-2023-0027-1998, at p. 3. SFAA is a non-profit, If finalized, this Proposed Rule will exacerbate the very "problem" it seeks to

B. Taxpayers and Small Businesses Pay a Steep Price for the Proposed Rule's Limited Protection

The fallout of the Proposed Rule will not be limited to the small businesses singularly burdened with enormous and unjustified compliance costs; it will also harm American taxpayers. BOEM readily acknowledges these consequences:

> This action, which is a significant regulatory action under Executive Order 12866 is likely to have a *significant effect on the supply, distribution, and use of energy.*
>
> * * * * *
>
> BOEM recognizes that this action may 'adversely affect' *in a material way the productivity, competition, or prices in the energy sector.*' By increasing industry compliance costs, the regulations could adversely make the U.S. offshore oil and gas sector less attractive than regions with lower operating costs. *Additionally, increased costs may depress the value of offshore assets or cause continuing production to become uneconomic sooner, leading to shorter-than-otherwise useful life and potentially a loss of production.*[21]

An informed evaluation of any proposed regulation is not possible without a fulsome cost- benefit analysis. If BOEM had conducted an adequate and accurate cost-benefit analysis for the Proposed Rule, it is incomprehensible that BOEM could have concluded that the Proposed Rule is justified.

Opportune LLP, a leading global energy advisory firm, conducted its own cost-benefit analysis of Proposed Rule and concluded that it would, if implemented:

- Result in a decrease in production of approximately fifty-five million barrels of oil equivalent from the GOM;
- Destroy 36,000 high paying jobs, mostly in disadvantaged areas;

national trade association of more than four hundred companies that write 98 percent of surety and fidelity bonds in the United States. SFAA is licensed

resolve. That is, it will accelerate the default of current lease owners in the GOM, increasing risk to taxpayers and destroying small businesses all along the Gulf Coast in its wake.

[21] 88 Fed. Reg. at 42168 (emphasis added).

as a rating or advisory organization in all states and is designated by state insurance departments as a statistical agent for the reporting of surety and fidelity experience. SFAA members provide most bonds that secure regulatory obligations.

- Forfeit $573 million in royalties to the U.S. Treasury;
- Reduce Gross Domestic Product (GDP) (particularly in the Gulf Coast states) by as much as $9.9 billion.[22]

In addition, finalizing the Proposed Rule will (1) increase overall emissions as the U.S. replaces the cleanest barrels in the world produced in the GOM with foreign, dirtier barrels, (2) reduce competition in the offshore oil and gas industry by forcing out smaller independents, and (3) weaken an already tenuous supply chain supporting the offshore oil and gas industry, (shipyards, steel manufacturers, rig owners, equipment manufacturers, construction workers and welders, seaman and vessel providers, aviation contractors, etc.), which could permanently harm the industry's ability to continue production and support future decommissioning activities.[23]

IV. BOEM's Proposed "Solution" is Not Workable, as $9.2 Billion of AdditionalSurety Market Capacity for Bonds Does Not Exist

Despite repeated inquiries from the surety industry, independent offshore producers, and other stakeholders, BOEM has failed to answer a critical threshold question underpinning the rulemaking: Is it the government's intent to treat these new bonds as "the last dollar out" protecting only the taxpayer, or is it the government's intent to transfer risk from predecessors to the surety industry?

As the SFAA noted in its comment letter:

> BOEM is silent as to how and when the required financial assurance will be calledupon. Without knowing the order of attribution of when the financial assurance isrequired to be utilized to fund decommissioning activity it provides an extreme amount of uncertainty to [sureties] being

[22] *See* Opportune Study at p. 7.

[23] For example, in 2014, there were thirty-six active jack-up drilling rigs in the GOM; today, there are only six active drilling rigs. *See* IHS Petrodata.

> asked to provide the financial assuranceon behalf of lessees...This is critically important for the sureties to understand when underwriting lease operators and the Gulf of Mexico lease marketplace generally, specifically whether to continue to write in this space or how to change the terms under which they will continue to write bonds for lease operators.[24]

The answer to this question determines whether (1) a surety market exists for BOEM's "solution," (2) whether the supplemental bonds will priced in a manner that does not create an undue burden, (3) whether the sureties providing the supplemental bonds will require cash collateral to provide new capacity, (4) whether there is a capital market that is willing to provide that funding for cash collateral, and (5) ultimately, who will benefit from the new bonding: major oil and gas companies or American taxpayers.

CAC Specialty, a broker that places approximately $1.5 billion of energy surety bonds, noted in its comment letter, "[t]his massive bond increase the BOEM seeks comes on the heels of some of the largest OCS related surety losses in history. Markets have withdrawn, capacity is low, reinsurance expenses and losses have driven up rates."[25]

If the new supplemental bonds are intended to protect major oil and gas companies from their joint and several liabilities, then there will be little to no capacity in the surety market. We understand that the current offshore surety market guarantees approximately $5 billion of obligations and that recent losses exceeding $1.9 billion associated with two bankruptcies have crippled the industry's ability to support current (and new) risks for the benefit of predecessors in title.[26]

CAC Specialty noted that "only about a dozen carriers remain, which means that, if distributed evenly, each carrier would take on an average of $750 million in new bonds, a potentially futile exercise."[27] The SFAA echoed this sentiment, stating that "it is unlikely that whatever appetite would exist [for these new bonds] will be in the aggregate amounts contemplated in the proposed rule, making commercial implementation of the Proposed Regulations unlikely through surety bonds alone."[28]

---

[24] SFAA Comment Letter at p. 7.
[25] CAC Specialty Comment Letter, https://www.regulations.gov/comment/BOEM-2023-0027-1201, at p. 2.
[26] *See* CAC Specialty's Comment Letter.
[27] *Id.* at p. 2.
[28] SFAA Comment Letter at p. 2.

What is more, the more financially sound independent operators will be impaired by the substantial cost of complying with the Proposed Rule, which as written does not allow operators to receive credit for previously posted private-party security. As the SBA pointed out, "[t]he proposed rule would require small businesses to pay twice to protect against decommissioning liabilities, once through the sales contract and again to the federal government."[29]

In other words, the Proposed Rule is unimplementable if BOEM intends for the new supplemental bonding to transfer risks from predecessors to sureties given the recent losses sustained by the surety market. Moreover, if BOEM does not clearly state its intent for whom the new bonds will benefit, sureties must assume that the new bonds may protect predecessors, and capacity will not materialize.

V. Congressional Oversight is Needed to Ensure that the Department of InteriorArrives at a Workable Solution to This Limited Problem

BOEM's Proposed Rule seeks to impose a $6 billion "solution" to a $58 million "problem." This is equivalent to "spend[ing] a dollar to save less than a nickel."[30] The ostensible provider of BOEM's "solution"—the surety industry—has made clear that there is no capacity in the surety market to support the Proposed Rule. At a more fundamental level, the fact that this rulemaking directly targets the viability of small businesses in the GOM necessitates a closer look at the true intent underpinning this rulemaking exercise. There is a path forward on this issue, but Congressional oversight is necessary to ensure BOEM's solution is commensurate with the actual risk to taxpayers.

Arena and the GEA have expended considerable effort and resources seeking resolution of this issue for over a decade, and we remain committed to working with BOEM and other stakeholders to develop a financial assurance framework focused on actual taxpayer risk. The offshore industry sorely needs finality on these regulations. But instead of protecting major oil and gas companies by ignoring security already in place and re-trading private, commercial transactions between buyers and sellers, any bonding framework

[29] SBA Comment Letter at p. 3.

[30] State of Louisiana Comment Letter at p. 2.

must recognize the joint and several liability of all current and former owners for decommissioning as a fundamental and grounding principle, which would be consistent with BOEM's actual practice and the stated intent of the Proposed Rule.[31]

I certainly recognize that there are many other regulations I could speak to today in the context of this hearing—whether already in place, proposed, or anticipated—negatively impacting small businesses in the GOM. But from my perspective, there is no issue of greater importance to small businesses in the GOM than this one.

Thank you for the opportunity to testify this morning and I look forward to your questions.

Testimony of Paul Danos
Hearing of the Subcommittee on Rural Development, Energy, and Supply Chains
House Committee on Small Business
"Energy Independence: How Burdensome Regulations Are Crushing Small Offshore Energy Producers"
Owner, President, and CEO, Danos Company
September 28, 2023

Chairman Hunt, Ranking Member Gluesenkamp, and members of the Committee, thank you for inviting me to testify today. My name is Paul Danos, and I am Owner, President, and CEO of Danos, a 76-year-old, family-owned company out of Houma, Louisiana, that provides labor and project services for energy companies throughout the country. I am also the current chairman of the National Ocean Industries Association, or NOIA. For more than 50 years, NOIA has represented the interests of all segments of the offshore energy industry, including offshore oil and gas, offshore wind, offshore minerals, and offshore carbon sequestration. The membership of NOIA includes energy project leaseholders and developers and the entire supply chain of companies – like Danos – that make up an innovative energy system contributing to the safe and responsible exploration, development, and production of energy for the American people.

---

[31] The Proposed Rule states that "[t]his proposed rule would retain the authority to pursue predecessor lessees for the performance of decommissioning." 88 Fed. Reg. at 42141.

In 1947, my grandfather started Danos as a small tugboat company supporting Gulf of Mexico oil and gas operations while conducting business from my grandmother's kitchen table. Today, Danos has grown into a global presence and a trusted strategic partner for energy project developers around the globe.

Danos offers onshore and offshore customers an extensive range of integrated services, including workforce, construction, fabrication, project management, supply chain, shorebase and logistics, mechanical maintenance, coastal restoration, power generation, and various other services. Danos has 2,700 employees, with nine offices in Louisiana and Texas and operations across the Gulf of Mexico and North American shale plays. Looking forward, we plan to continue adding service lines, expanding into renewable energy, and increasing operations to meet the needs of our customers. Throughout the past seven decades, our company has maintained an unfaltering commitment to values, safety, and overall results for our clients. This is the ethos we will continue to embrace as we rise to meet the energy needs of tomorrow.

A vital part of Danos' business profile has been the ability of our company to diversify our work in the energy sector and positively contribute to communities through environmental stewardship. Our environmental stewardship includes efforts taken to sustain and preserve the natural environment as well as actions to reduce or mitigate impacts to the environment.

Our commitment to preserving and protecting the environment is a natural extension of our company purpose, "to solve big challenges for our customers and communities." We have a great opportunity to leverage our expertise and competencies to help customers meet today's energy demands without compromising the environment for tomorrow. We aim to reduce our carbon footprint while supporting customers as they strive to meet global environmental standards. We have identified three key pillars of focus regarding the environment:

- Protect, preserve, and restore the natural environment.
- Reduce the environmental impact of Danos' operations.
- Support renewable energy business opportunities.

Danos is leading the way in wetlands and coastal environmental restoration and protection through several key initiatives:

- Danos has embarked on a collaborative partnership with 3D-printing technology company Natrx to positively impact coastal resiliency and restoration by designing, manufacturing, and installing innovative nature-based infrastructure solutions that reduce carbon emissions. The revolutionary process cuts material usage by up to 70%, generates up to 300% higher protective biomass, and increases habitat by 650% per linear foot of infrastructure vs. rock or solid concrete structure. Our first joint project involved the placement of "Cajun Coral," an innovative 3D-printed infrastructure, to establish a new reef in Catfish Lake, part of the Golden Meadow marshland area. The installation has provided more substantial protection from erosion for this vital coastal wetland and home to a growing population of oysters and other sea life.
- To that end, Danos and Natrx have signed a letter of intent to partner on coastal issues, and we have continued to work on projects deploying nature-based technologies. Danos has exclusive rights to manufacture Natrx's proprietary Oysterbreak and ExoForm technology across the Gulf South region. Through this partnership, we are able to execute projects for coastal restoration, artificial reef creation, and pipeline protection.

Danos actively participates in various wetland conservation groups, including Partnership for our Working Coast (POWC), Coalition to Restore Coastal Louisiana, Coastal Conservation Association, Louisiana Sea Grant, Restore or Retreat, and the LSU College of the Coast and Environment. As a member of POWC, an alliance of industry and environmental entities led by the Water Institute of the Gulf, Danos supports efforts to protect vital infrastructure in Port Fourchon. With planned improvements to the Port expected to produce millions of cubic yards of dredged materials, POWC has identified the most beneficial ways to use this material to contribute to Louisiana's coastal sustainability efforts, protect coastal communities, and support America's Working Coast.

The conversation around energy often focuses on the major oil and gas companies, but it is important to recognize and appreciate that it is companies like Danos that do much of the work and employ a substantial portion of the workforce that develops U.S. energy projects. For instance, there are thousands of companies and hundreds of thousands of U.S. workers that support oil and gas production out of the U.S. Gulf of Mexico. So, when it comes to energy policy, decisions in Washington have a massive impact on the employees of Danos in Louisiana, throughout the Gulf Coast, and in communities all across America. I can also personally attest that companies throughout the energy supply chain are taking tangible steps to reduce the environmental and emissions impact of operations.

The offshore energy sector is a proven leader in solving energy challenges and delivering diverse sources of energy to the global economy. The offshore industry brings together the companies that produce foundational energy sources such as oil and gas, while leading innovation and investment in energy sources and technologies that will drive decarbonization efforts well into the future. The offshore energy sector has unparalleled expertise and experience deploying and scaling technologies at levels necessary to achieve decarbonization objectives.

Companies throughout the offshore industry continue to lead the way in innovating low emission solutions that include offshore wind, carbon capture and storage, hydrogen, and geothermal, among others.

For the foreseeable future, the offshore industry will play an integral role in shaping an energy system that promotes the production of affordable and reliable energy while continuing to reduce environmental impacts, including emissions. Importantly, for the coming decades, oil and gas supplies will remain a vital energy source for Americans and our allies around the globe, while we simultaneously integrate and add low carbon sources into the mix.

Energy production in my backyard – the U.S. Gulf of Mexico – demonstrates that it is possible to develop offshore resources while adhering to the highest safety and environmental standards. A multitude of companies involved in offshore energy development are working collaboratively to shrink an already small carbon footprint. From electrifying operations to deploying innovative solutions that reduce the size, weight, and part count of offshore infrastructure – thus increasing safety and decreasing emissions – the U.S. Gulf of Mexico hosts a high-tech revolution.

Currently, global oil consumption is approximately 100 million barrels per day. Various scenarios forecast global oil consumption volumes through 2050 and beyond, and nearly all of them predict similarly high levels of oil production will be necessary through at least 2050. The facts, data, and our experience make clear that we should focus on the U.S. offshore region, and the Gulf of Mexico in particular, for securing those vital resources.

Oil produced from the U.S. Gulf of Mexico has a carbon intensity one-half that of other producing regions.[32]1 The technologies used in deepwater production – which represents 92 percent of the oil produced in the U.S. Gulf of Mexico – place this region among the lowest carbon intensity oil-producing regions in the world.[33]2 Moreover, a recent study by ICF International, and commissioned by NOIA, found that that U.S. Gulf of Mexico has a carbon intensity 46% lower than the global average outside of the U.S. and Canada, outperforming other nations like Russia, China, Brazil, Iran, Iraq, and Nigeria.[34]

Policies that restrict domestic offshore development require imports to make up the shortfall, and that supplemental production comes from higher-emitting operations in other countries. Foreign providers generally employ less environmentally conscientious production methods,[35] which when combined with the added emissions from transporting oil over great distances by tanker, increases the amount of carbon released into the atmosphere rather than decreasing it.

Emissions reduction is a global challenge. As analysts at Wood Mackenzie explain, "Removing or handicapping a low emitter hurts the collective global average."[36] Removing a proven, stable supplier such as the U.S. Gulf of Mexico would be a poor choice with devasting consequences. The better choice is to institute government policies that promote cleaner and safer domestic production, less reliance on higher-emitting foreign suppliers like Russia and China, and the preservation of hundreds of thousands of American jobs.

Efforts to restrict U.S. energy development could eventually lead to Americans of every walk of life having to contend with the issues Europe has been experiencing as a result of disrupted supply from Russia, including potential industrial curtailment and families having to make difficult choices between heat and food. Our energy reality makes it clear that U.S. energy policy should support U.S. energy production of all types, including offshore oil and gas and wind. Government policies play a substantial role in the ability to develop energy in the U.S., whether onshore or offshore, and whether the

---

32 Motiwala, and Ismail, "Statistical Study of Carbon Intensities in the GOM and PB," ChemRxiv, April 13, 2020.

33 https://www.woodmac.com/news/the-challenge-of-negative-emissions/.

34 https://www.noia.org/new-report-u-s-gulf-of-mexico-oil-gas-production-leads-with-lower-emissions-including- methane/.

35 https://epi.yale.edu/epi-results/2022/component/epi.

36 https://www.woodmac.com/news/opinion/could-restricting-oil-production-in-the-us-gulf-of-mexico-lead-to- carbon-leakage/.

energy source is oil and gas, wind, hydrogen, or another resource. Obstructive government policies inevitably lead to adverse consequences for our energy security, national security, economic security, and decarbonization efforts.

We are fortunate in the United States that our Gulf of Mexico region is up to the task of delivering the oil and gas the economy needs. Production numbers from the U.S. Gulf of Mexico place it in the company of some of the largest oil producing countries. If the Gulf of Mexico were its own country, it would be one of the top eleven oil producing countries:

Source: U.S. Energy Information Administration.

Offshore energy is truly a story of accomplishing more with less – creating more energy with less environmental impact. Offshore production platforms are incredible edifices of continuously evolving technology that allow enormous amounts of energy to be produced through a relatively small footprint. Incredibly, 18 deepwater facilities, which equate to about the size of only nine city blocks, produce about the same amount of oil as the entire state of North Dakota.[37]

---

[37] Director Scott Angelle, BSEE Director, BSEE Presentation to the Deepwater Technical Symposium, November 13, 2020.

From a regulatory standpoint, federal government policy must serve to eliminate potential roadblocks to investment in energy projects, including offshore wind. The recent debt ceiling agreement included important changes that will hopefully help streamline the permitting process. The National Environmental Policy Act (NEPA) is a bedrock law for guiding the federal decision-making process with due consideration of the potential environmental impacts.

However, as with any rule or regulation, it is important that we take the time to review and improve rules and regulations as necessary to promote efficiency and effectiveness in regulation. The inclusion of many aspects of Congressman Graves BUILDER Act in the debt ceiling agreement was a very positive step toward streamlining the NEPA process. We remain hopeful that Congress will continue to work together to refine and improve all aspects of permitting. In this vein, there are currently multiple legislative efforts, primarily within Chairman Bruce Westerman's House Natural Resources Committee and Chairman Joe Manchin's Senate Energy and Natural Resources Committee, that, if passed, would take significant and salient steps to streamline NEPA and related permitting processes, and provide a true statutory foundation for many of the critical energy development processes that the nation relies on, like offshore federal leasing.

We also remain extremely concerned about potential delays to investment in American energy projects as a result of the actions of the Administration. As the Administration reviews and reworks regulations and energy programs, it will be important to ensure changes to the regulatory framework are conducted in a way that promotes the development of all forms of American energy. Environmental stewardship and energy progress are not mutually exclusive; for example, Danos and members of NOIA have consistently been leaders in both arenas. Promulgating rules that balance the need for energy development with effective environmental stewardship will provide the certainty these massive investments require.

In order to fully unleash American energy potential, it is vital that federal policy promotes consistency and predictability in leasing, permitting, and regulation. In an unprecedented fashion, the Administration has paused and delayed offshore oil and gas leasing and has failed to timely develop a new leasing program for U.S. federal waters, putting into jeopardy U.S. energy production, major capital investments, and thousands of jobs.

Since its inception, offshore oil and gas production has created hundreds of thousands of jobs and generated billions in royalties for the U.S. Treasury, boosting our nation's energy independence and national security – all while yielding approximately half of the carbon intensity per barrel of other producers worldwide. The offshore industry has also worked with the federal government and conservation partners, such as the Coastal Conservation Association (CCA), to collaborate on innovative efforts like the Rigs-to-Reef program, which repurposes obsolete platforms into habitats for marine life and further helps create a national recreational fishing economy. Additionally, legislation and programs like the Great American Outdoors Act, the Gulf of Mexico Energy Security Act (GOMESA), and the Land and Water Conservation Fund ensure that billions of dollars from federal offshore oil and gas leasing are dedicated to long-term coastal conservation and restoration, environmental protection, and urban recreation programs. Without continued reliable offshore oil and gas leasing this funding is at risk.

The employees of companies like Danos, throughout the Gulf Coast and beyond, rely upon a steady stream of lease sales through a continuously maintained national leasing program so that our companies can thrive and grow. Importantly, based upon what we have seen with prior legislative proposals, Congress could consider legislation that sets a deadline for the completion of the next federal offshore oil and gas leasing program and mandates a minimum number of region-wide Gulf of Mexico lease sales. While the Inflation Reduction Act reinstates cancelled lease sales, it does not address the lack of an active federal offshore oil and gas leasing program. Interior is legally required to maintain a leasing program and to schedule and hold lease sales, yet a federal offshore leasing program is currently going through an unprecedented lapse. The long-term success of the Gulf of Mexico as a premier energy region is dependent on the ability of companies to continuously secure acreage through new lease opportunities. Contractors like Danos then have the opportunity to compete for the work in constructing and maintaining these innovative projects. With a heightened level of uncertainty in the Gulf of Mexico, investment dollars could naturally leave the U.S. to be spent in regions with weaker environmental oversight and weaken our energy security.

Some critics audaciously claim that the industry has enough leases, and that it is unnecessary to offer more. However, this ignores fundamental realities of the oil and gas market, particularly in the offshore region where hundreds of millions of dollars may often be spent to simply determine if oil exists in commercial quantities within a lease block. One way to think about leasing is through the analogy of a conveyor belt. So long as leases are continuously placed on the conveyor belt, the industry has the ability to continuously take the steps necessary to explore for, discover, develop, and then produce the resources that may be found within the lease. As the leases move along the conveyor belt, companies are continuously analyzing the geology, acquiring, and processing seismic data, contracting for drilling rigs and workers, drilling exploratory wells, evaluating drilling results, drilling additional wells, determining whether the field contains commercial quantities of oil and gas, and finally, designing and procuring production facilities and associated infrastructure. During each stage, companies must apply for various plans, permits and approvals. In many cases, companies determine that oil and gas is not commercially recoverable, and the company ultimately relinquishes the lease back to the federal government. It is also important to recognize that companies pay bonus bids to obtain a lease, rentals to continue to hold the lease, and then royalties if the lease is producing oil or gas. All told, it costs companies significant resources, in terms of capital investment as well as time and man-hours, to explore for and potentially develop these resources.

As the U.S. and its allies attempt to overcome mounting geopolitical instability provoked by the Russian Federation, the Chinese Communist Party, and other adversaries around the world, the importance of the Gulf of Mexico in providing energy and national security for our nation and our allies will only grow. With a five-year offshore leasing program and uninterrupted lease sales, energy experts predict that the Gulf of Mexico will continue as the backbone of U.S. energy production by producing an estimated average of 2.6 million barrels equivalent per day from 2022-2040. Conversely, experts also project that a delay in the program could translate to nearly 500,000 barrels equivalent per day less over that period. The reason for this delta in future production under different leasing program scenarios is simple; without new leasing, companies cannot replenish their energy portfolios with new lease blocks. Having a robust and diversified exploration portfolio is critical to business health and delivering energy, as most leases do not contain commercially viable amounts of oil or natural gas. Put simply, continued lease sales in the U.S. offshore region means continued U.S. oil and gas production for years to come. With the Gulf of Mexico recognized as a region that produces some of the lowest-carbon intensity barrels in the world, more lease sales are good for us, our allies, and global emissions.

It is also important to highlight the significant impacts that may occur to U.S. energy production in the Gulf of Mexico as a result of regulatory activity and litigation. While the courts have struck down the Administration's decisions to pause leasing and, more recently, to remove millions of acres and impose operational restrictions under the lease terms for Lease Sale 261, the

Administration's policies continue to ignore legal requirements and the vital importance of American energy production.

In July 2023, the Biden administration reached a voluntary settlement agreement with activist organizations over an expanded Rice's whale protected habitat area that had the potential to reshape the future of energy production in the Gulf of Mexico and disrupt the flow of commerce throughout the region. The actions of the Administration were claimed to be in support of the Rice's Whale, a species already protected under the Endangered Species Act and the Marine Mammal Protection Act.

Fortunately, just last week, the United States District Court for the Western District of Louisiana issued a preliminary injunction the prohibits the Administration from implementing acreage withdrawals and vessel restrictions in the Final Notice of Sale for Lease Sale 261 that the Administration agreed to in their "sue and settle" arrangement with activists.

The Stipulated Stay agreement aimed to achieve two main objectives. First, it intended to exclude millions of acres from Lease Sale 261 in the Gulf of Mexico and impose unwarranted restrictions within the lease terms for those obtained in the sale. These restrictions were designed to impact the operations of companies as they transited the expanded protected habitat.

Secondly, the federal government introduced new "recommendations" concerning mitigation measures, such as vessel speed and night travel in the Gulf of Mexico, through a new Notice to Lessees (NTL). While BOEM consistently refers to these as "recommendations," the practical application of these recommendations and how companies should incorporate them appeared to be far from clear and seemed to be, in practice, actual requirements.

The litigation is not over and the issues at play remain important for the industry. To begin with, the areas that had been removed for consideration for Lease Sale 261 are highly prospective areas across some of the most resource-attractive areas in the Gulf of Mexico and the federal government proposed designating these areas as critical habitat for the Rice's whale. This area holds immense potential for responsible domestic energy development yet is now inaccessible due to the settlement's restrictions. Taking millions of acres of attractive oil and gas lease areas off the table would create strategic repercussions to say the least. Based upon numerous empirical studies, the U.S. Gulf of Mexico is recognized for its low carbon intensity barrels. By hampering production in this region, we needlessly risk importing higher carbon intensity alternatives from abroad, undermining both environmental progress and domestic energy security.

In addition, the settlement's inclusion of stipulations like vessel speed restrictions and limitations on night travel and times of low visibility would further affect the offshore oil and gas industry by imposing unwarranted constraints. These measures target normal and proven operations and processes and would hinder our ability to create jobs, stimulate economic growth, and maintain our energy independence, all while lacking scientific evidence to justify such extensive bans. There is also a plethora of unanswered questions regarding the feasibility of complying with these types of mitigations in a manner that is conducive to vessel and mariner safety. Many operations that are critical to rig and platform safety, like well rebalancing, must happen on short notice including at night. It is unclear how the vast range of spur of the moment operations requiring vessel transit, from materials management to catering, would be interpreted under these restrictive mitigations.

Importantly, the new Notice to Lessees (NTL) opens up a pandora's box of questions. The NTL includes "recommendations" similar to the lease sale stipulations that impose knot vessel speed restrictions – which are well below safe, normal, and typical maneuverability speeds for vessels, new vessel strike avoidance protocols, new multi-year record keeping requirements, limitations on night travel, and more. The NTL directs companies to include these mitigation measures in plans for future offshore operations in the region, but how the federal government will interpret the recommendations in the NTL is uncertain at best. The NTL circumvented the normal regulatory process and injected ambiguity into the system for energy producers and vessel operators in the region.

Decisions like this are shortsighted and can result in needless adverse consequences to offshore safety, emissions, energy security, energy affordability, and national security. For example, making areas off-limits, imposing speed restrictions, and limiting transit at night and in times of low visibility significantly impacts the ability of the industry to explore, construct, and develop energy projects in the Gulf of Mexico. The development of offshore energy projects depends upon a diverse ecosystem of companies providing support, supplies, and services. This work further relies upon a vast network of vessels for the construction, servicing, and maintenance of projects and facilities.

To help maximize logistical efficiency and safety of operations, vessels often transit at night so that support activities can occur during the day. The proposed restrictions would have flipped this option on its head and potentially eliminate or hamper an efficient approach for safely conducting support operations at offshore facilities.

The NTL will likely lead to an increase in the number of vessels required to support offshore projects as operators try to mitigate impacts from this drastic shift in how vessels operate in support of oil and gas activities. Limiting transit at night and in periods of low visibility would idle vessels offshore and increase traffic in daylight and high visibility periods. The associated increase in the vessel miles traveled would correspondingly serve to increase the safety risk associated with offshore operations. A basic risk assessment will demonstrate that more activity carries greater risk than less activity.

Moreover, an increase in the number of vessels and the associated increase in vessel miles traveled would also lead to an increase in overall emissions – at a time when the industry has continued to successfully improve safety and decrease emissions through continued enhancements in efficiency and applications of new technology. And this is before accounting for potential emissions from vessels idling outside the restricted zone for hours to wait for daylight or changes in weather conditions.

Importantly, these adverse consequences would likely accrue without any appreciable benefit to the conservation of the Rice's whale, which is already afforded protections under the Endangered Species Act and the Marine Mammal Protection Act – specifically in the Eastern Gulf habitat area where there is evidence that the species inhabits the area.

The government is also moving forward to expand these protections to other ocean users through the proposed designation of critical habitat for the Rice's whale, greatly expanding the adverse impacts that will ripple throughout our entire economy. Everything from cruise ships to cargo vessels to fishing boats that are working in the Gulf of Mexico could be impacted.

NOAA is accepting comments through October 6th on its proposed expansion of the Rice's whale critical habitat in the waters from the 100-meter isobath to the 400-meter isobath in the Gulf of Mexico, which are the areas which bisect the Western and Central Gulf of Mexico.

The government is thus erecting an arbitrary barrier between vessel transit and the shoreline that runs the full length of the U.S. Gulf of Mexico from the Mexico border all the way through the Florida coast. Relatedly, the National Oceanic and Atmospheric Administration (NOAA) recently closed the comment period on a petition to establish a 10-knot mandatory speed limit, banning night travel, and other mitigation measures which would apply to all vessels in the area in the Eastern Gulf of Mexico that has been referred to as the core habitat.

The Gulf of Mexico is home to vital American port infrastructure, with shipments transversing the area that flow throughout the entire U.S. economy. The imposition of a restricted zone that runs throughout the entirety of a critical zone of commerce in the Gulf has the potential to further inflationary impacts and drive up the cost of goods for all Americans.

According to one estimate, approximately sixty-five percent of the length of the proposed expanded critical habitat for the Rice's whale is traversed by international shipping fairways.[38] As vessels queue and all wait for the same transit windows through the protected areas, it is safe to assume that if these vessels all transit the protected areas at the same time, they will likely reach port near the same time. Bottlenecks and delays will continue as vessels try to unload cargo onto trucks and trains all at the same time. The result will surely be cascading delays impacting the flow of critical goods and commerce at the start of the onshore supply chain at a time when American families are already besieged by inflation.

While environmental preservation is a shared goal of Gulf Coast residents and businesses, the approach chosen to safeguard this species bypasses the appropriate channels for public and congressional engagement. Expanding the Rice's whale critical habitat to include areas where there is only a negligible or no presence at all will dilute conservation resources that should be going towards protecting actual core habitat areas. As the National Marine Fisheries Service has noted, only a single Rice's whale has been observed in the western Gulf of Mexico off the coast of Texas. This is not enough evidence to warrant a massive expansion of critical habitat areas without the chance for experts and the local Gulf Coast maritime sector to offer input.

There are also significant concerns with the inputs and process used by federal agencies in arriving at the decision to designate this area as critical habitat. To date, various scientific literature and a modeling study used by the Administration to justify these proposals has not been released publicly in peer-reviewed, native academic form. This has prevented businesses and Gulf communities from accurately analyzing the critical habitat proposal in a full and open way that lends to the federal government's commitment to transparency and due process.

All told, the decision by the Biden administration to circumvent Congress, the normal regulatory process, the public, and stakeholders and experts puts at risk a national energy asset, as well as other critical industries, without any

---

38 https://www.noia.org/wp-content/uploads/2023/09/Rices-Whale-Expanded-Area-and-Shipping-Fairways.pptx.

tangible benefits for the species they claimed they were protecting, and it took a last-minute injunction from a court to halt those actions in the interim.

The need for a healthy federal leasing program is compounded by the impact of the energy provisions passed within the Inflation Reduction Act (IRA). With periodic oil and gas lease sales in the Gulf of Mexico now required in order for the Department of the Interior to issue offshore wind leases, the urgency for leasing for both sectors is now tied together by Congressional mandate. Many of the same companies that built the offshore oil and gas sector in the Gulf of Mexico are now participating in the build-out of the offshore wind sector in the Atlantic. This includes many service and supply companies, like Danos, who have expertise in areas such as marine construction, fabrication, subsea engineering and design, and offshore vessel services. A steady stream of offshore oil and gas and offshore wind lease sales is needed for the supply chain to fully realize these incredible opportunities before us.

The U.S. and global economies continue to depend upon reliable and affordable supplies of all forms of energy – and specifically oil and natural gas – to maintain a high standard of living. Continued U.S. domestic oil and gas development, particularly offshore production, provides vast benefits and a sensible pathway for energy security for the next few decades. At the same time, the U.S. energy industry is contributing to the development of low and zero carbon energy options, including wind, hydrogen, and carbon removal technologies. Danos, the members of NOIA, energy companies, and small businesses across the nation stand ready to work with policy makers to advance policies to ensure that Americans can rely upon an affordable and reliable energy system built upon strong pillars of energy, economic, national, and environmental security.

It is an honor for me to testify before the committee. Thank you for this opportunity.

How Today's Global Anti-Fossil-Fuel Agenda Is Crippling American Small Businesses
By Alex Epstein, founder of the Center for Industrial Progress and author of the Moral Case for Fossil Fuels (2014) and Fossil Future (2022).

Testimony to the House Committee on Small Business, Subcommittee on Rural Development, Energy, and Supply Chains

Because energy is a major input in every business, the lower cost and more reliable energy is, the lower cost and more reliable everything is. The higher cost and less reliable energy is, the less efficient and less competitive US businesses, including small businesses, are.

Today we're hearing how irrational policies are crippling one crucial area of the energy business: offshore oil production, which provides almost 1/6th of our domestic oil production—production without which gasoline would be far more expensive and far, far less secure.[39]

What I would like to do today is put the irrational attack on offshore drilling in its broader context, which is that it is just a tiny piece of today's rabidly irrational global anti-fossil-fuel agenda—an agenda that poses as rational and scientific but is actually the opposite.[40]

The dozens of discrete attacks of the Biden Administration on the fossil fuel industry are not isolated incidents; they are all attempts to impose today's global anti-fossil-fuel agenda, which says that fossil fuel use must be essentially eliminated by the year 2050.

At the beginning of his administration, President Biden announced a US "whole of government" approach to "climate change"—which in practice means an overarching commitment to rapidly eliminate fossil fuels. Every agency was told to figure out new ways to attack fossil fuels.[41]

---

[39] U.S. Energy Information Administration - Oil and petroleum products explained, Where our oil comesfrom.

[40] Alex Epstein - How a fake climate emergency created a real energy emergency.

[41] "The order establishes the National Climate Task Force, assembling leaders from across 21 federal agencies and departments to enable a whole-of-government approach to combatting the climate crisis." The White House - FACT SHEET: President Biden Takes Executive Actions to Tackle the Climate Crisis atHome and Abroad, Create Jobs, and Restore Scientific Integrity Across Federal Government, January 27,2021.

The Biden administration is systematically committed to attacking fossil fuels on 4 fronts, every one of which drives higher cost and lower reliability:

1) Fossil fuel investment
2) Fossil fuel production
3) Fossil fuel transport
4) Fossil fuel use, e.g., electricity generation

The Costs of Punishing Fossil Fuel Investment

Fossil fuels investments are long-term commitments that require confidence in a payoff.

Our government, often colluding with activists, incessantly discourage fossil fuel investments directly and by threatening the industry's future.

The Federal government has been a major driver of ESG anti-fossil-fuel climate policies, with regulators pressuring financial institutions to declare their commitment to getting off fossil fuels. The SEC's recent "climate disclosure rules" are taking the damage to the next level.[42]

U.S. SECURITIES AND EXCHANGE COMMISSION

Search SEC.gov

COMPANY FILINGS

ABOUT | DIVISIONS & OFFICES | ENFORCEMENT | REGULATION | EDUCATION | FILINGS | NEWS

Newsroom
Press Releases
Speeches and Statements
SEC Stories
Securities Topics
Media Kit
Press Contacts
Events
Webcasts

Press Release

SEC Proposes Rules to Enhance and Standardize Climate-Related Disclosures for Investors

FOR IMMEDIATE RELEASE

2022-46

Washington D.C., March 21, 2022 — The Securities and Exchange Commission today proposed rule changes that would require registrants to include certain climate-related disclosures in their registration statements and periodic reports, including information about climate-related risks that are reasonably likely to have a material impact on their business, results of operations, or financial condition, and certain climate-related financial statement metrics in a note to their audited financial statements. The required information about climate-related risks also would include disclosure of a registrant's greenhouse gas emissions, which have become a commonly used metric to assess a registrant's exposure to such risks.

Related Materials

- Proposed Rule
- Fact Sheet
- Comments Received

---

[42] SEC - SEC Proposes Rules to Enhance and Standardize Climate-Related Disclosures for Investors.

The Costs of Punishing Fossil Fuel Production

Punishing restrictions and onerous taxes specifically on fossil fuel production are common in developed countries, suppressing supply from the freest nations on the globe.

One of the many punishments President Biden has inflicted on the supposedly-preferred fossil fuel industry was his early moratorium on issuing oil and gas leases on Federal lands. And the Biden admin continues to delay and decline holding lease sales.[43]

THE WALL STREET JOURNAL.

POLITICS

Federal Oil Leases Slow to a Trickle Under Biden

President, citing climate change, spurns resources his predecessors relied on to boost U.S. energy production

By Timothy Puko Follow and Anthony DeBarros Follow

Sept. 4, 2022 5:30 am ET

Gift unlocked article Listen (9 min)

President Biden has promoted a transition to clean energy and away from fossil fuels. PHOTO: MICHAEL M. SANTIAGO/GETTY IMAGES

WASHINGTON—The Biden administration has leased fewer acres for oil-and-gas drilling offshore and on federal land than any other administration in its early stages dating back to the end of World War II, according to a Wall Street Journal analysis.

[43] WSJ - Federal Oil Leases Slow to a Trickle Under Biden. The White House - Executive Order on Protecting Public Health and the Environment and Restoring Science to Tackle the Climate Crisis.

The Costs of Punishing Fossil Fuel Transport

One of the most effective ways of punishing the fossil fuel industry has been attacking critical transport infrastructure, including pipelines and export/import facilities, with delays, regulations, and lawsuits.

Biden's destruction of the Keystone XL pipeline inhibited Canada from bringing oil to market, which prevents Canada from using its vast oil deposits to their full potential—meaning lower global supply and higher prices for oil.[44]

Perhaps our most destructive way of singling out fossil fuel transport for punishment hasbeen killing natural gas pipelines.

We have a virtually limitless supply of gas and an incredible ability to ramp upproduction.

But we're blocking pipeline after pipeline.[45]

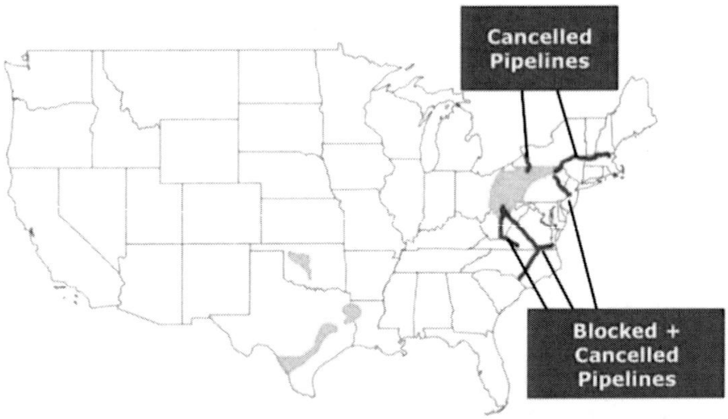

Project	Status	Gas Volumes (Bcf/d)
Constitution	Cancelled	0.7
Penn East	Cancelled	1.1
Northern Access	Opposed	0.5
MVP	Opposed	2.0
Atlantic Coast	Cancelled	1.5
Northeast Direct	Cancelled	1.2
Total		7.0 Bcf/d

---

[44] CNBC - Keystone pipeline officially canceled after Biden revokes key permit Competitive Enterprise Institute - Lessons from the Demise of ANWR and Keystone XL.

[45] EQT - Nationwide Poll Shows Nearly 70% of All Voters Support Increasing Natural Gas Production.

The Costs of Punishing Fossil Fuel Use

Our government's "whole of government" attack on fossil fuels seeks out every way it can to punish fossil fuel use.

Perhaps the most destructive form of this is punishing reliable, fossil fueled electricity generation and thereby destroying our grid.

As a Californian, I am well aware of the costs of punishing fossil fuel electricity generation. Prematurely shutting down vital fossil fuel plants led to devastating statewide blackouts in 2020.[46]8

And 5 days after pledging to go all EVs, Newsom told us there wasn't enough power to charge our EVs.[47]9

Flex Alert Conservation Actions

Before3 p.m.:

- **Pre-cool home by setting the thermostat to as low as 72 degrees**
- **Use major appliances, including:**
    - **Washer and dryer**
    - **Dishwasher**
    - **Oven and stove for pre-cooking and preparing meals**
- **Adjust blinds and drapes to cover windows**

From 3 p.m. to 10 p.m.:

- **Set thermostat to 78 degrees or higher, if health permits**
- **Avoid using major appliances and charging electric vehicles**
- **Turn off all unnecessary lights**

Nationally, we're in an electricity crisis, with reliable power plants shutting down far faster than they are being built.[48]10

And yet the Biden EPA plans to make things much worse with 7 policies that gravely threaten 10-20% of our reliable capacity in the next 7 years.[49]11

Every one of the hundreds of actions that the Biden Administration takes in the name of its "whole of government" anti-fossil-fuel agenda is harmful to American businesses, including small businesses, who have to struggle with higher energy costs and lower reliability.

---

[46] Alex Epstein - Talking Points on California Blackouts.

[47] The Babylon Bee - State With No Electricity Orders Everyone To Drive Cars That Run On ElectricityCAISO - California ISO issues Flex Alert for today, Aug. 31 FoxWeather.com - California asks residents to avoid charging electric cars amid intense heat wave.

[48] Alex Epstein - Electricity Emergency.

[49] Alex Epstein - The EPA vs. the grid.

The Lie That Solar and Wind Can Rapidly Replace Fossil Fuels

The anti-fossil-fuel agenda pretends it's not harming us by claiming, in contradiction to all evidence, that rapidly eliminating fossil fuels won't make energy more expensive and less reliable—because unreliable solar and wind can somehow replace fossil fuels.

If solar and wind could actually replace fossil fuels, they wouldn't need massive government punishments of fossil fuels plus massive government preferences of their own—mandates, subsidies, and no penalty for unreliability.[50]12

If solar/wind could actually replace fossil fuels, China, which dominates the production of solar/wind components, wouldn't be using coal to produce these components—and wouldn't have > 300 planned new coal plants designed to last over 40 years.[51]13

How to Fight Back Against the Global Anti-Fossil-Fuel Agenda

To counter the global anti-fossil-fuel agenda, American businesses need to use their powerful voices to do two things above all:

1) Refute the falsehood that fossil fuels are bringing about climate apocalypse
2) Offer a pro-freedom energy agenda

In my supplemental testimony, I have included a document, "Countering the world's rigged conversation about energy and climate," that thoroughly exposes the faulty thinking methods and abuse of science behind the falsehood that fossil fuels are bringing about climate apocalypse.

In my supplemental testimony I have also included a document, "The Energy Freedom Platform," that lays out a positive, pro-human, pro-science, pro-freedom energy agenda based on liberating our ability to empower the world using fossil fuels and to develop truly cost-effective alternatives.

My goal in testifying and answering your questions today, as well as in my work more broadly, is to give pro-freedom legislators, businesses, and

---

[50] Alex Epstein - The Myth of Fossil Fuel Subsidies.
[51] Global Energy Monitor - Global Coal Plant Tracker.

other influencers the resources they need to fight for the energy freedom that all Americans and our businesses need to flourish.

Countering the World's Rigged Conversation About Energy and Climate

Today's conversation about energy and climate is "rigged" with bad thinking methods, misleading terms, false assumptions, and anti-human values—all of which serve to promote deadly policies

JUL 13, 2023 · PAID

79 4 6 Share ...

I've been in Washington, DC this week speaking to major groups of legislators about how to effectively champion energy freedom, including the freedom to use fossil fuels. (Here's a story in Energy & Environment News covering one of my presentations.)

In preparation for my visit I put together this guide for how to discuss energy and climate issues in the face of a cultural conversation that is incredibly biased against fossil fuels. I think it's one of the most valuable things I've ever written. I hope you enjoy it.

Our Rigged Conversation About Energy and Climate

Just as Legal Systems Can Be Rigged, So Can Cultural Conversations

We are all familiar with the idea of a legal system that is rigged against certain types of people. For example, in the classic To Kill a Mockingbird, the legal system of (fictional) Maycomb County, AL, has a deep racist bias against black individuals that dismisses strong evidence of their innocence and embraces pseudo-evidence of their guilt.

A rigged legal system inevitably leads to immoral results—as captured by the saga of Tom Robinson in To Kill a Mockingbird, a good man who, afier resisting the sexual advances of a white woman, was arrested, prosecuted, and convicted for rape.

Just as it is possible for a legal system to be rigged, so it is also possible for a culture's intellectual conversation to be rigged. To continue with the example of racism, it is unfortunately commonplace throughout history for the conversation about particular racial minorities to be rigged. One element this almost always involves is ignoring the positives of individuals in the disfavored group and exaggerating or fabricating negatives.

For example, in (the unfortunately numerous) anti-Semitic cultures it is commonplace to ignore any positive attributes and contributions of individuals of Jewish descent, while fabricating the idea that all Jews are miserly and uncaring.

4 Common Ways in Which Cultural Conversations Are Rigged

4 common dimensions in which a culture's intellectual conversation can be rigged are:

1) Bad thinking methods. For example, with racist conversations, the aforementioned examples of ignoring positives and exaggerating or fabricating negatives.
2) Misleading terminology. For example, criticisms of Jews as "greedy" misleadingly associate 1) financial success earned by productive achievement, a good thing, with 2) getting money by uncaringly exploiting others, a bad thing.
3) False assumptions. For example, racist cultural conversations falsely assume that an individual's ideas and character are determined by their skin color.
4) Anti-human values. For example, racist cultural conversations treat some categories of human beings as intrinsically non-valuable.

Rigged Conversations Are Common—No Conspiracy Required

To say that a conversation is "rigged" is not to assert a conspiracy in which a few people covertly decide to crafi a cultural conversation with bad thinking methods, misleading terminology, false assumptions, and/or anti-human values. (Although this can happen.)

It is to recognize that very frequently, for whatever set of reasons, cultural conversations operate on bad thinking methods, misleading terminology, false assumptions, and anti- human values that rig them against coming to true and pro-human conclusions.

And the cultural conversation that I study, the conversation around energy and climate, is rigged to a degree that almost no one can imagine.

To Counter the Rigging You Must First Understand It

To help you counter the rigged nature of this conversation, I will identify 12 distortions that rig our global energy and climate conversation to reach the deadly conclusion that we should rapidly eliminate fossil fuel use to prevent climate catastrophe.

By making you aware of these distortions, I hope to

1) Help you point them out explicitly whenever they occur (which is all the time).
2) Help you lead and have energy/climate conversations without these distortions.

Afier explaining the 12 distortions I'll share some of my favorite "talking points" that reframe—de-rig—the conversation, so that we can make others see the truth.

12 Distortions Around Which the Energy and Climate Conversation Is Rigged

1) (Bad thinking method) *Looking only at the negative side-effects of fossil fuels,* while ignoring the massive and unique benefits of fossil fuels.
2) (Bad thinking method) *Only looking at the positives of solar and wind* while ignoring obvious negatives. E.g., praising solar and wind as "secure" because they don't depend on Russia like oil and gas do, when in fact they depend on China far more than oil and gas depend on Russia.
3) (Bad thinking method) *Only looking at the negatives of CO2 emissions* while ignoring the positives (such as greater plant

growth and the prevention of cold- related deaths—which far outnumber heat-related deaths).

4) (Bad thinking method) *Engaging in "partial cost accounting" for solar and wind*— claiming they are cheap by only looking at some of their costs (e.g., solar panels, wind turbines) while ignoring other huge costs (e.g., the cost of 24/7 life support for an unreliable input that can easily go near-0).
5) (Bad thinking method) *Ignoring the massive climate-related benefits of fossil fuels* —their benefits in helping us master climate danger—even though these benefits have thus far overwhelmed any negative climate side-effects of fossil fuels.
6) (Misleading terminology) *Using the vague term "climate change,"* which conflates *some* human impact on climate (which the vast majority of climate scientists agree with) with *catastrophic human impact on climate* (which is not supported by climate science and economics).
7) (Misleading terminology) *Using "climate crisis" or "climate emergency" as the basic noun to refer to the state of today's climate*—thereby asserting a catastrophe without needing to provide any evidence.
8) (Misleading terminology) *Using the terms "energy" and "electricity" interchangeably*, even though the vast majority of the energy that powers our machines is not electricity but the direct burning of fossil fuels for transportation, industrial heat, or residential heat. This (along with "partial cost accounting," helps promote the false idea that solar and wind electricity can rapidly replace all fossil fuel energy.
9) (False assumption) *Treating climate* (and, more broadly unimpacted nature) *as a "delicate nurturer"*: a stable, sumcient, safe phenomenon that human impact ruins, when in fact climate (and more broadly unimpacted nature) is dynamic, deficient, and dangerous—and human impact makes it a lot safer (e.g., irrigation radically reduces drought-related deaths).
10) (Anti-human value) *Treating today's global energy use as suncient or even excessive*, when in fact most of the world is desperately lacking in energy. E.g., 3 billion people use less electricity than a typical American refrigerator.
11) (Anti-human value) *Treating human impact on climate, and more broadly human impact on nature, as intrinsically bad.* E.g.,

assuming all "climate change" is bad even though rising CO_2 clearly leads to beneficial greening and warming will clearly save many lives in many places (far more people die of cold than of heat).

12) (Anti-human value) *Making eliminating human impact on climate at all costs (e.g., "net zero") our number one global climate, energy, and political goal*—instead of embracing the proper, pro-human goals of maximizing climate livability, human empowerment, and human flourishing.

Countering Our Rigged Conversation About Energy and Climate

Understanding the distortions around which our energy/climate conversation is rigged is one key to countering them, because once you understand these distortions you can explicitly and effectively point them out. 3 other keys are:

- Explaining what you think is the right way to think about energy and climate issues —not just criticizing the wrong ways.
- Explaining the essential facts about energy and climate that are relevant to policy- making—not just counters to various myths.
- Advocating a positive energy and climate policy—not just negatively reacting to bad ones.

Explaining What You Think Is The Right Way to Think About Energy and Climate Issues—Not Just Criticizing the Wrong Ways

Here are some ways I explain what I think is the right approach to thinking about energy and climate—including my thinking methods, assumptions, and values—all using precise terminology.

Thinking Methods

Summary: I believe that we should think about fossil fuels the way we should think about every other product or technology: we should carefully weigh the benefits and the side-effects. And that includes the many climate-related benefits of fossil fuels that no one talks about, such as their ability to power irrigation systems that alleviate drought.

- When we're evaluating what to do about any technology we must factor in not only its negative side-effects but also its benefits. E.g., oil-powered equipment and natural gas fertilizer are crucial to feeding 8 billion people.
- Even though we obviously need to factor in fossil fuels' benefits, not just their negative side-effects, most designated experts totally fail to do this. E.g., "expert" climate scientist Michael Mann 100% ignores fossil fuels' unique agricultural benefits in his book on fossil fuels and climate [1].
- One huge benefit we get from fossil fuels is the ability to master climate danger— e.g., fossil-fueled cooling, heating, irrigation— which can potentially neutralize fossil fuels' negative climate impacts.
- Even though we obviously need to factor in fossil fuels' climate mastery benefits, our designated experts totally fail to do this. E.g., the UN IPCC's multi-thousand- page reports totally omit fossil-fueled climate mastery! That's like a polio report omitting the polio vaccine [2].
- When we're carefully weighing the climate side-effects of fossil fuels we must consider both negatives (more heatwaves) and positives (fewer cold deaths). And we must be precise, not equating some impact with huge impact.
- Even though we obviously need to factor in both negative and positive impacts of rising CO2 with precision, most designated experts ignore big positives (e.g., global greening) while catastrophizing negatives (e.g., Al Gore portrays 20 fi sea level rise as imminent when extreme UN projections are 3fi/100yrs).

Assumptions

Summary: I believe that Earth is not a "delicate nurturer" that human impact inevitably ruins, but rather "wild potential" that human impact generally improves, transforming Earth's naturally deficient and dangerous environment into an unnaturally abundant and safe environment.

- Much of our energy and climate conversation operates on the false assumption I call "the delicate nurturer": Earth (including climate) exists in a delicate, nurturing balance—that's stable, suncient, and safe—but human beings are "parasite- polluters" whose impact inevitably destroys the delicate balance and us with it.
- Insofar as you believe that Earth is a "delicate nurturer" and humans are "parasite polluters," you expect that continuing human impact on Earth will inevitably lead to disaster. That's why catastrophists keep thinking their next catastrophe prediction will be the one that's right.
- In reality, Earth is not a "delicate nurturer" but "wild potential"—it's dynamic, deficient, and dangerous—and human beings are not "parasite-polluters" but "producer-improvers" whose impact generally produces new value and therefore makes the world much more livable for us.

Values

Summary: When we are evaluating what to do about global issues such as energy and climate, I believe our primary moral goal should be advancing human flourishing on Earth—which means a lot of impact, since Earth and its climate are not naturally very livable. I reject the idea that our primary goal should be to eliminate our impact on Earth.

- Many "experts," when evaluating what to do about global issues such as energy and climate, are implicitly or explicitly operating on the goal of minimizing or eliminating human impact on Earth. For example, the leading climate-related goal today is "net-zero emissions," which means "eliminate human impact on climate."
- If your primary goal is to eliminate human impact on Earth, then you will inevitably regard today's highly-impacted Earth as bad, even though it's never been more livable for human beings. Note how many "experts" say we have "destroyed" the Earth and its climate,

even though humans have never flourished more—including being safer than ever from climate danger.

- I believe that our goal for the Earth should not be to eliminate human impact on Earth—which, taken consistently, means mass poverty and ultimately death—but to advance human flourishing on Earth. That means we need to impact Earth a lot, just to do so intelligently such that the benefits to human flourishing of our impacts far outweigh the negative side-effects.
- I believe that our goal for climate should not be to eliminate human impact but to maximize climate livability. We want a climate that's as livable as possible for human beings. If we use fossil fuels and the planet is a little warmer overall as a side-effect, but the benefit of all the energy is that we have made ourselves 50 times less likely to die from climate-related disasters like temperature extremes and storms and floods, that's a great thing [3]
- Over the last century, fossil fuels have made us far better off climate-wise, driving down climate disaster deaths by a factor of 50. Compare today's fossil-fueled "climate mastery" to the real "climate crisis" of the past, when we were having very little impact on climate but were terrorized by climate—since climate is naturally so dangerous and we had so little energy and technology to master it.

Explaining the Essential Facts About Energy and Climate That Are Relevant to Policy-Making—Not Just Counters to Various Myths

Benefits of Fossil Fuels

Summary: If 8 billion people are going to have the cost-effective energy they need to flourish— including to master our naturally dangerous climate—in the far greater quantities needed, fossil fuel use needs to increase. Rapidly restricting fossil fuel use, as many experts advocate, is deadly.

- Undeniable energy fact 1: Cost-effective energy is essential to human flourishing. Cost-effective energy—affordable, reliable, versatile, scalable energy—is essential to human flourishing because it gives us the ability to use machines to become productive and prosperous.
- Undeniable energy fact 2: The world needs much more energy. Billions of people lack the cost-effective energy they need to flourish.

3 billion use less electricity than a typical American refrigerator. 1/3 of the world uses wood/dung for heating/cooking. Much more energy is needed [4]

- Undeniable energy fact 3: Fossil fuels are uniquely cost-effective Despite 100+ years of aggressive competition, fossil fuels provide 80%+ of the world's energy and they are still growing—especially in the countries most concerned with cost-effective energy. E.g., China.
- Undeniable energy fact 4: Unreliable solar/wind are failing to replace fossil fuels Despite claims that solar + wind are rapidly replacing fossil fuels, they provide < 5% of world energy—only electricity, ⅕ of energy—and even that depends on huge subsidies and reliable (mostly fossil-fueled) power plants [5].
- Undeniable energy fact 5: Fossil fuel energy gives us an incredible climate mastery ability Fossil fuels have helped drive down climate disaster deaths by 98% over the last century by powering the amazing machines that protect us against storms, extreme temperatures, and drought [6].

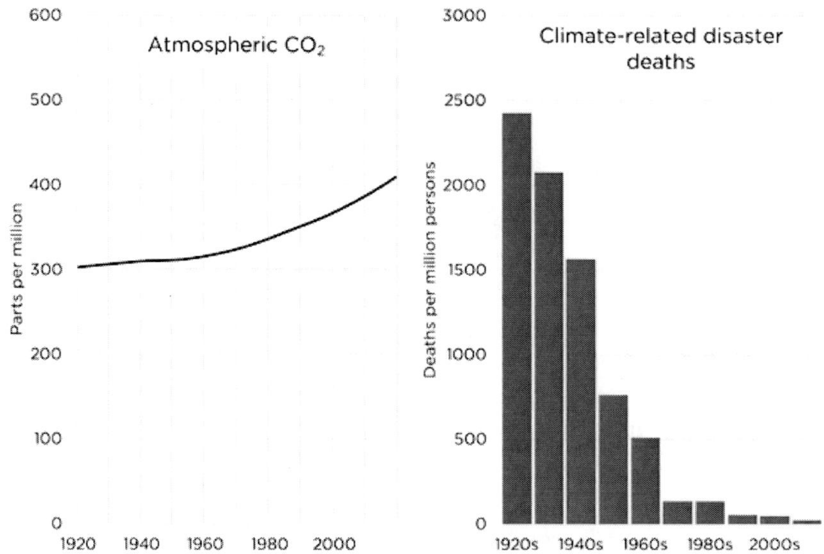

The Climate Side-Effects of Fossil Fuel Use

Summary: If we're free to use fossil fuels, we'll continue to have a warming impact that we can master and flourish with. If we follow "net zero" policies we'll have a less- impacted climate in the short-term, but the climate and the world as a whole will be incomparably less livable, with billions plunging into poverty and premature death.

- *Undeniable climate fact: Fossil fuel energy gives us an incredible climate mastery ability. Fossil fuels have helped drive down climate disaster deaths by 98% over the last century* by powering the amazing machines that protect us against storms, extremetemperatures, and drought [7].

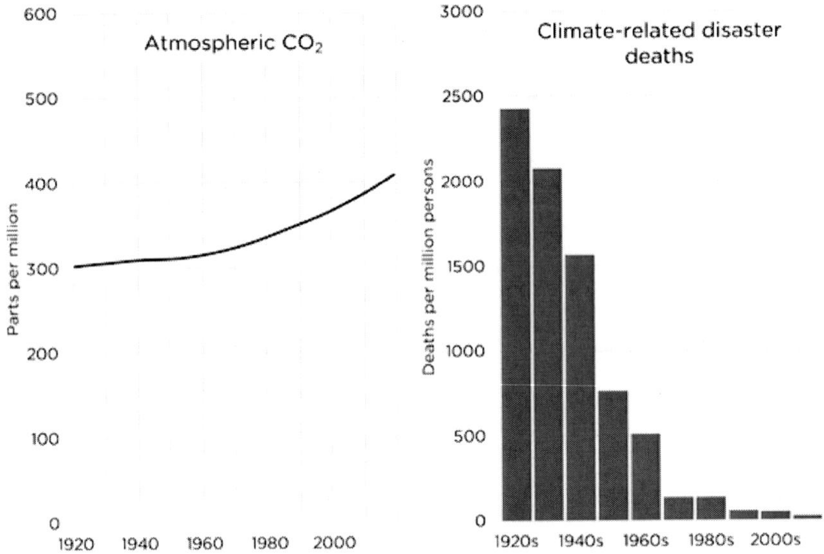

- *Undeniable climate fact: CO2 emissions correlate with 1°C warming, + greening* Fossil fuels' CO2 emissions have contributed to the warming of the last 170 years, but that warming has been mild and manageable—1° C. Here's what that lookslike compared to normal temp changes [8].
- *Undeniable climate fact: Deaths from cold far exceed deaths from heat.* While leading institutions portray a world as increasingly riddled with heat-related death, the fact is that even though Earth has gotten 1°C warmer far more peopledie from cold than heat (even in India!) [9].

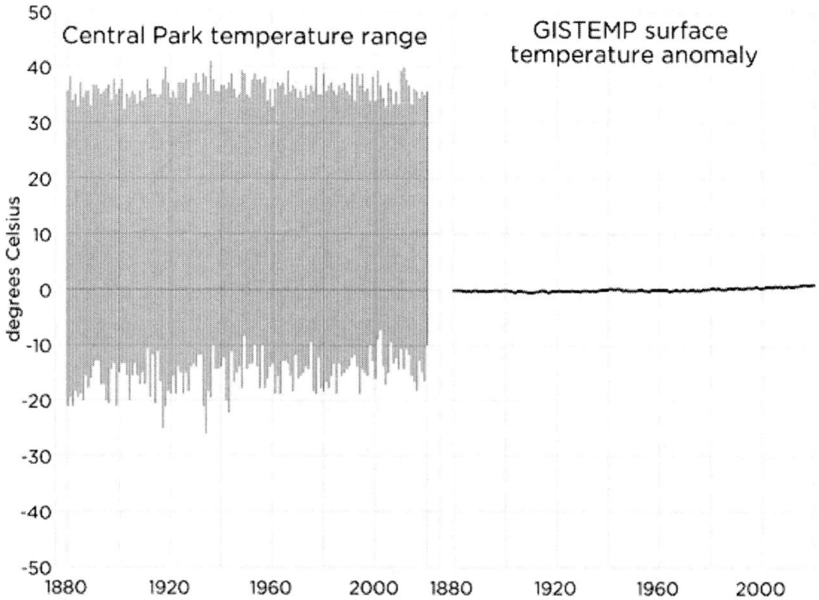

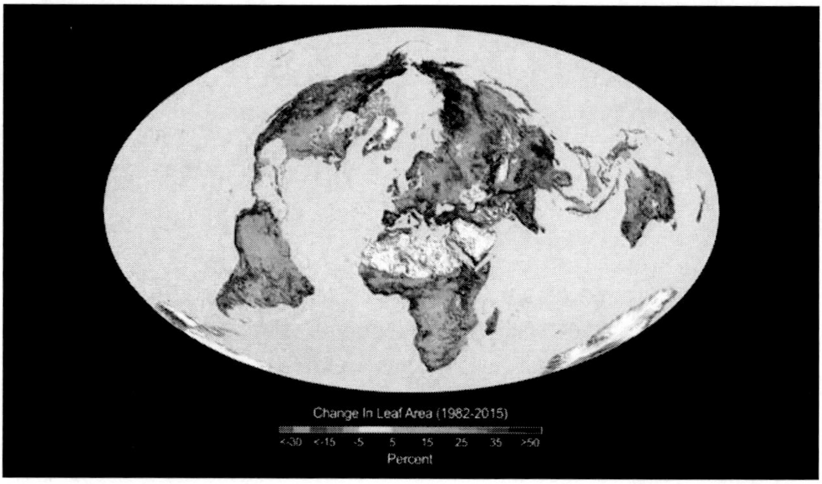

- *Undeniable climate fact: Warming from CO2 occurs more in colder places.* The mainstream view in climate science is that more warming will be concentrated in colder places (Northern latitudes) and at colder times (nighttime) and during colder seasons (winter). This is good news [10].

Global temperature anomalies, 2000-2009

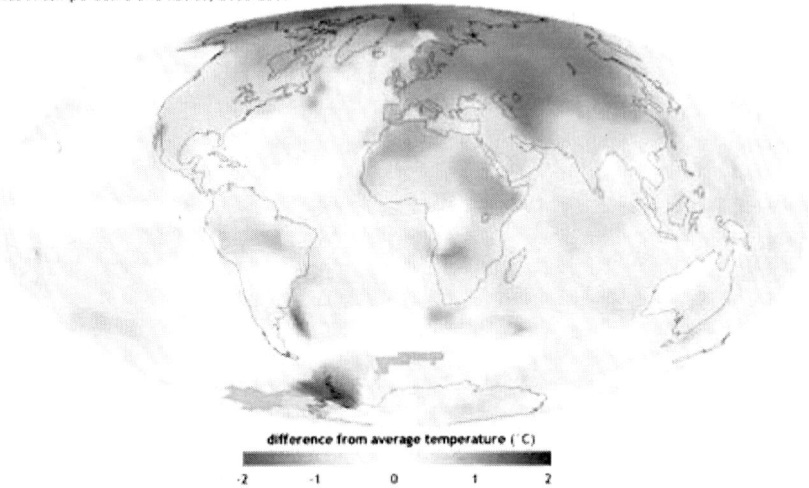

Arctic versus global temperatures anomalies, 1900-2014

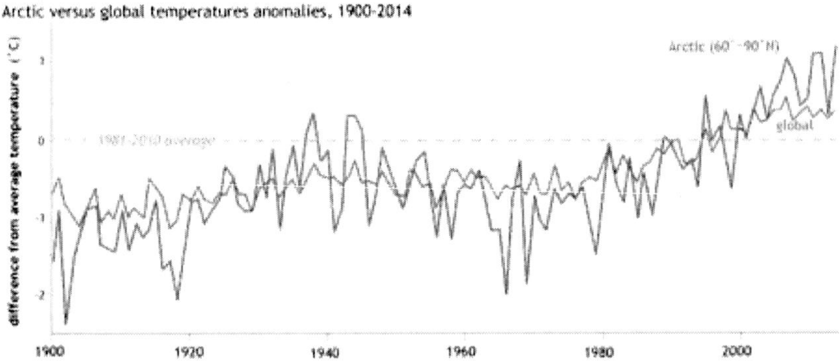

- *Undeniable climate fact: Rising CO2 leads to diminishing warming.* Mainstream climate science is unanimous about a conclusion that the public is, shamefully, not made aware of: the "greenhouse effect" of CO2 is a diminishing effect, with additional CO2 leading to less warming.
- The only moral and practical way to reduce CO2 emissions is innovation that makes low-carbon energy globally cost-competitive. So long as fossil fuels are the most cost-competitive option for people, especially in developing nations, they will (rightly) choose to emit CO2 vs. plunge even further into poverty.
- So long as America and other wealthy nations follow the anti-development "green energy" movement and the "climate emergency" narrative, they will continue to adopt senseless policies that harm their

economies and security while doing nothing to bring about globally cost-competitive low-carbon energy.

The Truth About Alternatives

Summary: No alternative or combination of alternatives to fossil fuels have any near- term hope of replacing fossil fuels' unique combination of affordability, reliability, versatility, and scalability in a world that needs far more energy. We should, however, liberate alternatives from any and all restrictions that are preventing them from reaching their full potential.

- *Myth: We can rapidly reduce fossil fuels at very low costs.*
  Truth: Fossil fuels are a uniquely cost-effective form of energy, which is why they are 80% of global energy and still growing. Rapidly reducing fossil fuels, in a worldthat needs far more energy, is catastrophic [11].
- *Myth: Solar and wind are cheap.*
  Truth: Solar and wind are unreliable, parasitical sources of energy that add coststo the grid.
  Claims of "cheapness" are based on ignoring the full costs of solar + wind—aboveall the cost of a reliable grid that gives them 24/7 life support.
- *Myth: Solar/wind is cheaper than fossil fuels because Lazard's "Levelized Cost of Energy"(LCOE) is lower for solar/wind.*
  Truth: LCOE, by Lazard's own admission, doesn't include many costs of solar/wind—above all the cost of a reliable grid needed for 24/7 life support [12].
- *Myth: Solar and wind are "winning in the marketplace," out-competing fossil fuels andnuclear with superior economics.*
  Truth: Unreliable, parasitical solar and wind are only "winning" when givenmassive preferences—mandates, subsidies, and no penalty for unreliability.
- *Myth: Nuclear is too expensive, so we should use solar/wind instead.*
  Truth: Solar/wind can't provide reliable energy; nuclear can. And nuclear is onlyexpensive because it has, with the help of many "green" activists, been falsely labeled unsafe and effectively criminalized.

- *Myth: Solar and wind will reduce our dependence on adversaries for energy.*
  Truth: If Europe's level of dependence on Russia for natural gas scares you, know this: America is even more dependent on China for many of the key componentsof solar panels, wind turbines, and batteries than Europe is on Russia for natural gas [13].

Advocating a Positive Energy and Climate Policy—Not Just Negatively Reacting to Bad Ones

Overall Energy Policy

- What's your policy on energy, environment, and climate?

I believe in energy freedom: the freedom to use all forms of energy, with lawsagainst emissions that are truly harmful and reasonably preventable.
5 key energy freedom policies are:

1. Liberate responsible development
2. End preferences for unreliable electricity
3. Reform air and water emissions standards to incorporate cost-benefit analysis
4. Reduce long-term CO_2 emissions via liberating innovation
5. Decriminalize nuclear

For a detailed positive energy policy, see The Energy Freedom Platform.

A Pro-Human, Pro-Energy, Pro-Freedom Policy Toward GHG/CO_2 Emissions

Summary: America is taking a "punish America" approach to CO_2 emissions, making our energy more expensive and less reliable while China, Russia, and others increase their emissions.

We need a "liberate American innovation" policy instead.

- The only moral and practical way to reduce CO2 emissions is innovation that makes low-carbon energy globally cost-competitive. So long as fossil fuels are themost cost-competitive option for people, especially in developing nations, they will (rightly) choose to emit CO2.
- The US causes < 1/6 of global CO2 emissions—and falling. The main reason global CO2 emissions are rising is because billions of people in the developing world are bringing themselves out of poverty by using fossil fuels to power factories, farms, vehicles, and appliances [14].

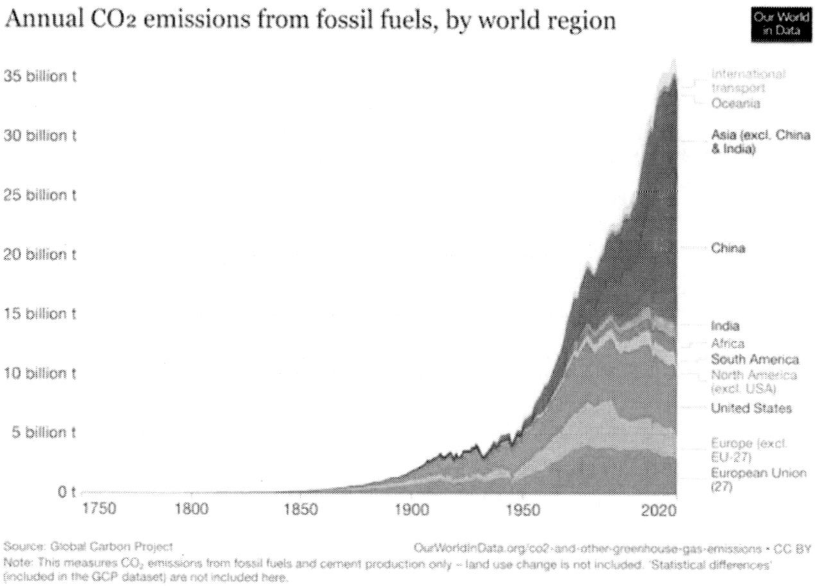

- The developing world overwhelmingly uses fossil fuels because that is by far the lowest-cost way for them to get reliable energy. Unreliable solar and wind can't
- come close. That's why China and India have hundreds of new coal plants in development [15]
- The only way to lower CO2 emissions and benefit America is to promote innovation that makes low-carbon energy truly reliable and low-cost. Are China and India going to stop using fossil fuels so long as they are the lowest-cost option? They won't and they shouldn't.

- The only moral and practical way to reduce emissions long-term is liberating innovation that makes low-carbon energy globally cost-competitive—while ending all policies that punish America via rapid short-term emissions reduction.

Some key policies

- Reject the false idea of "climate emergency."Our government's disastrous anti-fossil fuel policies are justified by the disastrous conflation of "climate impact," which is real, with "climate emergency," which is not, given today's unprecedented safety from climate danger.
  Withdraw from the Paris Agreement and encourage others to do the same. The Paris Agreement is an immoral agreement that calls for rapidly eliminating fossil fuels, which are the only near-term way to provide reliable energy for billions of people at prices they can afford.

- Reject and eliminate all carbon taxes. Carbon taxes increase our energy costs based on the false premise that the "negative externalities" of fossil fuels' CO2 emissions outweigh the "positive externalities" of the uniquely low-cost, reliable energy they provide for billions.
- Amend the Clean Air Act to explicitly reject the bogus "endangerment finding." Much of today's "punish America" CO2 policy is rooted in EPA's "endangerment finding," which treats fossil fuel use as a net harm to "public health and welfare" even though it radically improves both.
- Decriminalize nuclear energy. The overregulation of low-carbon nuclear verges on criminalization, making nuclear costs 10X higher than they need to be. Decriminalizing nuclear, including radical reform of the Nuclear Regulatory Commission and Environmental Protection Agency, will make energy far cheaper, safer, and cleaner.
- End all preferences for unreliable electricity. Today's electric grids are being ruined by systemic preferences for unreliable electricity, which causes prices to rise and reliability to decline.

Eliminating them can help make America a leader in low-cost, reliable electricity.

- Allow free-market competition for EVs. The proper policy toward EVs, which are promising but not cost-effective for the vast majority of Americans, is 1) let them compete on a free market and 2) make sure we have plenty of low-cost, reliable electricity.
- Summary: The pro-human CO2 policy is to reduce CO2 emissions long-term through liberating innovation, not punishing America. This will ensure plenty of energy for the foreseeable future and enable truly promising alternatives such as nuclear to be globally cost-competitive.

Bonus: Concise Answers to Common Questions, Including Loaded Questions

- How to answer loaded climate questions
- Snappy answers to energy questions

Popular Links

- EnergyTalkingPoints.com: Hundreds of concise, powerful, well-referenced talking points on energy, environmental, and climate issues.
- My new book Fossil Future: Why Global Human Flourishing Requires More Oil, Coal, and Natural Gas—Not Less.
- Speaking and media inquiries

"Energy Talking Points by Alex Epstein" is my free Substack newsletter designed to give as many people as possible access to concise, powerful, well-referenced talking points on the latest energy, environmental, and climate issues from a pro-human, pro-energy perspective.

References

[1] Michael E. Mann, Tom Toles - The Madhouse Effect

[2] Alex Epstein - The IPCC's perversion of science

[3] For every million people on earth, annual deaths from climate-related causes (extreme temperature, drought, flood, storms, wildfires) declined 98%--from an average of 247 per year during the 1920s to 2.5 in per year during the 2010s.
Data on disaster deaths come from EM-DAT, CRED / UCLouvain, Brussels, Belgium – www.emdat.be (D. Guha-Sapir).
Population estimates for the 1920s from the Maddison Database 2010 come from the Groningen Growth and Development Centre, Faculty of Economics and Business a University of Groningen. For years not shown, population is assumed to have grown at a steady rate.
Population estimates for the 2010s come from World Bank Data.

[4] IEA - Access to affordable, reliable, sustainable and modern energy for all Robert Bryce - A Question of Power: Electricity and the Wealth of Nations

[5] BP - Statistical Review of World Energy

[6] UC San Diego - The Keeling Curve
For every million people on earth, annual deaths from climate-related causes (extreme temperature, drought, flood, storms, wildfires) declined 98%--from an average of 247 per year during the 1920s to 2.5 in per year during the 2010s.
Data on disaster deaths come from EM-DAT, CRED / UCLouvain, Brussels, Belgium – www.emdat.be (D. Guha-Sapir).
Population estimates for the 1920s from the Maddison Database 2010, the Groningen Growth and Development Centre, Faculty of Economics and Business at University of Groningen. For years not shown, population is assumed to have grown at a steady rate.
Population estimates for the 2010s come from World Bank Data.

[7] UC San Diego - The Keeling Curve
For every million people on earth, annual deaths from climate-related causes (extreme temperature, drought, flood, storms, wildfires) declined 98%--from an average of 247 per year during the 1920s to 2.5 in per year during the 2010s.
Data on disaster deaths come from EM-DAT, CRED / UCLouvain, Brussels, Belgium – www.emdat.be (D. Guha-Sapir).
Population estimates for the 1920s from the Maddison Database 2010, the Groningen Growth and Development Centre, Faculty of Economics and Business at University of Groningen. For years not shown, population is assumed to have grown at a steady rate.
Population estimates for the 2010s come from World Bank Data.

[8] NASA - Carbon Dioxide Fertilization Greening Earth, Study Finds
Regional trends vary, but overall the world's leaf area increased by 5.4 million square kilometers, or an amazon rainforest worth of greening, between 2000 and 2017 alone with over 1/3 of vegetated land showing an increase while only 5% showed a loss of green vegetation.

"Long-term satellite records reveal a significant global greening of vegetated areas since the 1980s, which recent data suggest has continued past 2010. … Global vegetation models suggest that CO2 fertilization is the main driver of global vegetation greening."

Piao, S., Wang, X., Park, T. et al. Characteristics, drivers and feedbacks of global greening. Nat Rev Earth Environ 1, 14–27 (2020)

NASA - GISS Surface Temperature Analysis

NOAA National Centers for Environmental Information - Daily Summaries Station Details, NY, Central Park

[9] Zhao et al. (2021)

[10] NOAA - Climate change rule of thumb: cold "things" warming faster than warm things

[11] BP - Statistical Review of World Energy

[12] "Other factors would also have a potentially significant effect on the results contained herein, but have not been examined in the scope of this current analysis. These additional factors, among others, could include: capacity value vs. energy value; network upgrades, transmission, congestion or other integration-related costs…" "This analysis does not take into account potential social and environmental externalities or reliability-related considerations" Lazard's Levelized Cost of Energy Analysis — Version 14.0.

[13] Alex Epstein - How the Biden Administration threatens energy security

[14] Our World in Data - Annual CO2 emissions by world region

[15] Global energy Monitor - Global Coal Plant Tracker

Alex Epstein

Energy Talking Points

The Energy Freedom Platform

By Alex Epstein

The Root of Our Global Energy Crisis Is Green Energy Fascism. The Solution Is Energy Freedom.

Originally published August 10, 2022

The root of our global energy crisis is Green Energy Fascism—the idea that government should have total control over the energy industry (fascism) used for the "green" goal of rapidly eliminating fossil fuels and nuclear.

The solution is Energy Freedom—the idea that individuals should be free to produce and consume energy how they judge best, provided they follow laws protecting everyone from truly harmful emissions and dangerous practices.

- There is no inevitability whatsoever to today's energy crisis.

In fact, if we replace Green Energy Fascism with Energy Freedom policies, America can lead the world in producing low-cost, reliable, plentiful, and cleaner energy—enriching ourselves and billions around the world.

- The Opportunity

Human beings' knowledge of how to produce energy in a way that is low-cost, reliable, and plentiful has never been greater. And we can do it in ever-cleaner ways.

As the world's leading economy and energy producer, America can lead a global energy renaissance.

- The Problem

Even though the world could be in an energy renaissance, it is instead in an energy crisis—with skyrocketing costs and declining reliability around the world.

Europe fears industrial collapse and citizens freezing this winter. Poorer regions fear literal starvation [1].

- The Cause

The direct cause of our unnecessary energy crisis is simple: governments around the world have restricted the supply of energy through anti-fossil-fuel, anti-nuclear policies. When demand outstrips supply, energy prices rise—which causes all other prices to rise.

- The Root Cause

The reason that governments have been able to get away with anti-fossil-fuel, anti-nuclear policies is that they have used the pretext of climate apocalypse to justify Green Energy Fascism—total government control over the energy industry to achieve "green" goals.

- Today's energy policy is fascist.

  Economically, "fascism" means: unlimited control of govt over private industry.

  Our leaders exercise such control by using mandates, subsidies, prohibitions, and ESG rules to dictate the behavior of energy producers and energy consumers.

  Using energy fascism to achieve "green" goals makes a terrible policy far worse.

  "Green" = "minimal impact." Since all energy impacts nature, all energy can be opposed as not "green."

  In practice, "green" energy means, disastrously: oppose everything but unreliable solar and wind.

- The pretext of Green Energy Fascism is "climate emergency." But:

  1) While humans impact climate, there is no emergency; in fact we're safer than ever from climate.
  2) Trying to lower CO2 emissions by favoring unreliable solar/wind and punishing nuclear is wildly ineffective.

- While "climate change"—humans impacting climate—is real, "climate emergency" is not. The world is slowly becoming warmer—at a cold point in geological history, when many more people die of cold than of heat. This doesn't at all justify rapidly restricting global fossil fuel use [2].
- Fossil fuels actually overall make us far safer from climate byproviding low-cost energy for the amazing machines that protect us against storms, protect us against extreme temperatures, and alleviate drought. Climate disaster deaths have decreased 98% over the last century [3].

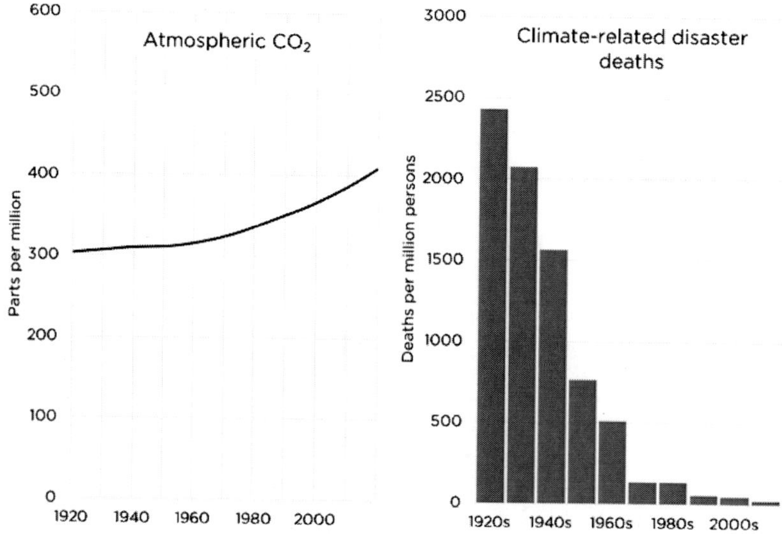

- The only practical way to lower CO2 emissions long-term is to make low-carbon energy globally cost-competitive.
- Green Energy Fascism policies of dictating the energy industry in favor of unreliables against nuclear—the most proven low-carbon form of energy—are unjustifiable.
- Green Energy Fascism not only ruins energy via dictatorial, anti-fossil-fuel/anti-nuclear control of industry, but also via "green" hostility to all development. Every aspect of energy is held back by "green" policies—including the massive mining and development solar and wind require [4].
- America and the world need an energy policy that ensures plenty of fossil fuel energy and fosters cost-competitive low- carbon alternatives like nuclear—making energy increasingly lower-cost, more reliable, more plentiful, cleaner, and lower- carbon.

That policy is Energy Freedom.

- The Energy Freedom Solution

America can enrich itself and empower billions with policies that free human ingenuity to produce energy in as low-cost, reliable, plentiful, and clean way as possible—while protecting us against harmful emissions and dangerous practices.

- The 5 steps to Energy Freedom are:

    1) Liberate responsible development
    2) End preferences for unreliable electricity
    3) Reform air/water emissions standards to incorporate cost-benefit analysis
    4) Reduce long-term CO_2 emissions via liberating innovation
    5) Decriminalize nuclear

1) Liberate responsible energy development

Anti-development policies prevent every form of energy from reaching its potential—from natural gas to nuclear to solar. Liberating responsible development throughout America will create unprecedented energy abundance and progress.

2) End preferences for unreliable electricity

Today's electric grids are being ruined by systemic preferences for unreliable electricity, which causes prices to rise and reliability to decline.

Eliminating them can help make America a leader in low- cost, reliable electricity [5].

3) Reform air/water emissions standards based on proper cost-benekt analysis.

Today's Environmental Protection Agency does huge harm to prosperity and health via energy-crippling emissions standards that fail any reasonable cost-benefit analysis.

Such analysis will lead to more energy and better health.

4) Reduce CO_2 emissions long-term through liberating innovation, not punishing America.

The only reasonable way to reduce CO_2 emissions long-term is innovation that makes low-carbon energy globally cost- competitive. Punishing US emissions is senseless self- destruction.

5) Decriminalize nuclear energy

The overregulation of low-carbon nuclear verges on criminalization, making nuclear costs 10X higher than they need to be. Decriminalizing nuclear, including radical reform of the Nuclear Regulatory Commission and Environmental Protection Agency, will make energy far cheaper, safer, and cleaner [6].

- Despite the fact that Green Energy Fascism policies are causing a global crisis, our leaders just passed a bill that should be called the Green Energy Fascism Act because it increased government control over energy, further punished fossil fuels, and did nothing to liberate nuclear [7].
- This Fall, Congressional and gubernatorial candidates have the opportunity to do the right thing and win over voters by rejecting this Administration and Congress's green energy fascism, and embracing Energy Freedom.
  Americans rightly want cheap energy and prosperity back.

References

[1] Reuters - BASF says it would stop output if gas supplies fell to half its needs Politico - Europe's looming coal crisis

[2] Zhao et al. (2021)

[3] For every million people on earth, annual deaths from climate- related causes (extreme temperature, drought, hood, storms, wildfires) declined 98%--from an average of 247 per year during the 1920s to 2.5 in per year during the 2010s.
Data on disaster deaths come from EM-DAT, CRED / UCLouvain, Brussels, Belgium – www.emdat.be (D. Guha-Sapir).
Population estimates for the 1920s from the Maddison Database 2010 come from the Groningen Growth and Development Centre, Faculty of Economics and Business at University of Groningen. For years not shown, population is assumed to have grown at a steady rate.
Population estimates for the 2010s come from World Bank Data.

[4] E&E News - Biden EPA deals major blow to Pebble mine

[5] Alex Epstein - An Open Letter to Senator Joe Manchin

[6] Lovering et al. (2016) - Historical construction costs of global nuclear power reactors

[7] Alex Epstein - Stop The Manchin Green New Deal

Next up: Manchin GND

September 26, 2023

Chairman Hunt, Ranking Member Gluesenkamp Perez, and members of the Subcommittee, thank you for the opportunity to testify today. My name is Kathleen Nisbet Moncy and I am the Chief Organizing Officer of Goose Point Oysters.

Our company is located in the heart of Willapa Bay Washington where we farm 2,000 acres of privately owned tidelands. Willapa Bay is one of 2 coastal estuaries in Washington that produces 25% of the nation's oysters with thriving export markets. The shellfish industry from Willapa Bay and the surrounding region generates tremendous economic impact. As the largest private employer in Pacific County, local shellfish farmers here and in Grays Harbor are responsible for nearly 2,000, rural, family- wage jobs, and contribute $102 million in economic output to our region. While shellfish farmers are the largest private employer in Pacific County, we still have one of the highest unemployment rates in our state. Shellfish farming in Washington State predates statehood. The tidelands where we farm are locally owned and passed down from generation to generation, one of the families has been farming for 150 years. We serve as ecological stewards keeping the water clean and protecting the sustainability of our farms. We are the backbone of our rural communities from jobs to kids in our local schools, shellfish farming is not just a livelihood, it's our way of life.

The ecosystem in which we farm is of upmost importance to us, as shellfish farmers depend on clean water and a healthy habitat. A single submerged oyster will filter seven gallons of water an hour which means farmers must protect the water while their oysters help to purify it. We are dedicated to managing the health of the ecosystem we farm in and as a result are regarded as one of the healthiest estuaries in the U.S. The impact to our delicate ecosystem with one oil spill would be devastating to our industry.

I would promote stricter regulations on offshore energy no matter the company size. Our oceans are delicate environments and ocean acidification is a real issue shellfish farmers face. We have been struggling with ocean acidification since 2008 and it will continue to worsen for the foreseeable future based on the amount of carbon dioxide we are putting into the atmosphere. The negative impacts due to ocean acidification are pH shifts, lack of calcium in the water and harmful algal blooms. We must fight to protect our waters and learn to evolve with the changes without adding anymore complications with offshore energy. Along with more stringent regulations for offshore energy there must be adequate funds set aside for any disasters that may arise. The Gulf was not prepared for the oil spill and the shellfish industry is still impaired due to it, we should learn from the mistakes made and ensure that they don't happen again.

Besides the biological issues that would arise with offshore energy our location does not have adequate services to respond to an oil spill. Our bay's waterways have not been properly managed by the Army Corps of Engineers and pose a danger for any type of response team. There is no positive scenario for our delicate ecosystem if there were to be a spill.

Along with the impacts to shellfish in our coastal estuaries we must also consider the Dungeness crab and salmon fishing communities. Shellfish tidelands are the rearing habitat for Dungeness crab and salmon. Dungeness crab fishing grounds are directly located off the Washington and Oregon coast. The access to current fishing grounds is already limited and further limitations would risk the livelihood of the fleet and the communities that process and sell the seafood. The cumulative impacts to the fishing grounds make it near impossible for future generations to be successful in the fishing industry. We should be focused on increasing seafood produced in the U.S. instead of polluting the very waters that provide it.

In conclusion, I do not support reduced regulations of offshore energy at the risk of the livelihoods of our rural communities. Shellfish is a sustainable seafood that has a huge benefit to the ecosystem by purifying the waters, reducing carbon and providing essential ecosystem services to the estuaries in which they are farmed. Our fishing communities must retain the available fishing grounds for the success of their industry. We must stand up and protect our waters from contamination and be responsible stewards of these delicate ecosystems for future generations.

Thank you for your time and considering the impact on rural communities.

In the United States District Court for the Western District of Louisiana Lake Charles Division

THE STATE OF LOUISIANA, by and through its Attorney General, JEFF LANDRY; AMERICAN PETROLEUM INSTITUTE; and CHEVRON U.S.A. INC.,

Plaintiffs,

v.

DEB HAALAND, in her official capacity as Secretary of the Interior; LAURA DANIEL-DAVIS, in her official capacity as Principal Deputy Assistant Secretary of the Interior for Land and Minerals Management; ELIZABETH KLEIN, in her official capacity as Director of the Bureau of Ocean Energy Management; JAMES KENDALL, in his official capacity as Director of the Gulf of Mexico Regional Office of the Bureau of Ocean Energy Management; U.S. DEPARTMENT OF THE INTERIOR; and BUREAU OF OCEAN ENERGY MANAGEMENT,

Defendants.

Hon. Judge James D. Cain, Jr.
Hon. Magistrate Judge Kathleen Kay

Consolidated cases:
Case No. 2:23-cv-1157 (Lead)
Case No. 2:23-cv-1167 (Member)

Unopposed Motion and Memorandum Requesting Leave to FileBrief of *Amici Curiae* 10 Members of the United States Congressin Support of Plaintiffs' Motion for Preliminary Injunction

Representative Bruce Westerman, Representative Garret Graves, Representative Pete Stauber, Representative Jerry Carl, Representative August Pfluger, Representative Wesley Hunt, Senator John A. Barrasso, Senator Ted Cruz, Senator Bill Cassidy, and Senator Cindy Hyde-Smith ("Amici") respectfully request that the Court allow the filing of an amici curiae brief (attached as Exhibit 1) in support of Plaintiffs' Motion for Preliminary Injunction [Dkt. No. 14].

In support of this motion, Amici state the following:

1) Amici are members of the United States Congress.
2) This case presents a recurring issue of paramount significance to the Amici: the Executive Branch's obligation to faithfully interpret and execute statutes that the Legislative Branch enacts. It is a foundational principle of our constitutional system that the three branches of government operate via checks and balances. Congress carries responsibility to draft the Nation's laws. Executive agencies, in turn, have a duty to faithfully execute those laws—neither adding to nor subtracting from what Congress has prescribed. When agencies in the Executive Branch modify or selectively apply statutes, they usurp the role that the Constitution assigned to Congress and undermine that balance.
3) Amici also have an interest in this case's specific subject matter. Amici include Members of Congress in leadership positions on the relevant U.S. House and Senate Committees with jurisdiction over offshore oil and gas development, and these Members have extensive knowledge of and expertise regarding the laws in question. Amici also represent States and congressional districts that will be directly affected by the outcome of this lawsuit.
4) The Inflation Reduction Act ("IRA") requires the Bureau of Ocean Energy Management ("Bureau") to conduct "Lease Sale 261 in accordance with the Record of Decision approved by the Secretary

on January 17, 2017." Pub. L. No. 117-169, § 50264(e), 136 Stat. 1818, 2060 (Aug. 16, 2022) (emphasis added). The IRA was the result of considerable deliberation concerning the economic, energy, environmental, and strategic interests of the United States:

- *Economic Interests:* Mineral leases provide a significant source of revenue for state and federal governments, and they also stimulate economic activity by creating jobs and spurring technological advancements.
- Energy Security: Minerals produced from our territorial waters reduce America's dependence on foreign sources. This bolsters our Nation's energy security and increases our resilience to fluctuations in the global market.
- *Environmental Oversight:* The Bureau's responsibility to oversee lease sales flows only from power that Congress has delegated. In the IRA, Congress recognized the years-long environmental evaluation that the Bureau has already undertaken as part of the process that led to the 2017 Record of Decision.
- *Strategic Positioning:* Leasing mineral rights from our territorial waters allows the United States to maintain a competitive edge globally, ensuring that we are not left behind in the race for resources.

5) Congress expressly struck a balance between these interests when it directed the Bureau to conduct the sale "in accordance with" the 2017 Record of Decision. Id.
6) The Bureau has no authority to revise that balance.
7) "District courts frequently welcome amicus briefs from non-parties concerning legal issues that have potential ramifications beyond the parties directly involved or if the amicus has unique information or perspective that can help the court beyond the help that the lawyers for the parties are able to provide." NGV Gaming, Ltd. v. Upstream Point Molate, LLC, 355 F. Supp. 2d 1061, 1067 (N.D. Cal. 2005) (internal quotation marks and citations omitted). Courts in this district have routinely granted requests to file amicus briefs. See, e.g., Louisiana v. Biden, 543 F. Supp. 3d 388, 397 (W.D. La., Lake Charles Division, June 15, 2021) (noting amicus briefs that were filed by local governments and by environmental associations); United States v. Catahoula Par. Sch. Bd., No. 1:69-CV-14430, 2020 WL 5261198, at *2 (W.D. La., Alexandria Division, Sept. 3, 2020)

(same for civil-rights association); Sweet Lake Land & Oil Co., LLC v. Exxon Mobil Corp., No. 2:09-CV-01100, 2011 WL 6300343, at *1 (W.D. La., Lake Charles Division, Dec. 16, 2011) (same for industry associations); Pel-Star Energy, Inc. v. U.S. Dep't of Energy, 890 F. Supp. 532, 535 (W.D. La., Shreveport Division, June 12, 1995) (same for shareholders).

8) Local Rule 7.4.1 Certificate. Pursuant to Local Rule 7.4.1, counsel for Amici sought consent from counsel for the parties to file this motion. Counsel for Plaintiffs stated that Plaintiffs are unopposed to the motion. Counsel for Defendants stated that Defendants take no position on the motion. Counsel for Intervenor-Defendants stated that Intervenor-Defendants take no position on the motion.
9) A proposed order is attached as Exhibit 2.
10) For the foregoing reasons, Amici respectfully request that this Court grant leave to file the attached brief as amici curiae.

Dated: September 15, 2023

Steven P. Lehotsky steve@lkcfirm.com
Michael B. Schon mike@lkcfirm.com
LEHOTSKY KELLER COHN LLP
200 Massachusetts Ave., NW Washington, DC 20001

Joshua P. Morrow josh@lkcfirm.com
LEHOTSKY KELLER COHN LLP
408 W. 11th Street, 5th Floor Austin, TX 78701

Respectfully submitted,

/s/ Kenneth H. Laborde
Kenneth H. Laborde (No. 8067)
Victoria E. Emmerling (No. 33117)
GIEGER, LABORDE & LAPEROUSE, LLC
701 Poydras Street, Suite 4800 New Orleans, Louisiana 70139 Telephone: (504) 561-0400
Facsimile: (504) 561-1011
Email: klaborde@glllaw.com
temmerling@glllaw.com

Certificate of Service

I hereby certify that a copy of the above and foregoing has been served on all counsel of record via the Court's electronic case management system on September 15, 2023.

/s/ Kenneth H. Laborde
KENNETH H. LABORDE

In the United States District Court for the Western District of Louisiana Lake Charles Division

THE STATE OF LOUISIANA, by and through its Attorney General, JEFF LANDRY; AMERICAN PETROLEUM INSTITUTE; and CHEVRON U.S.A. INC.,

Plaintiffs,

v

DEB HAALAND, in her official capacity as Secretary of the Interior; LAURA DANIEL-DAVIS, in her official capacity as Principal Deputy Assistant Secretary of the Interior for Land and Minerals Management; ELIZABETH KLEIN, in her official capacity as Director of the Bureau of Ocean Energy Management; JAMES KENDALL, in his official capacity as Director of the Gulf of Mexico Regional Office of the Bureau of Ocean Energy Management; U.S. DEPARTMENT OF THE INTERIOR;and BUREAU OF OCEAN ENERGY MANAGEMENT,

Defendants.

Hon. Judge James D. Cain, Jr.
Hon. Magistrate Judge Kathleen Kay

Consolidated cases:
Case No. 2:23-cv-1157 (Lead) Case No. 2:23-cv-1167 (Member)

Brief of Amici Curiae
10 Members of the United States Congress
in Support of Plaintiffs' Motion for Preliminary Injunction

Interest of Amici Curiae

Representative Bruce Westerman, Representative Garret Graves, Representative PeteStauber, Representative Jerry Carl, Representative August Pfluger, Representative Wesley Hunt, Senator John A. Barrasso, Senator Ted Cruz, Senator Bill Cassidy, and Senator Cindy Hyde-Smith("Amici ") are Members of the United States Congress.

This case presents a recurring issue of paramount significance to the Amici: the ExecutiveBranch's obligation to faithfully interpret and execute statutes that the Legislative Branch enacts.It is a foundational principle of our constitutional system that the three branches of government operate via checks and balances. Congress carries responsibility to draft the Nation's laws. Executive agencies, in turn, have a duty to faithfully execute those laws—neither adding to nor subtracting from what Congress has prescribed. When agencies in the Executive Branch modify or selectively apply statutes, they usurp the role that the Constitution assigned to Congress and undermine that balance.

Amici also have an interest in this case's specific subject matter. Amici include Members of Congress in leadership positions on the relevant U.S. House and Senate Committees with jurisdiction over offshore oil and gas development, and these Members have extensive knowledgeof and expertise regarding the laws in question. Amici also represent States and congressional districts that will be directly affected by the outcome of this lawsuit.

The Inflation Reduction Act ("IRA") requires the Bureau of Ocean Energy Management ("Bureau") to conduct "Lease Sale 261 in accordance with the Record of Decision approved by the Secretary on January 17, 2017." Pub. L. No. 117-169, § 502f64(e), 136 Stat. 1818, 2060 (Aug. 16, 2022) (emphasis added). The IRA was the result of considerable deliberation concerning the economic, energy, environmental, and strategic interests of the United States.

Mineral leases provide a significant source of revenue for state and federal governments,and they also stimulate economic activity by creating jobs and spurring technological advancements. Production from our territorial waters also reduces America's dependence onforeign sources, which bolsters our Nation's energy security and increases our resilience to fluctuations in the global market. Leasing minerals from our territorial waters also allows the United States to keep a competitive edge, ensuring we are not left behind in the race for resources.

The Bureau's authority to conduct lease sales flows only from the power that Congress vested through the Outer Continental Shelf Lands Act (OCSLA). In passing the IRA, Congress recognized the years-long environmental evaluation that the Bureau has already undertaken as partof creating the 2017 Record of Decision pursuant to the OCSLA's directions.

Congress expressly struck a balance between all of these interests when it directed the Bureau to conduct the sale "in accordance with" the 2017 Record of Decision. Id. Amici have an institutional interest in preventing the Bureau from attempting to revise that balance.

Introduction & Summary of Argument

The Bureau made a last-minute decision to remove millions of acres from Lease Sale 261 and to impose additional and burdensome vessel-travel restrictions on the remaining acreage (together, the "challenged provisions"). *Amici* respectfully urge the Court to grant the Plaintiffs' motion for preliminary injunction, for three reasons.

First, the challenged provisions violate the IRA's plain text. That statute requires the Bureau to conduct Lease Sale 261 "in accordance with" the 2017 Record of Decision. This requires more than bare attention or passing consideration—it requires conformity. The challenged provisions fail that requirement. Any deviations from the terms of sale that appear in the 2017 Record of Decision (whether as added restrictions on vessels or as subtractions from acreage) by definition *do not* conform to the sale process that Congress specifically prescribed. By attempting to circumvent Congress's express instructions, the Bureau not only sidesteps its immediate obligations to follow IRA to the letter, but it also undermines the foundational and delicate balanceof power that protects each Branch of government from the other's encroachment.

Second, even beyond IRA's text, there are considerable concerns regarding the proceduralapproach the Bureau took with regard to Lease Sale 261. The Bureau's unexpected pivot—without adequate explanation or consideration for stakeholders' reliance interests—exemplifies what the Fifth Circuit and other courts have dubbed an unlawful "surprise switcheroo." This procedural flaw is particularly problematic given the stakes. Louisiana and other major stakeholders like API, Chevron, and Shell risk substantial and harmful economic consequences as a result of the challenged provisions. Yet the Bureau did not explain the abrupt-about face in its reasoning for adopting the challenged provisions, nor did it exhibit awareness of the affected reliance interests.Congress has an interest in ensuring that agency decisionmaking is explained and supported by an administrative record and is not the result of a sweetheart settlement in litigation in which an executive agency, such as the Bureau, was not even a party.

Third, Congress has already addressed the factors that make Lease Sale 261 so significant: bolstering the economy, incentivizing technological progress, creating jobs, maintaining geopolitical strength and energy security, reducing dependence on foreign energy, preserving theGulf of Mexico's fragile ecosystems, and stewarding the Nation's resources by preventing them from going derelict. The challenged provisions are the Bureau's attempt to reassess and realign thebalance that Congress established. The Bureau lacks any such authority.

Argument

1) The Challenged Provisions Violate The IRA's Plain Text Because They Are Not "in accordance with" The 2017 Record of Decision.

The IRA requires that the Bureau "shall" conduct Lease Sale 261 "in accordance with" the 2017 Record of Decision. IRA § 50264(e). "The traditional, commonly repeated rule is that shall is mandatory" Antonin Scalia & Bryan A. Garner, Reading Law 112 (2012). So it is here. The same instance of "shall" dictates both the sale's timing and its process. The timing is mandatory ("not later than September 30"), and thus the process is likewise mandatory ("in accordance with the [2017] Record of Decision"). IRA § 50264(e). The Bureau argues that the IRA relates to timing only, but this view leaves no role for the words "in accordance with the [2017] Record of Decision." Contra Federal Defendants' Opposition to Plaintiff Louisiana Et Al.'s and Plaintiff Shell Offshore Oil Inc.'s Motions for a Preliminary Injunction [Dkt. No. 55] at 15, id. at 18 (arguing that this language merely "parrot[s]" OCSLA). But courts are "obliged to give effect, if possible, to every word Congress used." Nat'l Ass'n of Mfrs. v. Dep't of Def., 583 U.S. 109, 128– 29 (2018) (internal quotation marks and citation omitted). The only remaining interpretive question is what it means to conduct the Lease Sale "in accordance with" the 2017 Record of Decision.

"Accordance" means "conformity." The American Heritage Dictionary of the English Language 11 (5th ed. 2011); Accordance, Oxford English Dictionary Online. Congress chooses its words carefully. That care ensures that citizens, businesses, and agencies can understand the law—and can rely on it. Here, Congress chose the word "accordance," and that word's plain meaning carries Congress's intent. "In accordance with" is precise language that leaves no wiggle- room for the Bureau to improvise. Had Congress wished to allow the Bureau latitude to alter the 2017 Record of Decision, or to add or subtract from its terms of sale, then Congress would have used different language. But instead, Congress expressly dictated the sale process for Lease Sale 261. The Bureau cannot add terms to the process that Congress dictated any more than it can add days to the deadline that Congress specified.

The challenged provisions do not conform to—and therefore are not "in accordance with"—the process that Congress chose. The Bureau's brand-new restriction on vessel activity is not required under the 2017 Record of Decision. It makes no difference whether the restriction is viewed as an addition to the 2017 Record of Decision (a new stipulation) or as a subtraction from that Decision (removal of otherwise-available transport avenues). Neither view conforms to the 2017 Record of Decision, and thus both views deviate from the process that the IRA mandated. The Bureau's last-minute acreage withdrawal is likewise defective. It is a clear attempt to make an end-run around Congress's determination—plain in the IRA's text—that the 2017 Record of Decision is sufficient and that Lease Sale 261 should proceed in conformity with that Decision's terms. But, of course, the acreage withdrawal is not part of the 2017 Record of Decision.

The Bureau must take Congress at its word. Here that word is "accordance." As an agency in the Executive Branch, the Bureau's authority begins and ends with its responsibility to carry out the laws that Congress has enacted—all of them. By circumventing the IRA's directives regarding Lease Sale 261, the Bureau not only dodges its statutory obligations in this particular instance, but it also arrogates power to itself and undermines the balance of power between the branches of government. The Bureau's course of action regarding Lease Sale 261 also sends the distressing message that agencies can pick and choose which laws to follow, disregarding those that do not align with their immediate interests or preferences. The IRA's clear and unambiguous language requires the Bureau to conduct Lease Sale 261 exactly as Congress prescribed, without adding to or subtracting from the process that Congress specifically identified.

2) The Challenged Provisions Are A Procedurally Unlawful "Surprise Switcheroo."

Of parallel importance to IRA's text are the procedural requirements that Congress enacted as part of the Administrative Procedure Act, 5 U.S.C. §§ 551–559, 701–06. As the Plaintiffs have explained, the Bureau's actions concerning Lease Sale 261 raise significant issues of procedural irregularity. See Plaintiffs' Memorandum in Support of Motion for Preliminary Injunction ("Plaintiffs' Memorandum") [Dkt. No. 14-1] at 18–27. The Bureau's missteps are gravely concerning. In particular, Amici call the Court's attention to the Bureau's sudden about-face.

The Bureau acted unlawfully when it "pull[ed] a surprise switcheroo on" the states and the other stakeholders in Lease Sale 261. Env't Integrity Project v. EPA, 425 F.3d 992, 996 (D.C. Cir. 2005) (Sentelle, J.); accord Azar v. Allina Health Servs., 139 S. Ct. 1804, 1810 (2019) (citing the "surprise switcheroo" doctrine); Wages & White Lion Invs., LLC v. FDA, 16 F.4th 1130, 1138 (5th Cir. 2021) (granting motion for stay after concluding that plaintiffs were likely to succeed on the merits of their "surprise switcheroo" argument). "When an agency changes course, . . . it must be cognizant that longstanding policies may have engendered serious reliance interests that must be taken into account." DHS v. Regents of the Univ. of California, 140 S. Ct. 1891, 1913 (2020) (citing Encino Motorcars, LLC v. Navarro, 579 U.S. 211, 222 (2016)). Here, the Bureau neither explained its 180-degree turn nor adequately addressed the reliance interests that it was disrupting.

The Bureau did not explain its "change[d] course." Id. Despite a long-held position that existing protections were sufficient, the Bureau's new, stricter measures seem to have appeared from thin air. See Plaintiffs' Memorandum [Dkt. 14-1] at 22 (explaining that, "[u]ntil days ago, [the Bureau] has consistently found it unnecessary to adopt more protections for the Rice's whale outside of their 'core' habitat in the eastern Gulf that is already excluded from lease sales."). The Bureau's sudden decision to impose the challenged provisions is puzzling and unaccounted for given the Bureau's previous stance deeming additional protections unnecessary. This abrupt shift is a strong indication of arbitrary-and-capricious decision-making. Indeed, as the Plaintiffs have explained, the nature of the Bureau's decision appears to have more to do with placating parties in separate litigation than with addressing any genuinely new evidence. All the same, it is not the Plaintiffs' burden to provide convincing evidence that explains the Bureau's actions. Instead, it is the Bureau's burden to point to a reasonable explanation justifying its reversal. But despite the Lease Sale's significance, no such explanation is evident in the Bureau's decision.

The Bureau also failed to demonstrate that it was "cognizant" of the "serious reliance interests" that its decision affected. Id. The Plaintiffs, including Louisiana and various stakeholders like API, Chevron, and Shell, assert that they face serious and irreparable economic harm due to the challenged provisions related to Lease Sale 261. See, e.g., Plaintiffs' Memorandum [Dkt. No. 14] at 27–29. For Louisiana, the harm lies in potential lost revenues, as the State is entitled to a portion of the proceeds from the Lease Sale—proceeds that the challenged provisions threaten to diminish. The Bureau has not explained whether or how its decision to implement the challenged provisions accounted for Louisiana's reasonable reliance interest in those revenues.

Similarly, API, Chevron, and Shell face repercussions due to the challenged provisions. API's members stand to lose opportunities to purchase vast tracts of acreage. And even if theseacres were to be made available in the future, the delay's financial implications are significant and likely non-recoverable. Plaintiffs also point out that the challenged provisions, especially the stipulation concerning vessels, would generate hefty compliance costs and impact bidding strategy. This would alter the bidding dynamics, potentially affecting not just the prices but also the very identity of the winning bidder. If the challenged provisions remain, they threaten to distort the entire bidding process. Yet again, however, the Bureau has offered no explanation for how it treated these reliance interests.

For the foregoing procedural reasons, too, the challenged provisions are unlawful— separate and apart from the Bureau's failure to comply with the IRA's plain text.

3) Congress Already Struck The Balance Between The Various Policy Concerns That Lease Sale 261 Implicates.

The IRA balances diverse, complex, and overlapping considerations including growth and conservation, domestic needs and global positioning, and security and diplomacy. Of crucial importance here, however, is that Congress struck the balance between those interests when it dictated the timing and terms of sale for Lease Sale 261. Congress rarely intervenes in individual Lease Sales, nor does it make a habit of dictating the when and how for a specific sale. But here it has. That makes the Bureau's attempted re-balancing all the more stark—and all the more startling. Congress's special attention reflects the special significance that attaches to Lease Sale 261, and indeed to offshore leases in general. These leases represent critical investments in our national economy—as well as in state and local communities. Revenue from Lease Sale 261 will bolster federal and state coffers and will also drive economic expansion across the various sectors that service and supply operations at the Lease. Likewise, offshore leasing spurs innovation, fosters technological advancements, and paves the way for job creation. All these benefits create a ripple effect that reinforces secondary industries and local businesses, and that enriches regional and local economies while also easing the energy costs that everyday Americans must face.

Offshore leasing, and Lease Sale 261 in particular, also has heightened significance because using mineral resources from our own territorial waters allows us to rely less on foreign energy suppliers. In turn, this helps insulate the American people from unpredictable geopolitical shifts and potential market manipulations. Strategic self-reliance ensures a continuous and stable energy supply, reduces our vulnerability to international disruptions, and solidifies our position, enabling us to navigate international relations from a position of self-sufficiency. Likewise, by ensuring that Lease Sale 261 complies with the 2017 Record of Decision's rigorous environmental standards, Congress has voiced its commitment to protecting the Gulf of Mexico's delicate marine ecosystems while also ensuring that the Nation's energy supply does not sit unused.

By enacting IRA, Congress has already weighed each of these interests.

The Bureau does not have authority to second-guess the balance that Congress chose. But the challenged provisions do just that. The acreage-withdrawal removes millions of acres from sale, and the vessel restrictions add a last-minute stipulation. Neither term is "in accordance with" the 2017 Record of Decision. Congress was aware of the potential tradeoffs between efficiency and stewardship when it enacted the IRA. Yet Congress deemed the 2017 Record of Decision to strike a proper balance between those interests. The Bureau must honor that choice.

Conclusion

For the reasons set forth above, Amici respectfully urge that the Court grant the Plaintiffs' motion for preliminary injunction.

Dated: September 15, 2023

Steven P. Lehotsky steve@lkcfirm.com
Michael B. Schon mike@lkcfirm.com
LEHOTSKY KELLER COHN LLP
200 Massachusetts Ave., NW Washington, DC 20001

Joshua P. Morrow josh@lkcfirm.com
LEHOTSKY KELLER COHN LLP
408 W. 11th Street, 5th Floor Austin, TX 78701

Respectfully submitted,

/s/ Kenneth H. Laborde

Kenneth H. Laborde (No. 8067)
Victoria E. Emmerling (No. 33117)
GIEGER, LABORDE & LAPEROUSE, LLC
701 Poydras Street, Suite 4800 New Orleans, Louisiana 70139
Telephone: (504) 561-0400
Facsimile: (504) 561-1011
Email: klaborde@glllaw.com;
temmerling@glllaw.com

Certificate of Service

I hereby certify that a copy of the above and foregoing has been served on all counsel of record via the Court's electronic case management system on September 15, 2023.

/s/ Kenneth H. Laborde
KENNETH H. LABORDE

In the United States District Court for the Western District of Louisiana Lake Charles Division

THE STATE OF LOUISIANA, by and through its Attorney General, JEFF LANDRY; AMERICAN PETROLEUM INSTITUTE; and CHEVRON U.S.A. INC.,

Plaintiffs,

v

DEB HAALAND, in her official capacity as Secretary of the Interior; LAURA DANIEL-DAVIS, in her official capacity as Principal Deputy Assistant Secretary of the Interior for Land and Minerals Management; ELIZABETH KLEIN, in her official capacity as Director of the Bureau of Ocean Energy Management; JAMES KENDALL, in his official capacity as Director of the Gulf of Mexico Regional Office of the Bureau of Ocean Energy Management; U.S. DEPARTMENT OF THE INTERIOR;and BUREAU OF OCEAN ENERGY MANAGEMENT,

Defendants.

Hon. Judge James D. Cain, Jr.
Hon. Magistrate Judge Kathleen Kay

Consolidated cases:
Case No. 2:23-cv-1157 (Lead)
Case No. 2:23-cv-1167 (Member)

[Proposed] Order

Upon consideration of the Motion and Memorandum Requesting Leave to File Brief of Amici Curiae 10 Members of the United States Congress, it is hereby ORDERED that the motion is GRANTED. The proposed brief that accompanied the motion shall be deemed to have been filed and served by ECF on the date of this Order.

Dated: ______________________________

James D. Cain, Jr.
United States District Judge

United States District Court Western District of Louisiana Lake Charles Division

STATE OF LOUISIANA ET AL	CASE NO. 2:23-CV-01157
VERSUS	JUDGE JAMES D. CAIN, JR.
DEB HAALAND ET AL	MAGISTRATE JUDGE KAY

Memorandum Order

Before the court are Motions for Preliminary Injunction[52] filed, respectively, by plaintiffs the State of Louisiana, American Petroleum Institute, and Chevron USA Inc., and by plaintiff Shell Offshore Inc. Plaintiffs seek to halt the addition of a term to Lease Sale 261 by the Bureau of Ocean Energy Management and the withdrawal of six million acres from that sale. The motion is opposed by both the government defendants originally named in the suits as well as intervenor-defendants Center for Biological Diversity, Friends of the Earth, Sierra Club, and Turtle Island Restoration Network. Docs. 55, 70. The matter came before the court for hearing on September 21, 2023, and the undersigned now issues this decision.

I. Introduction

This suit arises from the Bureau of Ocean Energy Management ("BOEM")'s plans regarding "Lease Sale 261," an oil and gas lease on the Outer Continental Shelf in the Gulf of Mexico. Lease Sale 261 is proposed for sale on September 27, 2023, in accordance with deadlines set under the Inflation Reduction Act ("IRA"). On August 23, 2023, BOEM issued a Final Notice of Sale that withdrew six million acres from the lease and inserted a new lease stipulation restricting vessel activity in the lease area. The agency justified

---

[52] The first motion was filed in lead case *Louisiana v. Haaland*, No. 23-cv-1157, at doc. 14. The second was filed in member case *Shell Offshore Inc. v. U.S. Department of the Interior*, No. 23-cv-1167 (W.D. La.), at doc. 4.

both measures as necessary for the protection of Rice's whale, a species of baleen whale native to the northeastern Gulf of Mexico that is protected under the Endangered Species Act.

Plaintiffs allege that the last-minute insertion of these provisions violates (1) the IRA, which directed BOEM to conduct Lease Sale 261 in accordance with the agency's previously adopted Five-Year Plan for oil and gas leasing, (2) the procedural requirements of the Outer Continental Shelf Lands Act, and (3) the Administrative Procedure Act ("APA"), insofar as they represent an arbitrary and capricious change in position by BOEM. Accordingly, plaintiffs have filed suit in this court seeking declaratory and injunctive relief. They also seek a preliminary injunction, asserting that delaying the sale or allowing it to proceed with the challenged provisions will result in irreparable harm. Defendants and intervenor-defendants[53] oppose the motion, arguing that (1) the challenged measures are well-supported and were properly implemented, (2) plaintiffs have not met the high bar for showing that preliminary injunctive relief is warranted, and (3) the court lacks the authority to grant the requested relief. Docs. 55, 70. An amicus brief has also been filed in support of plaintiffs by ten members of the United States Congress. Doc. 66.

II. Background

A. Statutory Background

I. Outer Continental Shelf Lands Act

The Outer Continental Shelf Lands Act of 1953 ("OCSLA") provides the framework through which the Department of the Interior leases areas of the Outer Continental Shelf for oil and gas exploration and development. Ctr. for Biological Diversity v. U.S. Dep't of Interior, 563 F.3d 466, 472 (D.C. Cir. 2009). This is accomplished through a four-stage process, "proceeding from broad-based planning to an increasingly narrower focus as actual development grows more imminent." California ex rel. Brown v. Watt, 668 F.2d 1290, 1297 (D.C. Cir. 1981). The steps are: (1) formulation of a five-year leasing plan, (2) lease sales, (3) exploration by the lessees, and (4) development and

[53] The court permitted the intervention of four environmental nonprofit organizations (Sierra Club, Center for Biological Diversity, Friends of the Earth, and Turtle Island Restoration Network) as defendants, based on their advocacy on behalf of Rice's whale in a suit in the District of Maryland, described infra, and the relation between that suit and the measures challenged here. *See* doc. 58.

production. Secretary of the Interior v. California, 464 U.S. 312, 337 (1984). The Secretary's delegated authority in this area is divided between BOEM and the Bureau of Safety and Environmental Enforcement ("BSEE"), with BOEM overseeing lease sales and BSEE bearing primary responsibility for drilling operation safety and environmental protection. See 76 Fed. Reg. 64,432 (Oct. 18, 2011).

At the first stage, formulation of the leasing plan, BOEM coordinates with other agencies to evaluate the impacts of the program and to balance the potential for discovery of oil and gas with the potential for environmental damage and adverse impacts to the coast. 43 U.S.C. §§ 1344(a)(1)–(3), (b)(3); 42 U.S.C. § 4332(C). OCSLA also requires that the Secretary consider the impact on affected states and communicate with the governors of those states regarding their concerns. 43 U.S.C. §§ 1344(a)(2)(F), 1345. This consultation results in the development of a "schedule of proposed lease sales . . . which [the Secretary] determines will best meet national energy needs" for the applicable period, consistent with the balancing of environmental and social factors with economic and energy needs. 43 U.S.C. § 1344(a).

At the second stage Interior solicits bids and issues leases. 43 U.S.C. § 1337. To prepare for an individual lease sale, BOEM issues a "Call for Information and Nominations" on an area proposed in the Five-Year Plan through publication in the Federal Register. 30 C.F.R. § 556.301. To this end it requests comments from industry and the public on the sale area's potential for drilling as well as its "socioeconomic, biological, and environmental information." Id. BOEM considers these factors in determining the scope and terms of the lease, which are published in a proposed notice of sale ("PNOS"). Id. at§§ 556.302(b), 556.304(a). Upon approval by the Secretary, BOEM also sends the PNOS to the governors of affected states and publishes notice of its availability in the Federal Register. Id. at § 556.304(c). Governors of affected states may submit comments and recommendations to BOEM on the size, timing, and location of the proposed sale. Id. at § 556.305(a). The Secretary will accept these recommendations if it determines that they "provide a reasonable balance between the national interest and the well-being of the citizens of the State[.]" Id. at § 556.307(b). Finally, BOEM considers comments received in response to the proposed notice before publishing a final notice of sale ("FNOS"). Id. At §§ 556.307(a); 556.308(a). The sale is then conducted by a sealed-bid auction held at least 30 days after publication of the FNOS. Id. at § 556.308(a)–(b).

"A lessee does not, however, acquire an immediate or absolute right" to exploration and production in the lease area upon securing the lease; "those activities require separate, subsequent federal authorization." Interior, 464 U.S. at 317. At the third stage, Interior reviews and determines whether to approve exploration plans. Ctr. for Biological Diversity, 563 F.3d at 473. The Secretary will reject a plan if he determines that it may result in, inter alia, serious harm to the environment and "cannot be modified to avoid such condition." 43 U.S.C. § 1340(c)(1) (citing 43 U.S.C. § 1334(a)(2)(A)(i)). Finally, at the fourth stage, more detailed plans are required and reviewed in relation to the construction of platforms, installation of processing equipment, and the laying of pipelines. See 43 U.S.C. § 1351. Here, again, the plan and consequently the leasing program may be terminated if Interior finds a risk of serious harm or damage "to the marine, coastal or human environments." Ctr. for Biological Diversity, 563 F.3d at 473 (quoting 43 U.S.C. § 1351(h)(1)(D)(i)).

2. Endangered Species Act

The Department must also comply with federal environmental laws in its oversight of offshore leasing. Section 7(a)(2) of the Endangered Species Act ("ESA") provides:

> Each Federal agency shall, in consultation with and with the assistance of the Secretary [of the Interior], insure that any action authorized, funded, or carried out by such agency . . . is not likely to jeopardize the continued existence of any endangered species or threatened species or result in the destruction or adverse modification of habitat of such species which is determined by the Secretary, after consultation as appropriate with affected States, to be critical, unless such agency has been granted an exemption for such action by the Committee pursuant to subsection (h) of this section. In fulfilling the requirements of this paragraph each agency shall use the best scientific and commercial data available.

16 U.S.C. § 1536(a)(2). If the proposed action may affect a listed species, the agency must formally consult with either the National Marine Fisheries Service ("NMFS") or the Fish and Wildlife Service ("FWS"), depending on whether the species is marine or terrestrial. Oceana v. Bureau of Ocean Energy Mgmt., 37 F.Supp.3d 147, 151 (D.D.C. 2014); see 50 C.F.R. §§ 402.01(b), 402.14(a).

NMFS, as applies here, then prepares a biological opinion using "the best scientific and commercial data available" to evaluate the proposed action's impact on the species. 16 U.S.C. § 1536(a)–(b); 50 C.F.R. § 402.12. Here it considers whether the proposed action is likely to violate the ESA by jeopardizing the continued existence of a listed species or resulting in the destruction or adverse modification of a critical habitat. 16 U.S.C. § 1536(a)(2); 50 C.F.R. § 402.14(h)(1)(iv). If NMFS concludes that such a violation is likely and issues a "jeopardy" opinion, it must provide "reasonable and prudent alternatives" ("RPAs"). 16 U.S.C. § 1536(b)(3)(A); 50 C.F.R. § 402.14(h)(1)(iv), (h)(2). RPAs are defined under the regulations as:

> alternative actions identified during formal consultation [1] that can be implemented in a manner consistent with the intended purpose of the action, [2] that can be implemented consistent with the scope of the Federal agency's legal authority and jurisdiction, [3] that is economically and technologically feasible, and [4] that the Director believes would avoid the likelihood of jeopardizing the continued existence of listed species or resulting in the destruction or adverse modification of critical habitat.

50 C.F.R. § 402.02. Following the issuance of a "jeopardy" opinion, Interior must either terminate the action, implement the proposed alternative, or seek an exemption from the Cabinet-level Endangered Species Committee. Nat'l Ass'n of Home Builders v. Defenders of Wildlife, 551 U.S. 644, 653 (2007).

B. Factual Background

1. Approval of the 2017-22 Leasing Program

The Secretary approved the 2017-22 Five-Year Plan on January 17, 2017, in a Record of Decision that directed BOEM to proceed with 10 scheduled leases in the Gulf and one in Alaska's Cook Inlet. Doc. 14, att. 22. The ten Gulf of Mexico sales were thereby scheduled to occur over the five years of the program, with one sale in 2017, two each in 2018–2021, and one (Lease Sale 261) in 2022. Id. at 4. The Gulf of Mexico sales were to be "region-wide and include unleased acreage not subject to moratorium or otherwise unavailable, in the Western, Central, and Eastern Gulf," with the goal of "provid[ing] greater flexibility to industry, including more frequent opportunities to bid on rejected, relinquished, or expired OCS lease blocks." Id. This decision adopted

BOEM's Proposed Final Program and Programmatic Environmental Impact Statement ("EIS"), the latter of which evaluated the environmental impact of sales of differing scopes. Id. at 2–4. In reference to Rice's whale (then called Bryde's whale), the Programmatic EIS concluded that the "biologically important area" for the whale off the Florida coast did not overlap with the Gulf of Mexico program areas. Doc. 14, att. 24, p. 73. The evaluation assumed the continued implementation "of protective measures required by statute, regulation, or current lease stipulations that would likely continue to be adopted in the future." Doc. 14, att. 24, p. 50. Among these were several measures relating to protection of endangered species. Doc. 14, att. 25, pp. 548–58.

2. Additional Measures Adopted for Protection of Rice's Whale

Following approval of the Five-Year Plan BOEM and BSEE consulted with NMFS to determine the potential impact of the plan on protected species over the next fifty years. This culminated with NMFS's issuance of a biological opinion in March 2020 ("2020 BiOp"). See doc. 16, att. 2, pp. 28–29. NMFS concluded that oil and gas activity posed a risk to Rice's whale in portions of the eastern Gulf, particularly from vessel strikes. Id. at 578–81. NMFS determined that the jeopardy could be avoided, however, with an RPA consisting of several vessel operating conditions within the purple area shown below:

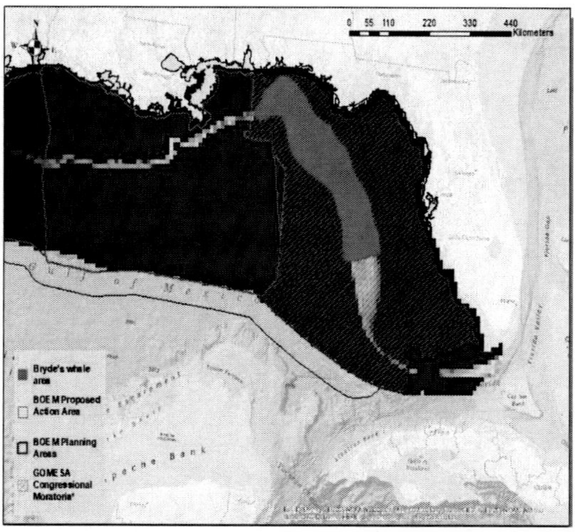

Id. at 624–26.

BOEM disagreed with NMFS's analysis and maintained that vessel traffic from the leases was unlikely to cross the identified habitat.[54] Doc. 16, att. 1, p. 232. Nevertheless, BOEM and BSEE formally accepted NMFS's RPA for Rice's whale and included it as a stipulation in future lease sales under the Five-Year Plan. *Id.*; *see* doc. 14, att. 20, p. 9 (FNOS for Lease Sale 256); doc. 14, att. 18, p. 9 (FNOS for Lease Sale 257); doc. 14, att. 10, p. 9 (PNOS for Lease Sale 261). Meanwhile, environmental groups filed suit against NMFS in the District of Maryland.[55] They argued that the 2020 BiOp understated the risks of oil and gas activity to various species, including Rice's whale, particularly through the threat of oil spills after the Deepwater Horizon disaster and the failure to consider alleged changes in habitat following that spill. *See Sierra Club v. NMFS*, No. 8:20-cv-3060 (D. Md.), at doc. 1.

3. Lease Sales Paused and Then Resumed Under Inflation Reduction Act

In January 2021, President Biden entered an executive order directing the Department of the Interior to pause new oil and gas lease sales pending a comprehensive review. 86 Fed. Reg. 7,619, 7,624 (Jan. 27, 2021). At this point seven of the 10 Gulf lease sales scheduled under the Five-Year Plan had been conducted, with Lease Sales 257, 259, and 261 still pending. See Louisiana v. Biden, 622 F.Supp.3d 267, 287–89 (W.D. La. 2022). Louisiana and several other states sued to enjoin the pause and the district court granted the preliminary injunction. *Louisiana v. Biden*, 543 F.Supp.3d 388 (W.D. La. 2021). BOEM then proceeded with Lease Sale 257 in November 2021. *See* doc. 14, att. 17 (Lease Sale 257 FNOS). Before the leases could be issued to the winning bidders, however, environmental groups challenged the sale through a lawsuit based on Interior's alleged failure to consider foreign oil consumption in its pre-sale EIS. *See Friends of the Earth v. Haaland*, 583 F.Supp.3d 113 (D.D.C. 2022). The court vacated BOEM's Record of Decision to conduct the lease sale and industry parties intervened and appealed. *Friends of the Earth v. Haaland*, 2023 WL 3144203 (D.C. Cir. Apr. 28, 2023).

---

[54] Specifically, it noted (as NMFS had in the 2020 BiOP) that the whale's habitat was largely in the area of the Gulf of Mexico which had been under a congressional moratorium from leasing and was now subject to presidential withdrawal. Doc. 16, att. 1, p. 232. BOEM determined that vessels expected to service the leases were more likely to use ports closer to the Western and Central Planning Areas, and thus "unlikely to transit across greater distances through the withdrawal area to get to the leases." *Id.*

[55] As noted above, the plaintiffs in that matter are the same groups as the defendant-intervenors in this case.

In August 2022, with the appeal of *Friends of the Earth* pending, Congress passed the Inflation Reduction Act ("IRA"), PL 117–169. There it ordered Interior to proceed with Lease Sale 257, in accordance with the terms and conditions in the FNOS. *Id.* at § 50264(b). It also ordered that the remaining lease sales be scheduled by specific dates, notwithstanding the expiration of the previous Five-Year Program, "in accordance with the Record of Decision approved by the Secretary on January 17, 2017": Lease Sale 258 (Cook Inlet) by December 31, 2022, Lease Sale 259 by March 31, 2023, and Lease Sale 261 by September 30, 2023. *Id.* at § 50264(c)–(e).

4. BOEM Conducts Further Environmental Review and Proceeds with Lease Sale 259

In October 2022 BOEM and BSEE sent a letter to NMFS requesting reinitiation of consultation on the 2020 BiOp, based on the issues raised in the District of Maryland suit. Doc. 55, att. 1. BOEM then issued a final supplemental EIS on the two remaining Gulf lease sales, 259 and 261, in January 2023.[56] Doc. 16, att. 1. In the public comment period on this document, environmental groups urged BOEM to remove blocks in the western and central Gulf from the leasing area for the protection of Rice's whale and to expand the areas where mitigation measures were applied. Id. at 469, 626–27. In support they referenced a study published in July 2022 by Melissa Soldevilla and others, positing broader distribution of Rice's whale throughout the Gulf based on passive acoustic data. Id. at 560.

Upon review of the study, BOEM determined that "not enough information is available at this time to confirm [the whales'] distribution or any seasonal movements outside of the core area that is already considered in this Supplemental EIS." Id. It also stated that its environmental analyses had not identified justifiable reasons to restrict the lease sale area and that it believed its stipulations and mitigations provided adequate environmental protection. Id. at 469. It further noted that "any individual lease sale could be scaled back during the prelease sale process to offer a smaller area should circumstances warrant." Id. BOEM also advised that "BOEM and BSEE's review of plans, permits, and/or authorizations at the post-lease stage includes review of any planned transits through the Rice's whale core habitat." It noted that "[a]t this time, critical habitat has not been identified for the Rice's whale"

---

[56] There it acknowledged that it had no discretion on whether to conduct these sales, but stated that it was preparing the supplemental EIS "to follow its normal leasing process to the fullest extent possible" *Id.* at 6.

and that it consulted with NMFS and FWS to determine the required mitigating measures. Id. at 626. Accordingly, BOEM stated in the final supplemental EIS:

> BOEM believes the potential for vessel strikes to sperm and Rice's whale is extremely unlikely to occur due to the generally slow vessel transiting and surveying speeds, limited vessel routes originating from the eastern GOM, and the additional mitigations on vessels within the Rice's whale core area (as defined by the 2020 GOM Biological Opinion [BiOp]) (Soldevilla et al. 2022). The core area has been changing over the years as baseline information becomes available (Rosel and Garrison 2022). *BOEM will continue to monitor current literature and work with NMFS as it relates to consultations, though the conclusions found in the 2017-2022 GOM Multisale EIS and 2018 GOM Supplemental EIS still remain valid.*

Id. at 170 (emphasis added). Lease Sale 259 was then conducted as scheduled in March 2023, as a regionwide sale of "all of the available unleased acreage in the GOM OCS" with unrelated minor adjustments to the lease stipulations.[57] Doc. 14, att. 14 (Lease Sale 259 FNOS). The sale was closed with industry bidding nearly $310 million for 313 tracts covering 1.6 million acres, and BOEM accepting bids and issuing leases for 295 tracts. See Oil and Gas Lease Sale 259, Final Bid Recap (Mar. 29, 2023), available at bit.ly/3EfOhqQ. That month, BOEM also issued a PNOS for Lease Sale 261 with the same stipulations. Doc. 14, atts. 8 & 10. BOEM indicated, however, that it was "considering removing the area comprising the northeastern Gulf of Mexico and continental shelf break between the 100 meters and 400 meters in depth isobaths to protect Rice's Whales that may transit through the area." Doc. 14, att. 8, p. 7. It also expressly "reserve[d] the right to modify the sale area in the Final NOS, including removing additional areas from Lease Sale 261." Id. As required by statute, the PNOS was published in the Federal Register and subject to comment from the governors of affected states. In response the governors of Alabama, Mississippi, and Texas all objected to the proposed withdrawal. Doc. 55, att. 3, p. 23.

---

[57] BOEM converted three stipulations (for Baldwin County, topographic features, and live bottom) into area exclusions and added a stipulation for royalties, as required by § 50263 of the IRA. *See* doc. 14, att. 14, pp. 14–15.

5. BOEM Shifts Approach to Rice's Whale and Adopts Challenged Provisions for Lease Sale 261's Final Notice of Sale

Despite the conclusions of the supplemental EIS, the government soon appeared to change its mind regarding the protections needed for Rice's whale. In a stipulation filed on July 21, 2023, in the Sierra Club suit, NMFS and plaintiffs agreed to stay proceedings while NMFS worked to update the challenged 2020 BiOP. No. 8:20-cv-3060 (D. Md.), at doc. 147. The stipulation indicated that this consultation had been prompted by the letter from BOEM and BSEE to NMFS, requesting to reopen consultation on the 2020 BiOp. Id. at doc. 147, p. 2. The stipulation also indicated that NMFS might designate a critical habitat for Rice's whale and that BOEM would issue notice to lessees and operators during the reinitiated consultation. Id. at doc. 147, pp. 2–3; see 88 Fed. Reg. 47,453, 47,460 (proposed designation). Finally, the parties agreed that a lease stipulation for the protection of Rice's whale would be added to Lease Sale 261 and any subsequent Gulf of Mexico oil and gas leases during the reinitiated consultation. Sierra Club, No. 8:20-cv-3060, at doc. 147, pp. 2–3; see id. at doc. 147, att. 2 (proposed lease stipulation). The proposed stipulation would add several vessel mitigation conditions between the 100- to 400-m -isobaths across the northern Gulf of Mexico on the OCS, eastward from the Mexican border with Texas and westward of the newly designated Rice's whale area shaded in blue below:

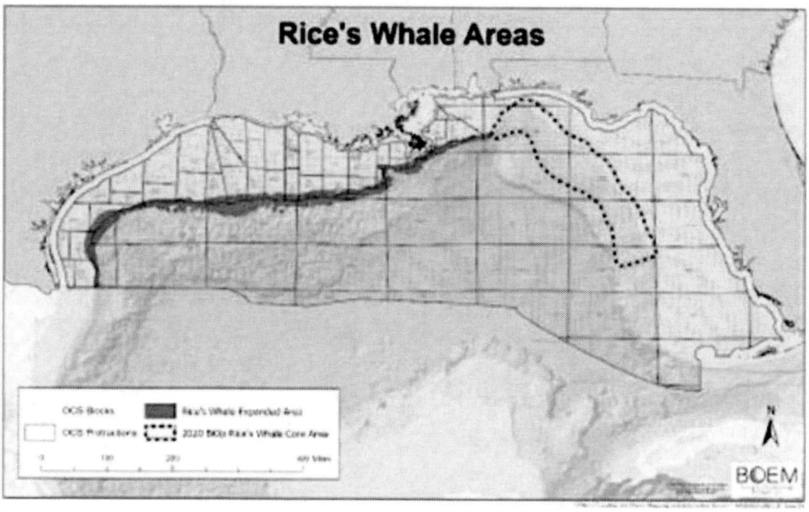

Id. at doc. 147, att. 2.

The new mitigation measures include the presence of visual observers on vessels, speed restrictions, limitations on transit "after dusk and before dawn, and during other times of low visibility," and use of Automatic Identification Systems on vessels. Id. The district court granted the motion and stayed proceedings. Id. at doc. 154.

The FNOS for Lease Sale 261 was issued on August 23, 2023, incorporating the new measures in Stipulation 4(B)(4) and covering a lease area of approximately 67.3 million acres. Doc. 14, att. 4, pp. 11–13; see doc. 16, att. 5, pp. 4–5. BOEM also confirmed that it was withholding six million acres in the expanded Rice's whale area from the sale, as alluded to in the PNOS. Doc. 14, att. 2, pp. 11–13. The Department of the Interior explained the actions as follows in a Record of Decision:

> The existing 2020 Biological Opinion, as amended, remains in effect until the reinitiated consultation is completed and a new or amended Biological Opinion becomes available. During the reinitiation process, BOEM will continue to implement the Reasonable and Prudent Alternative for the Rice's whale, and comply with all Reasonable and Prudent Measures and Terms and Conditions under the existing 2020 Biological Opinion, as amended. This includes continuing to request step-down reviews for the prescribed activities and implementing and adaptively managing the mitigation, monitoring, and reporting requirements (2020 Biological Opinion Appendices and/or Conditions of Approval, as amended) imposed by the Bureaus on plans and permits, and as coordinated with NMFS and industry. *In addition, based on a recent study that the endangered Rice's whale occurs in portions of the northern Gulf of Mexico between the 100-meter and 400-meter isobaths eastward from the Mexico border with Texas and westward of the Rice's Whale Core Area identified in the 2020 Biological Opinion (as amended in April 2021), removing this area from the lease sale could reduce risks to this species while reinitiated consultation with NMFS is ongoing.* Further, the updated Protected Species Stipulation includes interim measures to require certain speed restrictions and other measures between the 100- meter to 400-meter isobaths. These measures will remain in place while the reinitiated consultation is ongoing and until a new or amended biological opinion is issued by NMFS.

Doc. 14, att. 5, pp. 13–14 (emphasis added).

On August 24, 2023, plaintiffs filed their petition for declaratory and injunctive relief in this court. Doc. 1. Four days later they filed a motion for preliminary injunction. Docs. 9, 14. Plaintiffs ask the court to enjoin BOEM from implementing the changes in the FNOS for Lease Sale 261 and argue that these violate the APA and the IRA. Doc. 14; Shell Offshore Inc., No. 2:23-cv-1167, at doc. 4. The two suits have been consolidated for briefing purposes and the environmental group plaintiffs from the District of Maryland suit have been granted leave to intervene. Docs. 52, 58. Government defendants and intervenor defendants oppose the motion. Docs. 53, 55, 56.

III. Law & Application

To obtain a preliminary injunction, the movant must show (1) a substantial likelihood of success on the merits of its claims; (2) a substantial threat of irreparable injury if the injunction is not granted; (3) that the threatened injury to the movant outweighs the harm the injunction may do to the nonmovant; and (4) that granting the injunction will not disserve the public interest. Anderson v. Jackson, 556 F.3d 351, 360 (5th Cir. 2009). "The grant of injunctive relief is an extraordinary remedy which requires the movant to unequivocally show the need for its issuance." Valley v. Rapides Par. Sch. Bd., 118 F.3d 1047, 1050 (5th Cir. 1997). The decision lies within the district court's discretion. Allied Mktg. Grp., Inc. v. CDL Marketing, Inc., 878 F.2d 806, 809 (5th Cir. 1989).

A. Likelihood of Success on the Merits

Plaintiffs advance three legal grounds for voiding the challenged lease terms: (1) their implementation was procedurally invalid; (2) the decision amounted to arbitrary and capricious agency action; and (3) the implementation of the terms violated the IRA. Finding a sufficient showing of merit under the first two arguments, the court does not consider the third. First, however, the court considers procedural objections lodged by defendants.

1. Defendants' Procedural Objections

Both sets of defendants have asserted that plaintiffs are unlikely to succeed on the merits based on certain procedural objections. Intervenor defendants maintain that this court is an improper venue while government defendants contend that plaintiffs have violated OCSLA's requirement that potential litigants provide Interior with notice at least 60 days before filing suit. 43 U.S.C. § 1349(a)(2).

As to venue, a civil suit against a federal agency or official may only be brought in the judicial district where:

> (A) a defendant in the action resides, (B) a substantial part of the events or omissions giving rise to the claim occurred, or a substantial part of property that is the subject of the action is situated, or (C) the plaintiff resides if no real property is involved in the action.

28 U.S.C. § 1391(e)(1). Plaintiffs base venue on § 1391(e)(1)(B) and (C) in both the lead case, Louisiana v. Haaland, and member case Shell Offshore Inc. v. U.S. Department of Interior. Intervenor defendants argue that venue is improper in the latter suit because Shell Offshore is the lone plaintiff and has admitted that it is headquartered in Houston, though it maintains facilities in this district in support of its offshore leasing program. Shell Offshore Inc., No. 2:23-cv-1167, at doc. 1, ¶¶ 5, 13. Intervenor defendants also assert that venue is improper under § 1391(e)(1)(B) because the challenged decision-making occurred in Washington, D.C. and the stipulated agreement was signed in Maryland.

On the venue objection, the issue has not been properly raised through a motion to transfer or dismiss. The court declines to issue an advisory opinion on the issue. On government defendants' procedural objection, the court finds no merit. First, as defendants point out, these provisions apply only to citizen suits under OCSLA. They are irrelevant when plaintiffs seek review under the APA. Hornbeck Offshore Servs., LLC v. Salazar, 696 F.Supp.2d 627, 636 (E.D. La. 2010). And even if the notice requirements did apply, they are likely excused under § 1349(a)(3) because "the alleged violation constitutes an imminent threat to the public health or safety or would immediately affect a legal interest of the plaintiff." 43 U.S.C. § 1349(a)(3).[58] Plaintiffs' failure to

---

[58] Defendants maintain that plaintiffs' prospective interest in the parcels does not qualify as a legal interest and so they have no recourse for challenging any changes in the FNOS. This position is based on an unpublished decision from the D.C. Circuit, wherein the court held that a scheduled wind lease auction did not immediately affect plaintiffs' legal interests because the leases themselves did not authorize any activity in the leased area. *Fisheries Survival Fund v. Haaland*, 858 F. App'x 371, 374 (D.C. Cir. 2021). As explained in *Friends of the Earth*, however, a case involving Lease Sale 257, the lease sale stage for oil and gas leases authorizes certain ancillary activities including geological and geophysical exploration. 583 F.Supp.3d at 133. Additionally, the lease sale form grants the lessee exclusive rights to drilling and production and the applicable regulations provide "an irretrievable commitment of resources in the sense that once a lease is issued, BOEM

provide notice therefore has no impact on their ability to proceed with these claims.

Plaintiffs contend that the implementation of the challenged provisions and acreage withdrawal were procedurally invalid. Specifically, they assert that BOEM/Interior were obliged under OCSLA to provide notice and an opportunity for comment before altering the terms of Lease Sale 261 in the FNOS. Defendants maintain that the PNOS provided sufficient notice of the changes, and point out that governors of certain affected states already objected to the proposed acreage withdrawal.

As to those violations, the APA requires that the court hold unlawful and set aside any agency action found to be "without observance of procedure required by law." 5 U.S.C. 706(2)(D). OCSLA carves out a significant opportunity for state stakeholders to participate in the leasing process. Within sixty days of a PNOS, state and local governments may submit recommendations on "the size, timing, or location of a proposed lease sale." 43 U.S.C. § 1345(a)–(b). The Secretary must accept these recommendations if they "provide for a reasonable balance between the national interest and the well-being of the citizens of the affected State." Id. at § 1345(c). The Secretary must also communicate his or her decision on the recommendation in writing to the governor of the affected state. Id.

BOEM's regulations facilitate this input as well comment by the public. Specifically, BOEM is required to publish each proposed notice of sale in the Federal Register to facilitate the state and local governments' input. 30 C.F.R. § 556.304(c). The publication must include "the proposed lease terms and conditions of sale, and proposed stipulations to mitigate potential adverse impacts on the environment." *Id.* BOEM must "consider *all* comments and recommendations received in response to the proposed notice of sale." 30 C.F.R. § 556.307(a) (emphasis added).

The PNOS for Lease Sale 261 was issued in March 2023, soon after BOEM had publicly confirmed through its supplemental EIS that it did not believe additional measures were required to protect Rice's whale. The challenged lease term for the expanded Rice's whale area only arose in a July 2023 district court filing and then appeared in the FNOS for Lease Sale 261 on August 25, 2023—one month before the statutory deadline for the sale. BOEM failed to follow its own procedures by making significant changes to

---

cannot unilaterally undo that decision for at least five years and the government must pay a penalty if it does so." *Id.* at 136. Accordingly, the lease sale stage grants the lessee a legal right and this right is immediately impacted by changes to the FNOS.

the FNOS, thereby depriving both affected states and the public the opportunity for meaningful review and comment. The procedural error is particularly grave here, because of both the compressed timeline and BOEM's inexplicable about-face on the scientific record it had previously developed. As described below, Louisiana has a significant stake in offshore leasing. Because of BOEM's bait and switch tactics, however, it was deprived of its statutorily guaranteed right to address a significant modification to leasing protocols for a long-term lease sale off its coast. As emphasized at oral argument, the last-minute changes and challenged terms may also have sizable impacts on bidding dynamics and operations for the industry parties.[59]8 Accordingly, the insertion of the challenged provisions into the FNOS for Lease Sale 261 was procedurally invalid. Plaintiffs have carried their burden of showing a likelihood of success on the merits under this claim.

2. Whether the Challenged Provisions Are Arbitrary and Capricious

The APA also requires that a court "hold unlawful and set aside" an agency action if it is found to be "arbitrary, capricious, [or] an abuse of discretion[.]" 5 U.S.C. § 706(2)(A). This standard "requires that agency action be reasonable and reasonably explained." FCC v. Prometheus Radio Project, U.S. , 141 S.Ct. 1150, 1158 (2021). The court "must set aside any action premised on reasoning that fails to account for relevant factors or evinces a clear error of judgment." Texas v. United States, 40 F.4th 205, 226 (5th Cir. 2022) (internal quotations omitted). Put another way, the reviewing court must ensure that "the agency has acted within a zone of reasonableness and, in particular, has reasonably considered the relevant issues and reasonably explained the decision." Prometheus Radio Project, 141 S.Ct. at 1158. In reviewing an agency's action, the court may only consider the reasoning articulated by the agency itself and may not credit post hoc rationalizations. Wages and White Lion Investments, LLC v. USDA, 16 F.4th 1130, 1136 (5th Cir. 2021) (citing Motor Vehicle Mfrs. Ass'n of U.S., Inc. v. State Farm Mut. Auto. Ins. Co., 463 U.S. 29, 53 (1983)).

Plaintiffs assert that BOEM acted arbitrarily and capriciously by failing to (1) explain its change in position, (2) justify the challenged provisions, and (3) consider other important factors. They show a strong likelihood of success on the merits under all three prongs. Neither the Record of Decision announcing the new provisions nor the FNOS explains BOEM's shift in position from its supplemental EIS issued in January 2023. In the latter

[59] See, e.g., doc. 14, atts. 13 & 15.

document BOEM concluded on review of the Soldevilla study that "not enough information is available at this time to confirm [the whales'] distribution or any seasonal movements outside of the core area that is already considered in this Supplemental EIS." Doc. 16, att. 1, p. 560. Accordingly, it determined that "the potential for vessel strikes to sperm and Rice's whale is extremely unlikely to occur due to the generally slow vessel transiting and surveying speeds, limited vessel routes originating from the eastern GOM, and the additional mitigations on vessels within the Rice's whale core area" and "the conclusions found in the 2017-2022 GOM Multisale EIS and 2018 GOM Supplemental EIS still remain valid." Id. at 170. In the Record of Decision supporting the challenged provisions, however, the Secretary justified the change in position "based on a recent study that the endangered Rice's whale occurs in portions of the northern Gulf of Mexico between the 100-meter and 400-meter isobaths eastward from the Mexico border with Texas and westward of the Rice's Whale Core Area identified in the 2020 Biological Opinion," i.e., the Soldevilla study.

This is the only scientific justification offered for the challenged provisions, representing an unexplained about-face from a position taken by the agency just months before. BOEM offered no explanation for its new stance on the same body of evidence, let alone using this pivot to justify measures pending a final determination by NMFS. This leaves the impression that the 2023 Record of Decision is merely an attempt to provide scientific justification to a political reassessment of offshore drilling. Additionally, the 2023 Record of Decision does not show any reevaluation of the other factors BOEM is required to consider in recommending areas for leasing. See 43 U.S.C. § 1344(a)(2); 30 C.F.R. § 556.302(a)(1). Though the relevant paragraph discusses BOEM's obligations under the ESA and its other environmental obligations, it makes no reference to the impact of the measures on production or the availability of acreage for energy exploration. BOEM devoted only one paragraph to its decision and failed to explain how these alternatives were selected, a stark contrast to the prior environmental reviews conducted with respect to the leases authorized under this Five-Year Plan.

The agreed-upon RPAs for protection of Rice's whale were added after years of deliberation. After the leasing pause, BOEM conducted a supplemental EIS with notice and comment but found no basis for altering its measures with respect to the whale. The challenged provisions inserted into these leases at the eleventh hour, and the acreage withdrawal, are based only on an unexplained change in position by BOEM on a single study a few months after that supplemental EIS. The process followed here looks more like a weaponization of the Endangered Species Act than the collaborative, reasoned approach prescribed by the applicable laws and regulations. Even when an agency's decision is based on political considerations, it is not excused from justifying the position—particularly when the decision is a pivot from a prior policy. Failure to do so leads to "surprise switcheroo" by an agency against regulated entities. E.g., Wages and White Lion Investments, LLC, 16 F.4th at 1138 (quoting Env't Integrity Project v. EPA, 425 F.3d 992, 996 (D.C. Cir. 2005)). The law requires more, especially where the prior policies have engendered reliance interests. Id. at 1139. Here BOEM failed to justify its pivot or account for those reliance interests, reinforcing the arbitrariness of its decision. Accordingly, plaintiffs have shown a sufficient likelihood that they will succeed on this claim.

B. Substantial Threat of Irreparable Injury

Next, plaintiffs must show a substantial threat of irreparable harm. Here plaintiffs rely on their economic harm and lost business opportunities. Specifically, the acreage removal and the addition of burdensome lease stipulations will result in lost revenue from Lease Sale 261. Doc. 14, att. 27, ¶¶ 28–34. This could cause economic injury of up to $2.2 million to Louisiana, which is entitled to a share of offshore royalties. Id. The industry plaintiffs will also suffer economic injury and impairment to their business development. According to an affidavit from Shell's commercial manager, the new restrictions on vessel traffic apply to an area of the northern Gulf that separates Shell's existing offshore leases from the onshore infrastructure that supports them. Shell Offshore Inc., No. 2:23-cv-1167, at doc. 4, att. 2, ¶¶ 23–27. Imposition of restrictions in this barrier will undermine long- term planning for and execution of its lease operations by, for example, (1) reducing the available hours for supply and increasing the number of ships necessary to make supply runs, (2) restricting Shell's ability to obtain seismic survey data for areas obtained in other lease sales because a survey cannot be designed to exclude individual blocks that are subject to the new stipulation, and (3) creating "costly and unreasonable complications" for Shell as it loses the ability to consolidate operations across leases because of the new restrictions. Id. at ¶¶ 23–27. These complications will increase attendant costs for all leases issued from Lease Sale 261 as well as some issued from non-Lease Sale 261 sales. Id. at ¶

23. The same logistical complications apply to Chevron, whose operations manager posits that vessel delays and inefficiencies caused by the challenged stipulation will likely add to the amount of time needed to complete projects in the Lease Sale 261 and increase drilling costs on Lease Sale 261 leases "by a significant factor."[60]9 Doc. 14, att. 15. Finally, the industry plaintiffs show that the last-minute changes impact their ability to bid fairly on leases under Lease Sale 261 and to bid on the six million acres withdrawn from the sale. Doc. 14, att. 11, ¶¶ 12–16; doc. 14, att. 13, ¶¶ 14–16; Shell Offshore Inc., No. 2:23-cv-1167, at doc. 4, att. 2, ¶ 28.

In return defendants point out that there was low interest in the now-excluded area when it was offered during Lease Sale 259, with BOEM receiving bids on only 16 out of the circa 1200 blocks in that area. Doc. 55, att. 6, ¶ 8. They challenge Louisiana's estimate of lost revenues, arguing that the calculations are simplistic and overlook that most revenue is generated from older leases. They also argue that this harm is not irreparable because, if the court were to ultimately vacate Lease Sale 261 in a final adjudication on the merits of this case, Interior would then conduct a sale consistent with the court's orders. Finally, they maintain that the industry defendants' purported compliance costs are too speculative. They also assert that alleged harms based on bidding dynamics are based on the faulty premise that bidders are entitled to know all final conditions of sale at the time of the PNOS. A harm is irreparable when "there is no adequate remedy at law, such as monetary damages." Janvey v. Alguire, 647 F.3d 585, 600 (5th Cir. 2011). Economic loss therefore rarely qualifies as an irreparable harm. Optimus Steel, LLC v. U.S. Army Corps of Eng'rs, 492 F.Supp.3d 701, 725 (5th Cir. 2020). Because of the government's sovereign immunity, however, it will "be difficult, if not impossible" for plaintiffs to recover their funds.

Louisiana v. Biden, 543 F.Supp.3d at 417. "In determining whether costs are irreparable, the key inquiry is 'not so much the magnitude but the irreparability.'" Restaurant Law Ctr. v. U.S. Dep't of Labor, 66 F.4th 593, 597 (5th Cir. 2023) (quoting Texas v. EPA, 829 F.3d 405, 433–34 (5th Cir. 2016)). Accordingly, "complying with a regulation later held invalid almost always produces the irreparable harm of nonrecoverable compliance costs." Texas v. EPA, 829 F.3d at 433 (quoting Thunder Basin Coal Co. v. Reich, 510 U.S. 200, 220–21 (1994) (Scalia, J., concurring in part and concurring in the judgment)).

---

[60] Robichaux estimates that this cost could rise to a million dollars or more per vessel, per day if supply disruptions cause drilling to be suspended. Id. at ¶ 14.

The court observes that plaintiffs have demonstrated substantial potential costs resulting from the challenged provisions. While the government defendants largely focus on the acreage withdrawal and dynamics of the sale itself, many of plaintiffs' alleged hardships arise from the vessel restrictions. Industry plaintiffs have shown a likelihood that these will burden their operations on current and planned leases. The resulting costs would not be undone by the court's entry of a permanent injunction and order of another sale. Plaintiffs also show that no other mechanism exists for recovering the compliance costs because of the government's sovereign immunity. Accordingly, plaintiffs satisfy this step of the inquiry.

C. Whether Balance of Equities Supports Injunctive Relief

Finally, the party seeking a preliminary injunction must show that (1) the threatened injury to the movant outweighs the harm the injunction may do to the nonmovant and (2) granting the injunction will not disserve the public interest. Anderson v. Jackson, 556 F.3d 351, 360 (5th Cir. 2009). When the government is a party, these factors merge. Nken v. Holder, 556 U.S. 418, 435 (2009).

As described above, the industry plaintiffs and the state of Louisiana face a substantial risk of irreparable economic harm if the injunction is not granted. Additionally, the sale is currently under a statutory deadline. If it does not take place as scheduled, government defendants are no longer under any congressional mandate to proceed with the lease sale. Meanwhile, if the government finds justification for additional measures to protect Rice's whale, it may pursue these through the Department of the Interior's oversight of lessees' exploration and production plans. See 43 U.S.C. §§ 1340(c)(1), 1351(c) (describing Interior's ability to require modifications of exploration and production plans or cancel them to avoid environmental harm); see also Ctr. for Biological Diversity, 563 F.3d at 473 (same). Given the shaky justification offered by BOEM, the court cannot find that the challenged provisions are so necessary that withholding them even on a preliminary basis will outweigh the risk of irreparable economic harm shown by plaintiffs. Additionally, "there is generally no public interest in the perpetuation of unlawful agency action." Louisiana v. Biden, 55 F.4th 1017, 1035 (5th Cir. 2022) (internal quotations omitted).

The government defendants contend that last-minute changes to the sale will create administrative chaos for BOEM and potentially disadvantage non-party bidders who have been preparing for the sale on the announced terms. They have produced no evidence on behalf of these non-party bidders, however. Meanwhile, API, as representative of nearly 600 companies in the oil and gas industry, is one of the parties to this suit requesting preliminary injunctive relief. Additionally, the plaintiffs only request that certain terms be stricken from the existing FNOS. Government defendants fail to show that this necessitates the issuance of a new Notice of Sale.[61] The court must also balance the IRA's mandate that this sale proceed with the regulatory complications created by the unlawful agency action. Accordingly, the weight of equities supports injunctive relief.

D. Court's Authority to Grant Injunction

Finally, defendants challenge the court's ability to grant the requested injunctive relief. They argue that (1) plaintiffs seek an improper permanent injunction; (2) a mandatory injunction is not available on plaintiffs' claims under § 706(2) of the APA; (3) alternatively, plaintiffs have not met the heightened standard for a mandatory injunction; and (4) the court lacks authority to grant mandamus relief.

3. Whether Relief Sought Is Improper Permanent Injunction

Government defendants assert that plaintiffs' motion for preliminary injunction is per se a motion for permanent injunction because it is indistinguishable from the final relief sought under the complaint. But certain factors in this case, namely the eleventh-hour change to the FNOS and the looming statutory deadline for the sale, necessitated that plaintiffs seek injunctive relief before the court could issue a final adjudication on the merits.[62] Additionally, plaintiffs have shown a substantial likelihood of success on the merits of their claims. Given that the statutory deadline for the sale will expire before this matter can be brought to trial, and that plaintiffs

---

[61] Indeed, as plaintiffs point out, BOEM has changed actual terms less than a week prior to a sale in order to comply with the law. *See* News Releases, Fact Sheets, and Notices for Sale 196, BOEM, https://www.boem.gov/oil-gas-energy/leasing/news-releases-fact-sheets-and-notices-sale-196 (including "Change to Notice of Sale 196" on August 10, 2005, and sale results on August 16, 2005).

[62] Government defendants point out that the statutory mandate may remain in effect even if the deadline is missed. *E.g.*, *Brock v. Pierce Cnty.*, 476 U.S. 253 (1986). Given the lengthy political fight over these sales, however, and the resulting congressional mandate, the court views the statutory deadline as a critical matter.

have established a legal interest in seeing this sale proceed, the court can only preserve their right to relief in a final adjudication through the issuance of a preliminary injunction. Meanwhile, the government can see these preliminary measures undone if it does prevail on the merits by voiding the sales or proceeding with additional regulation through Interior's authority over further steps of offshore leasing, exploration, and production.

Availability of Mandatory Injunction for APA Violations

Next, the court turns to the availability of mandatory injunctive relief under the APA. Both of plaintiffs' claims arise under § 706(2) of that statute, based on government defendants' failure to follow procedures and the arbitrary and capricious agency action. Plaintiffs maintain that their requested relief is not an order for affirmative agency action but instead one prohibiting the agency from including the two challenged provisions in the sale. They therefore maintain that it is well within the scope of the court's authority. See, e.g., Cigar Ass'n of Am. v. FDA, 480 F.Supp.3d 256, 281 (D.D.C. 2020) (enjoining only one aspect of agency action).

Assuming that the requested relief amounts to a mandatory injunction, the court finds that such relief would be available in this matter. In "rare circumstances," remand to the agency is not the appropriate solution for a § 706(2) violation and the court may instead order affirmative agency action. O'Reilly, 477 F.3d at 238–39 (quoting Fla. Power & Light Co. v. Lorion, 470 U.S. 729, 744 (1985)). "Remand, not vacatur, is generally appropriate when there is at least a serious possibility that the agency will be able to substantiate its decision given the opportunity to do so" and where vacatur would be "disruptive." Tex. Assoc. of Manufacturers v. U.S. Consumer Product Safety Comm., 989 F.3d 368, 389–390 (5th Cir. 2021); Cent. & S.W. Servs., 220 F.3d 683, 692 (5th Cir. 2000). Here the government defendants have not shown such a possibility, even if the court credits their post hoc rationalizations. As for disruptiveness, government defendants have failed to substantiate their claims of the complications arising from changes to the sale. Additionally, the court must consider the statutory deadline imposed on this specific lease sale after multiple contentious legal battles over the remaining leases in the 2017–22 Five- Year Plan. Plaintiffs have shown that it would be significantly disruptive to their interests to allow Interior to proceed with Lease Sale 261 under the challenged terms, and that it is plaintiffs rather than defendants who lack any other recourse. Accordingly, remand is not the proper remedy here.

Finally, as defendants note, preliminary injunctions ordering a party to take a certain action must meet a higher standard. Mandatory preliminary injunctive relief, "which goes well beyond simply maintaining the status quo pendente lite, is particularly disfavored, and should not be issued unless the facts and law clearly favor the moving party." Martinez v. Mathews, 544 F.2d 1233, 1243 (5th Cir. 1976) (citations omitted). In this matter, plaintiffs have met the heightened standard. The imposition of the challenged terms right before a statutory sale deadline, the unavailability of any other recourse to plaintiffs, and the fact that defendants may see all this neatly undone if they prevail on the merits leave the court with only one option at this stage.

IV. Conclusion

For the reasons stated above, the court hereby ORDERS that the Motions for Preliminary Injunction be GRANTED. Accordingly, the government defendants are enjoined from implementing the acreage withdrawal and Stipulation 4(B)(4) as described in the Final Notice of Sale and Record of Decision for Lease Sale 261. Government defendants are ordered to proceed with Lease Sale 261, absent the challenged terms, by September 30, 2023.

THUS DONE AND SIGNED in Chambers on the 21st day of September, 2023.

JAMES D. CAIN, JR. UNITED STATES DISTRICT JUDGE

October 4, 2023

The Subcommittee on Rural Development, Energy, and Supply Chains The Committee on Small Business
U.S. House of Representatives Washington, D.C. 20515

Re: Subcommittee on Rural Development, Energy, and Supply Chains Hearing: "Energy Independence: How Burdensome Regulations are Crushing Small Offshore Energy Producers" (September 28, 2023)

Dear Members of the Subcommittee:

Oceana is the largest international conservation organization solely focused on protecting the world's oceans, with more than 1.7 million members and supporters in the United States. For twenty years, Oceana has campaigned to win strategic, directed campaigns that achieve measurable outcomes to help make our oceans more biodiverse and abundant.

For over 15 years, Oceana has been campaigning to oppose expanded offshore oil and gas exploration and development. Offshore drilling causes dangerous oil spills and perpetuates energy development based on fossil fuels. The United States must shift from fossil fuel-based energy sources to clean energy. Addressing climate change is important for oceans, wildlife, and our future. By shifting from fossil fuel energy to clean, renewable energy sources, the United States can help address this crisis.

Oceana thanks you for the opportunity to submit a letter for the record following the Subcommittee on Rural Development, Energy, and Supply Chains Hearing: "Energy Independence: How Burdensome Regulations are Crushing Small Offshore Energy Producers".

Offshore Drilling Is Dirty and Dangerous

No offshore oil and gas company can guarantee that oil spills will not happen. Despite repeated promises that toxic oil spill disasters are rare, the oil and gas industry proves again and again that offshore drilling is dirty and dangerous. Oil spill disasters are not isolated incidents. In the U.S. alone, there were over 6,800 oil spills between 2010 and 2021 — an average of almost two spills every day.[63] Toxic oil disasters poison marine wildlife, cause beach closures, and shut down lucrative fishing areas. Permanently protecting our coasts from offshore drilling will safeguard our coastal economies from the next oil disaster.

There are over 8,000 miles of active pipelines in the Gulf of Mexico and the government does not require subsea inspection of them.[64] There could be oil spewing below the surface that would not be visible from an aerial survey, especially given how deep oil companies are drilling in the Gulf of Mexico.

Offshore Drilling Fuels the Climate Crisis

The climate crisis has raged through our country this summer. Extreme weather has been front page news for months, with wildfire smoke billowing across the country, California hit by the first tropical storm in 80 years, and the hottest month ever on record.

Oceana's analysis found that permanently protecting all federal waters from new offshore drilling could prevent more than 19 billion tons of greenhouse gas emissions– nearly three times the total annual emissions of the U.S.[65] This would be equivalent to taking every car in the nation off the road for 15 years.

BOEM's Proposed Changes to Financial Assurances Regulation Are an Important Step in the Right Direction but Can and Should Be Stronger

The Bureau of Ocean Energy Management's (BOEM) proposed changes to financial assurances do not aim to impede fossil fuel development as claimed during this hearing, but are a step towards protecting taxpayers and coastal communities from the pollution of abandoned infrastructure.

---

[63] BSEE. (2022). Aggregated Data of OCS Oil and Gas Industry Activities (e.g. Production and drilling levels), Work Hours, Incidents, Inspections and Compliance: 2010-2021. https://www.bsee.gov/sites/default/files/2023-02/performance-data-table-2010-2021-10-26-2022.pdf.

[64] GAO. (2021). OFFSHORE OIL AND GAS: Updated Regulations Needed to Improve Pipeline Oversight andDecommissioning. https://www.gao.gov/assets/gao-21-293.pdf.

[65] Oceana. (2021, January 15). Offshore Drilling Fuels the Climate Crisis and Threatens the Economy. Oceana USA.https://usa.oceana.org/publications/reports/offshore-drilling-fuels-climate-crisis-and-threatens-economy.

BOEM has the obligation to be a responsible steward, both to the environment and to the U.S. taxpayer. While BOEM's proposed rule makes some positive changes, it requires further improvements to meet the agency's responsible stewardship obligation.

The only way to fully ensure that decommissioning costs are not borne by the federal government and the taxpayer is to require bonds from all lessees in the full amount of estimated decommissioning liabilities at the highest probabilistic estimate. In the absence of full bonding, however, BOEM should only exempt lessees from supplemental bonding for companies that show significant financial strength and have met all prior decommissioning obligations.

According to a recent study by Mark Agerton et al., industry is sitting on nearly 10,800 unplugged wells in federal waters alone, with an estimated decommissioning cost of $42 billion.[66] Of those wells, over 7,000 are inactive, with an estimated decommissioning cost of $28.65 billion.[67] And that only accounts for unplugged wells in federal waters. In state waters, the study estimated another 7,000 wells were inactive and unplugged, but the decommissioning cost estimate was much lower, at $2 billion.[68] Oil and gas can leak from these wells and cause damage to ecosystems, particularly closer to the coast.[69] To make matters worse, wells that were previously plugged and abandoned are still leaking oil and harmful gases, including methane, benzene, nitrogen oxides, and carbon dioxide,[70]8 due to vague and inadequate regulations when those wells were plugged.[71] And BOEM and the Bureau of Safety and Environmental Enforcement (BSEE) do not regularly monitor the state of the wells.

---

[66] Mark Agerton et al., *Financial liabilities and environmental implications of unplugged wells for the Gulf of Mexico and coastal waters*, NATURE ENERGY (May 8, 2023), https://doi.org/10.1038/s41560-023-01248-1.

[67] *Id.*

[68] *Id.*

[69] *Id.*

[70] Hannah Seo, *Unplugged: Abandoned oil and gas wells leave the ocean floor spewing methane*, ENV'T HEALTHNEWS (Dec. 8, 2020), https://www.ehn.org/oil-and-gas-wells-methane-oceans-2649126354.html.

[71] Torbjørn Vrålstad et al., *Plug & abandonment of offshore wells: Ensuring long-term integrity and cost-efficiency*, 173 J. PET. SCI. & ENG'G 478 (Feb. 2019), sciencedirect.com/science/article/pii/S0920410518309173.

Methane Pollution and Carbon Intensity of Gulf Of Mexico Offshore Drilling

During the hearing we heard untrue statements regarding fossil fuel production in the United States. Offshore drilling is dirty and dangerous in all locations. Claims were made about the carbon intensity of oil in the Gulf of Mexico, but the reality is that emissions are highly underreported and by some estimates are worse than average global emissions.

Ayasse et al. 2022[72] used an airplane to measure methane emissions in the Gulf of Mexico. They surveyed over 150 offshore platforms and surrounding infrastructure in shallow water. Tanks, pipelines, wells, and vents all released methane. Some vents were persistently releasing methane for days or months. The loss rates compared to production offshore varied from 10% to 66%, vastly higher than onshore estimates for the Permian Basin around 3.3%–3.7%.

Gorchov Negron et al. 2020[73] used methane measurements from aircrafts to show the Environmental Protection Agency Greenhouse Gas Inventory (GHGI) and the US Bureau of Ocean Energy Management Gulfwide Offshore Activities Data System (GOAD) underestimate emissions. Methane emissions from the largest shallow water facilities were underestimated by at least an order of magnitude compared to GOAD. The true emission factor for shallow oil platforms could be 80% higher than what is used by the GHGI. Under-sampling of facilities with disproportionately high emissions can lead to underestimates of basinwide emissions.

Gorchov Negron et al. 2023[74] collected airborne observations and combined them with previous surveys to evaluate the climate impact of offshore drilling in the Gulf of Mexico. The study included methane from losses and venting as well as CO2 from combustion. Based on their research, the climate impacts of drilling in the Gulf of Mexico are over twice what is reported in government inventories, and on average, shallow water platforms

---

[72] Alana K. Ayasse et al., *Methane remote sensing and emission quantification of offshore shallow water oil and gasplatforms in the Gulf of Mexico*, ENVTL. RES. LETTERS 17(8) (Aug. 11, 2022), https://doi.org/10.1088/1748- 9326/ac8566.

[73] Alan M. Gorchov Negron et al., *Airborne Assessment of Methane Emissions from Offshore Platforms in the U.S.Gulf of Mexico*, ENVTL. SCI. & TECH. 54(8) 5112–5120 (April 13, 2020), https://doi.org/10.1021/acs.est.0c00179.

[74] Alan M. Gorchov Negron AM et al., *Excess methane emissions from shallow water platforms elevate the carbonintensity of US Gulf of Mexico oil and gas production*, PROCEEDINGS OF THE NAT'L ACADEMY OF SCI.120, 10.1073/pnas.2215275120.

in the Gulf of Mexico are worse for the climate than typical oil production around the world.

Conclusion

Neither the federal government nor taxpayers should be held responsible for the decommissioning costs of offshore drilling infrastructure. Offshore drilling is a threat to coastal communities during construction and operation, and infrastructure can continue to pollute the ocean for decades. Oceana urges the Committee to prioritize protecting coastal communities and to recognize the need for improved financial assurances for the offshore drilling industry.

We appreciate the opportunity to provide input and thank you for your attention.

Sincerely,
Sarah Giltz, Ph.D. Marine Scientist Oceana
Andres Perotti Staff Attorney Oceana

August 21, 2023

Ms. Kelley Spence Office of Regulations
Department of the Interior, Bureau of Ocean Energy Management
45600 Woodland Road, Mailstop VAM–BOEM DIR
Sterling, VA 20166

Re: Bureau of Ocean Energy Management Docket No. 2023-0027 Risk Management and Financial Assurance for OCS Lease and Grant Obligations, RIN 1010-AE14

Dear Ms. Spence:

The Surety & Fidelity Association of America ("SFAA")[75] appreciates the opportunity to provide written comments, in addition to our virtual meeting

---

[75] SFAA is a non-profit, national trade association of more than 400 companies that write 98 percent of surety and fidelity bonds in the United States. SFAA is licensed as a rating or

two weeks ago, regarding the proposed Regulations entitled Risk Management and Financial Assurance for OCS Lease and Grant Obligations (the "Proposed Regulations").

1. The Entire Value of Decommissioning Obligations Should Be Subject to Financial Assurance Requirements for All Lessees

BOEM and the US taxpayers would benefit from requiring all lessees to provide financial assurance covering the entire value of decommissioning obligations without exception. Such should be the case regardless of credit rating or the point-in-time value of the reserves associated with a given lease.

Furthermore, bonds should be required whenever a lease transfers in both the ordinary course of business and as part of a bankruptcy matter. This would be a much more equitable, sustainable system.

As currently proposed, the financial assurance requirements continue to be a form of adverse selection against the financial assurance provider, including sureties. Granting exceptions to the posting of financial assurance for certain "creditworthy" lessees based on assessing which assets will likely be purchased and assessing the lessee's financial ability to perform its ultimate decommissioning requirements is a form of adverse selection against financial assurance providers. Further, these criteria, standing alone, are not reliable metrics for measuring the totality of the risk to BOEM – and, ultimately, the U.S. taxpayers – of whether the lessee will be able to meet its decommissioning obligations.

The problem is further exacerbated by the fact that, as noted in BOEM's commentary regarding the Proposed Rules, the financial position of lessees and commodity prices can change rapidly. It is not clear what the measurement date of the calculation of the value of reserves versus decommissioning costs will be, nor is it clear how frequently this measurement will be monitored. Though this is unclear from the Proposed Regulations, BOEM advised during our meeting it plans to review these two metrics annually at a minimum.

Further, it is unclear whether or how a lessee operating under a current permit that does not at some point meet the financial or proven reserve

---

advisory organization in allstates and is designated by state insurance departments as a statistical agent for the reporting of surety and fidelity experience. SFAA members provide the vast majority of bonds that secure regulatory obligations.

valuation requirements would be impacted. Would such a lessee be prohibited from continuing to operate until financial assurance is posted?

SFAA generally agrees that BOEM should eliminate Director discretion in determining which lessees in the Gulf of Mexico need to post supplemental financial assurance. As a result of the Director's discretion, BOEM's inconsistent enforcement of its existing regulations is improper and causes uncertainty for all stakeholders. For these reasons, SFAA recommends that BOEM require supplemental financial assurances of all lease operators.

2. SFAA Also Suggests That the Blanket Bond Should Be Increased

Based on the amount of time that has passed and inflation experienced since the last revisions, which required $3 million bonds, SFAA recommends an increase to the blanket bond amount. This will serve to provide blanket protection across all wells and can act as a partial backstop should decommissioning costsexceed supplemental bond amounts.

3. Surety Appetite Has Already Been Restricted Due to Recent Changes in the Marketplace. BOEM's Proposed Regulations Will Result in an Even More Limited Appetite Amongst the Industry and Potentially More Stringent Terms Where a Surety Appetite Remains

Given BOEM's current and proposed positions and the fact that BOEM has continued to allow some current lessees to operate without providing financial assurance, there will be even less of an appetite within the surety industry to provide financial assurances that lease operators will perform their decommissioning obligations. Moreover, it is unlikely that whatever appetite would exist will be in the aggregate amounts contemplated in the proposed rule, making commercial implementation of the Proposed Regulations unlikely through surety bonds alone.

Each surety member underwrites accounts independently and determines its terms and conditions for providing such financial assurance on an instrument that is non-cancellable unless replaced by acceptable alternative security or paid in full. Further, given BOEM's position that wells or fields that are subject to lease will only require financial assurance when the proven reserves are less than three times BOEM's estimated costs of decommissioning, SFAA would expect that lessees that are required to post bonds would be required to post collateral with the sureties to fund the newly required obligations as a result of BOEM's proposed rule and for both the existing BOEM bonds and private obligation bonds that have been written on behalf of a lessee to a named record chain title holder ("Private Bonds"); the percentage of collateral required would be subject to each individual surety's underwriting appetite for such obligations.

It is unknown whether the oil and gas operators will retain sufficient ability to provide such collateral to the sureties in support of the new and existing obligations. Given that "less" (relatively speaking) creditworthy lessees will necessarily have restricted access to capital and would be the ones subject to the proposed financial assurance requirements, having their capital tied up in collateral obligations will present a substantial burden on these operators and likely lead to more financial failures. Such financial failures will have cascading negative impacts that undermine BOEM's stated purpose of increasing access to these bonds and alleviating the burden on the taxpayer. For a sense of magnitude, the surety industry supports $2.93 billion of financial assurance with BOEM as beneficiary. This does not include Private Bonds, which are estimated to be at least as much, if not more, in amounts outstanding than those bonds written with BOEM as beneficiary.

a. Surety Industry Cannot Use Pricing to Improve the Economics of This Marketplace, Particularly Due to the Nature of the Proposed Regulations Being Steeped in Adverse Selection

The Proposed Regulations estimate surety bond pricing to range from 2% to 12% based on BOEM's new framework contemplated therein, but sureties do not price based on expected loss levels. Unlike insurance, where risk is pooled and premium calculated to cover the anticipated losses from the risk profile of the policyholders in the pool, surety companies underwrite risks based on an expectation of zero losses. This is the reason the Proposed Regulations are considered to be adverse selection against the surety, where only the riskiest accounts and leases are required to post security for decommissioning obligations. SFAA respectfully recommends all lessees be required to post security at any time the lease is transferred to that party, whether in the ordinary course of business or as part of a bankruptcy matter, as suggested earlier in these comments.

4. Respectfully, BOEM Evaluating a Lessee's Ability to Meet Their Decommissioning Obligations Using the Proposed Methodology Would Be Flawed and Outside the Scope of BOEM's Purpose as a Regulatory Body. Financial Assurance Providers Are Better Suited Than BOEM to Evaluate a Lessee's Risk of Default on Its Decommissioning Obligations

SFAA takes exception to certain aspects of the methodology described in the solicitation for comments of the Proposed Regulations, and there is one reference to Moody's ratings in the proposal, which is inconsistent. Specifically:

a) Regarding the proposed use of an Issuer Credit Rating (Page 42146 VI. A.), we agree that credit statistics support a lower probability of default with investment grade companies (sureties typically include third-party-provided credit ratings as part of the underwriting process). However, the inherent volatility in the oil & gas industry can, and has, resulted in circumstances where exploration and production companies received multiple notch downgrades at once and/or quickly moved into bankruptcy. This occurred in 2015 when we witnessed a dramatic increase in multi-notch (as many as five notches) downgrades, distressed bond exchanges, and bankruptcies. As both BOEM and the surety industry have experienced, bankruptcies can lead to defaults on existing and future decommissioning obligations that ultimately result in bond claims and/or losses borne by the taxpayers. Further, some companies, including those emerging from bankruptcy proceedings, will have an investment grade rating that nevertheless may be insufficient to demonstrate the financial wherewithal of the lessee to fulfill its decommissioning responsibilities. These aspects of credit ratings support the conclusion that any system where certain lessees are exempted from financial assurance requirements will result in extensive losses by US taxpayers and financial assurance providers.
Further, BOEM's use of NRSRO credit ratings and private company equivalent ratings as one of two criteria for determining exemption status is over-reliant upon NRSRO credit rating agencies. This overreliance, especially upon the private company equivalent ratings, may create strong incentives for market participants to game the

model, as they did with rating agency models in the lead-up to the 2007-2009 financial crisis (see John Soroushian's 2016 brief, "Credit Ratings in Financial Regulation: What's Changed Since the Dodd-Frank Act?"[76]; and see Gretchen Morgenson and Louise Story's 2010 New York Times article, "Rating Agency Data Aided Wall Street in Deals").[77]

To be clear, sureties utilize NRSRO credit ratings to augment their analyses. However, BOEM's significant reliance upon credit ratings as one of only two evaluation tools is improper and is a primary reason why SFAA maintains exemptions are inappropriate.

b) On page 42146 (part of Section VI.A) and elsewhere in the proposal, there is reference to utilizing a third-party credit model to generate proxy ratings for companies that lack an external credit rating. This proposal is another example of an effort to construct an adverse selection system that will result in expenditures by the US Taxpayer. The surety industry has and continues to use internally created and externally provided credit models for non-rated companies well. The general surety industry experience has been that implied credit ratings change frequently and dramatically. One factor contributing to the differences between external credit ratings and ratings provided through a credit model is the size of the company as measured in equity. If BOEM implements this proposal, we suggest that rules be modified such that a minimum net worth of some multiple of total decommissioning obligations be required, in addition to the credit rating.

c) In Section VI. and elsewhere in the proposed rulemaking, where BOEM provides for exemptions for certain lessees and leases based on Issuer Credit Rating or third-party credit model, the Government and ultimately US Taxpayers are unnecessarily taking on the risk of decommissioning obligations. SFAA contends that the purpose of BOEM is not to underwrite and assume risk on behalf of U.S. taxpayers in the normal course such as a surety company, but rather

---

[76] Soroushian, J. (2016, April 22). Credit Ratings in Financial Regulation: What's Changed Since the Dodd-Frank Act? FinancialResearch.gov. Retrieved August 24, 2023, from https://www.financialresearch.gov/briefs/2016/04/21/credit-ratings-in-financial-regulation/.

[77] Rating Agency Data Aided Wall Street in Mortgage Deals. GRETCHEN MORGENSON and LOUISE STORY. NYT April 23,2010. http://www.nytimes.com/2010/04/24/business/24rating.html.

to enforce the rules and regulations. BOEM's evaluation criteria for exemptions is too simplistic and thus will inevitably lead to taxpayers footing the bill when an exempt lessee cannot meet its decommissioning obligations. The oil and gas industry is so volatile that even the healthiest of operators one day can become financially distressed the next, and the credit ratings of an entity are only a small piece of its overall ability to meet the decommissioning obligations.
The oil and gas commodities markets have exhibited significant volatility in recent years. This volatility has driven quick and dramatic changes to both reserve values – which fluctuate daily with price – and credit ratings for operators. In pricing environments such as 2014 and 2020, pricing largely made the Gulf of Mexico basin non-economic. Were BOEM's proposed rules in effect during those times, most of the leases that would be afforded an exemption by BOEM under the Proposed Regulations would have been non-compliant on a 3:1 reserve value to decommissioning evaluation. Should BOEM have performed a point-in-time evaluation and determined that leases were non-compliant with the Proposed Regulations, it would have required financial assurance for those leases during a period where there would have been no appetite from any financial assurance provider to bring the lease into compliance. This would have placed the taxpayers at tremendous risk should pricing not have improved. Similarly, should BOEM have waited to evaluate the reserve value to decommissioning amounts on an annual basis, it would have put taxpayers at risk during the downcycle pricing event without having required financial assurance for the leases that would have been non-compliant on a point-in-time basis. As such, BOEM's proposed use of reserve valuations effectively puts the taxpayer at risk on a basis that is subject to an industry's volatile commodity pricing environment, backed by the taxpayer assuming risk.
Surety bonds transfer taxpayer risk to entities that provide stable credit strength independent from the sector that BOEM is regulating, and that has independent governmental regulations governing capital standards, pricing, and conduct. SFAA respectfully submits that financial assurance providers like sureties are better positioned

than BOEM to underwrite these risks to ensure that U.S. taxpayers are protected.

5. Many Open Questions Remain for SFAA and Its Members with the Proposed Regulations

a) It appears that BOEM will no longer require the initial lessee to provide financial assurance or guarantee once the lease is assumed by a new lessee that either qualifies to operate without financial assurance or posts its own financial assurance. Since several leases are being drilled withBOEM relying on the guarantee of current lessees and predecessor lessees, it seems some of the "private bonds" (where current lessees provide security to a predecessor lessee) would no longerbe required. What is the timeframe for reducing the guarantee of the original lessee? For example, would that guarantee by a predecessor be reduced over a three-year period, just as the proposed requirement is for the current lessee to provide financial assurance over a three-year period? Has BOEM factored "private bonds" written between the transferring parties where the new lessee provides assurances to the predecessor in its decommissioning liabilities calculations?

b) How will BOEM proceed if an operator fails to meet its decommissioning obligations? Will BOEMfirst seek recoveries from the predecessors on the lease and/or will it proceed against the bonds posted by the predecessors, or will it first pursue the current lessee? This is critically important for the sureties to understand when underwriting lease operators and the Gulf of Mexico lease marketplace generally, specifically whether to continue to write in this space or how to change theterms under which they will continue to write bonds for lease operators.

c) We further understand that existing lessees who need to post financial assurance can provide the financial assurance over a three-year period, but the proposal does not indicate what happens when a lessee who had previously qualified to operate without providing financial assurance at your measurement date ceases to qualify at some point in the future. Will there be a specific timeline for posting the additional assurance in the case of a lessee experiencing a downgrade at some point in the future? If a lessee experiences credit rating fluctuations while utilizing the three- year ramp-up, what happens to the financial assurances they provided?

d) What would happen to the lease if a lessee was unable or unwilling to post financial assurance as required by BOEM while operating under a current lease? Based on SFAA's meeting with BOEM, it is our understanding that BOEM will still have all of its current rights available to it where a lessee fails to meet its obligation to post supplemental financial assurance.
e) How will the financial assurance amount be calculated? Will financial assurance calculations consider any additional private security (e.g. Private Bonds, decommissioning escrow funds, letters of credit, etc.) guaranteeing decommissioning performance provided by a current lessee in favor of a predecessor lessee?
f) How frequently will BOEM require lessees and individual leases to be evaluated? Does BOEM intend to standardize reserve assessments and measurements? BOEM expects companies to have reserve information as a matter of general operations – will BOEM have any verification mechanism?
g) Will a "negative event" (Section 556.903) continue to be the bankruptcy, insolvency, or suspensionor revocation of the charter or license of the financial assurance provider?
h) If a lessee appeals BOEM's demand to require financial assurance, per Section VIII of the proposedrules (page 42148) BOEM would require an appeal bond (referred in the Proposed Rules as "appeals bond"), and the Proposed Regulation indicates "if the appeal is unsuccessful, the appealsbond could be replaced or converted into bonds to cover the supplemental financial assurance." Would such a conversion be required by the surety providing the appeal bond? We request that such a requirement be at the option of the surety, understanding that the lessee would need to provide financial assurance in some form.
i) In the event a surety allowed forfeiture of a bond in an amount demanded by BOEM (in accordance with 556.902(a)(3)), which is in an amount less than the amount of the bond, wouldBOEM be able to make a further "forfeiture demand" on the balance of the bond? We see reference to a surety being able to avoid forfeiture if the surety agrees to complete corrective action even if the amount of such corrective action exceeds the amount of the financial assurance.We

also see that any amount forfeited by the surety not utilized to fund the corrective action is returned to the entity that paid BOEM. The question is whether, when the amount of forfeiture is less than the bond amount, BOEM can make further claims?

j) BOEM's proposed second criterion to evaluate the requirement of supplemental financial assurance based on "a 3 to 1 ratio of the value of proved oil and gas reserves on a lease to the decommissioning liability associated with these reserves" leaves open the option to Lessee to abandon certain infrastructure and equipment in bankruptcy. This creates a decommissioning liability for either a predecessor, provider of financial assurance (such as a surety), or the government. Recent offshore bankruptcies show that sophisticated financial parties will use existing bankruptcy law to abandon obsolete equipment and sell proven reserves to new entities.To remove the incentive that exists in the current law, BOEM should implement an accountability program similar to the Applicant Violator System administered by the Office of Surface Mining (OSM) in the coal industry. Such a program should bar any officer from being employed in the oil and gas industry whose company does not fulfill a decommissioning obligation. An effective program that holds individual oil and gas company officers accountable for transferring oil decommissioning costs to predecessors, the government, or other third parties will reduce the incentive to abandon offshore infrastructure.

k) BOEM's Proposed Regulations are intended to provide greater protection to taxpayers – not predecessor lessees. In the Proposed Regulations, BOEM states, "BSEE and BOEM regulations hold predecessors and current co-lessees responsible for decommissioning when a current lessee is unable to perform." The Proposed Regulations further state that, "the proposed rule would retain the authority to pursue predecessor lessees for the performance of decommissioning." However, BOEM is silent as to how and when the required financial assurance will be called upon. Without knowing the order of attribution of when the financial assurance is required to be utilized to fund decommissioning activity it provides an extreme amount of uncertainty to those parties being asked to provide the financial assurance on behalf of lessees. When will the taxpayer, vis-à-vis BOEM and BSEE, perform the decommissioning activities when a lessee fails to do so? Will BOEM require available co-lessees and predecessor lessees to perform decommissioning prior to BOEM/BSEE performing the decommissioning? When will financial assurance be called upon to fund decommissioning? For sureties to adequately understand the risk they are underwriting, there needs to be consistency in application. While SFAA understands BOEM's desire to retain flexibility, this is not something that can be done on an ad hoc, case-by-case basis if BOEM wants these bonds to be reasonably available in the marketplace.

Secretary's Order 3299, which gives BOEM the authority to manage the Nation's offshore energyand mineral resources, also gives the Bureau of Safety and Environmental Enforcement (BSEE) authority to enforce lessee decommissioning obligations. How often does BOEM review BSEE's Incidents of Non-Compliance (INCs) records and its Increased Oversight List? Can BOEM provide documentation of BSEE referrals of unacceptable performance that resulted in BOEM action?

Surety evaluates bonding opportunities based, in-part, on the information BSEE and BOEM provide about well-operations and operators. Recent bankruptcy proceedings involving Cox Operating revealed a situation where Cox had received multiple warnings of decommissioning responsibility violations over an extended period of time. Without this kind of information being readily available, surety carriers could find themselves offering additional financial assurances to otherwise unworthy operators that they would not have otherwise bonded. Have BOEM and BSEEworked out a more responsive and transparent violations reporting paradigm?

6. SFAA Recommends Several Specific Clarifications and Revisions Within the Language of the Proposed Regulations

a) *Page 42147*. It is our understanding that BOEM will assess the need for a lessee to post financial security annually, though this is not specified in the Proposed Regulations. The Proposed Regulation should so specify.

b) *Page 42147*. On page 42147 (part of Section VI.A) indicates that non-publicly traded companies do not have credit ratings. In fact, some non-publicly traded companies do have credit ratings, and some publicly traded companies do not.

c) *Section 556.901(d)(2).* On page 42152 (part of Section X "Section-by-Section Analysis"), BOEM discusses Section 556.901(d)(2) and references credit ratings of BBB- from S&P or Ba3 from Moody's. This reference to Moody's should be changed to Baa3, which is the equivalent of S&P's BBB- credit rating and consistent with the other credit ratings references in the proposed rulemaking.

d) *Section 556.906.* We appreciate BOEM's effort to add clarity by replacing the word "terminated" with "canceled" in paragraph (b)(1) (As stated on page 42155 of the Proposed Regulations, the term "terminated" will be replaced by "canceled.") However, we believe the term "terminated" should also be replaced by "canceled" in paragraph (b)(3). This would make that section, which refers to "additional bonds" (as opposed to the base bonds), consistent with the wording referencing the base bonds.

e) *Appeal Bonds.* The Proposed Regulations currently specify that an appeal bond will automatically convert to a financial assurance obligation should the lease operator lose its appeal. However, appeal bonds do not operate in this manner. The appeal bond should provide a certain number of days for the lease operator to post its financial assurance to allow the surety to underwrite the operator at the time said bond is needed. Importantly, the obligations under the financial assurance bond relate to the operator's decommissioning obligations while the appeal bond guarantees that the operator will meet BOEM's requirements for it to post security for the decommissioning obligations under the lease.

7. Responses to Additional Comments Solicited by BOEM

SFAA submits the following responses to BOEM's requested comments on the topics below

a) "BOEM is considering the inclusion of offshore joint and several decommissioning liabilities (of the co- lessees that would otherwise have exempted the lessee from providing supplemental financial assurance) in the determination of a proxy credit rating when these liabilities are ''disproportionately high'' and may encumber that co-lessee's ability to carry out future obligations. BOEM is requesting comments on the appropriate criteria to determine what constitutes ''disproportionately high'' offshore liabilities, for example, a ratio of decommissioning liabilities to the net worth of the co-lessee above X times, or other financially significant and reasonable criteria on how these liabilities should best be incorporated into the proxy credit rating that BOEM will derive."

BOEM is seeking to modify credit ratings in a unilateral way utilizing an unregulated and undefined manner. SFAA discourages BOEM from doing so. Please refer to SFAA's broader comments to this effect in section 3 of its comments herein. SFAA reiterates that section in its entirety in response to this direct request for comment from BOEM.

b) "The use of End-of-Life (Years) in the evaluation of asset value as an alternative to using the decommissioning costs ratio. BOEM requests comments on the use of a minimum number of years of production remaining criterion to qualify for an exemption from supplemental financial assurance. Possibly, End-of-Life criteria could be an alternative to the 3:1 ratio of value of reserves to decommissioning costs."

BOEM is seeking to modify credit ratings in a unilateral way utilizing an unregulated and undefined manner. SFAA discourages BOEM from doing so. Please refer to SFAA's broader comments to this effect in section 3 of its comments herein. SFAA reiterates that section in its entirety in response to this direct request for comment from BOEM.

c) "The consideration of bond issuance ratings, in addition to issuer credit ratings, in determining the financial risk posed by lessees and grant holders. BOEM also invites comments on determining an appropriate threshold for bond issuance ratings, such as general unsecured debt ratings."

Bond issuance ratings are specifically for one tranche of security and are unique to the underlying offering associated therewith. It is therefore not applicable to the overall credit health of an entity. Therefore, it should be taken into consideration as an indicator of credit health, but not serve as a proxy to an overall credit rating.

BOEM is seeking to modify credit ratings in a unilateral way utilizing an unregulated and undefined manner. SFAA discourages BOEM from doing so. Please refer to SFAA's broader comments to this effect in section 3 of its comments herein. SFAA reiterates that section in its entirety in response to this direct request for comment from BOEM.

d) "Should BOEM exclude third-party guarantors from the requirement of § 556.902(a)(3) that guarantees must ''guarantee compliance with all obligations of all lessees, operating rights, owners and operators on the lease'' in addition to allowing a third- party guarantee to be limited in amount?"

No. If a third-party meets the U.S. government's regulatory requirements for conducting activity and assuming risk in the normal course then BOEM should not intercede with any limitation other than thoserequired by the applicable regulatory body whose responsibility it is to regulate such a third-party.

Though well intended, SFAA fears the Proposed Rule might have unforeseen and possibly negative consequences. Requiring additional financial assurances from those lessees least able to provide it mightresult in increased collateral demands and downward pressures leading to an uptick in bankruptcies among small, independent operators – ultimately worsening unfunded decommissioning liabilities.

Additionally, because the underlying asset – in this case well revenue – isn't apt to change, the Proposed Regulations would alter the environment such that not only would surety be unlikely to supply the additional coverage without significant, and in most cases, unavailable collateral from lessees, it would create a need for surety to reassess current contracts which could result in additional collateral demands.Given the increased risks in the market created by recent bankruptcy rulings, the Proposed Regulations could have an even stronger chilling effect on surety's appetite to participate. As is, players are already exiting the marketplace, further exacerbating capacity concerns.

Protecting taxpayers from shouldering the burden of decommissioning costs is a valid and worthwhile endeavor. SFAA remains open and willing to work with BOEM to find a way for the Proposed Regulationsto achieve that goal.

Thank you for your consideration of our comments. We look forward to hearing further from you.

Respectfully submitted,

Julie Alleyne
Vice President,
Policy & General Counsel
The Surety & Fidelity Association of America
JAlleyne@surety.org
202.778.3630

September 7, 2023

Via Electronic Submission

Kelley Spence
Department of the Interior,
Bureau of Ocean Energy Management,
Office of Regulations,
45600 Woodland Road, Mailstop
VAM–BOEM DIR, Sterling, VA 20166

Re: Risk Management and Financial Assurance for OCS Lease and Grant Obligations, RIN 1010–AE14; 88 Fed. Reg. 42136 (June 29, 2023); Docket ID: BOEM–2023–0027

Ms. Spence:

I write as the chief legal officer of the State of Louisiana. My State is committed to protecting the quality of our environment while at the same time promoting economic progress, developing natural resources, protecting our citizens' access to affordable energy, and ensuring that government agencies follow the law. I accordingly provide the following comments in response to the Bureau of Ocean Energy Management's Notice of Proposed Rulemaking, *Risk Management and Financial Assurance for OCS Lease and Grant Obligations, RIN 1010–AE14* ("Proposed Rule").

I. Introduction

Louisiana is, by every metric, an oil and gas state.[78] But more than that, Louisiana is an offshore drilling state. It was off the coast of Louisiana where the first offshore platform produced oil in paying quantities, and it was off the coast of Louisiana where the first platform was placed "out-of-site-of-land."[79] Indeed, Louisiana was spearheading the modern offshore drilling era long before OCSLA, MMS, and BOEM were even conceived. Louisiana's dominance in the offshore sector continues to present day, with vast quantities of offshore production flowing to onshore facilities in Louisiana, vast numbers of Louisiana citizens working to process that production, and vast quantities of oil and gas from the Gulf of Mexico ("GOM") flowing out of Louisiana to the nation's consumers. Given this unique position, BOEM's leasing decisions in the Gulf of Mexico have a significant impact on Louisiana citizens and the State's revenue.

II. Summary of Our Position

The Proposed Rule recites a salutary objective: protecting the American taxpayer from having to maintain and decommission orphan wells on the Outer Continental Shelf. However, the problem this Proposed Rule was purportedly designed to remedy is more imaginary than real. The Proposed Rule thus falls within the unpleasant category of a "solution in search of a problem."

In reality, there is no problem within taxpayer-funded decommissioning under the longstanding joint-and-solidary liability arrangement whereby current and predecessor lessees are each fully liable for well decommissioning costs. There certainly is no problem that would justify eviscerating the oil and gas industry. The Proposed Rule's (1) lack of meaningful cost-benefit analysis, (2) absence of sufficient authority, and (3) disparate treatment of renewables, all suggest this Proposed Rule is intended to burden industry in furtherance of the Administration's hostility to fossil fuels rather than serve as a genuine effort to "protect the taxpayer."

---

[78] *Louisiana,* ENERGY INFORMATION ADMINISTRATION, https://www.eia.gov/state/analysis.php?sid=LA (last visitedAug. 21, 2023).

[79] Austin, D., et al. *History of the Offshore Oil and Gas Industry in Southern Louisiana*, U.S. DEPT. OF THE INTERIOR,MINERALS MANAGEMENT SERVICE. OCS Study MMS 2004-049 at p.30.

III. The Proposed Rule Will Decrease Revenue to the Treasury and Louisiana by Increasing Costs and Limiting the Number and Decreasing the Amount of Lease-Sale Bids

The Outer Continental Shelf Lands Act ("OCSLA") charges BOEM with balancing costs and benefits in connection with the outer continental shelf leasing program.[80] Yet BOEM failed to do so here. That failure is readily explained, however, by pretext: no matter how creative its inputs, no such analysis would ever favor the Proposed Rule. It would be a non-starter. And while BOEM concedes the Proposed Rule may "adversely affect in a material way the productivity, competition, and prices in the energy sector," BOEM gives no concrete figure as to what the American people get in return for this great sacrifice.[81] In fact, OIRA directed BOEM back to the drawing board thrice for its failure to quantify numerous aspects of the cost-benefit analysis.[82] Cost-benefit analyses cannot be conducted with, or satisfied by, hunches and feelings.[83]

Thankfully, a private entity (Opportune LLC) conducted a robust cost-benefit analysis independent of BOEM, which points to only one (already obvious) conclusion—the Proposed Rule isn't worth it.[84] As expected, the figures make evident that BOEM has committed the cardinal error of regulators: spending a dollar to save a dime. Opportune's study concludes that BOEM's Proposed Rule will cost $10,500,000,000 to save $391,000,000 in potential Uncovered Risk. So really, BOEM wants to spend a dollar to save less than a nickel. This disconnect between means and ends, if not naturally wrong, is legally wrong. "No regulation is 'appropriate' if it does significantly more harm than good." Michigan v. EPA, 576 U.S. 743, 752 (2015). Of course, BOEM must know that the costs and benefits of its Proposed Rule are blatantly lopsided.

---

[80] 43 U.S.C. § 1344(a).

[81] 88 Fed. Reg. at 42168; *see* OMB Passback 1.1, BOEM-2023-0027-0018, at p.14 (explaining that BOEM has not quantified the benefit—the actual cost of taxpayer-borne decommissioning—saved by the Proposed Rule).

[82] *Id*; OMB Passback 1, BOEM-2023-0027-0017; OMB Passback 2, BOEM-2023-0027-0020; OMB Passback 3,BOEM-2023-0027-0022.

[83] *See Motor Vehicle Mfrs. Ass'n of U.S., Inc. v. State Farm Mut. Auto. Ins. Co.*, 463 U.S. 29, 43 (1983) ("the agencymust examine the relevant data and articulate a satisfactory explanation for its action including a rational connectionbetween the facts found and the choice made.")

[84] *A Cost-Benefit Analysis of Increased OCS Bonding*, OPPORTUNE LLP (July 2023), attached as Exhibit A.

Basic economics tells us that making the leasing process more costly will whittle away the potential bidding pool to a select few industry players with either sufficient credit ratings or sufficient market power to obtain bonds at a manageable cost. This is where Louisiana's interests are most-directly harmed. Under GOMESA, the GOM drilling states and their political subdivisions receive a share of lease proceeds for coastal restoration and other important projects, with Louisiana receiving the highest share of GOMESA distributions.[85] If BOEM runs bidders out of the market with onerous supplemental bonding requirements, competition will decrease, bid prices will decrease, and Louisiana will lose out for no good reason.

IV. BOEM's Purported Statutory Authority to Require the Supplemental Bonding Under 43 U.S.C. § 1338(A) or Any Other Statute Is Highly Dubious

Section 1338(a)—which BOEM contends authorizes it to require supplemental bonding— gives no authority whatsoever to BOEM. It merely allows forfeited bonds to accrete to BOEM rather than the Treasury. It doesn't authorize anything. Rather, Section 1338(a) merely creates an exception to the Miscellaneous Receipts statute.

The remainder of the statutes are interpreted by BOEM to mean it has a general authority to develop "the Nation's offshore energy and mineral resources in an economically responsible way."[86] Assuming arguendo BOEM's interpretation is correct, BOEM fails by its own measure because the supplemental bonding requirements in the Proposed Rule are economically irrational and infeasible. While economics can sometimes be imprecise, BOEM must be objectively economically responsible, not subjectively adherent to musings about moral hazard. If BOEM had quantitative studies establishing either (1) that the current joint and solidary liability regime dis-incentivizes current operators from maintaining their wells, or (2) that this alleged disincentive has produced any negative economic or environmental effect to date, BOEM might be entitled to propose a rule remedying that problem. But because the problem BOEM cites is imaginary, and because its authority in attempting to solve it is also imaginary, the Proposed Rule does not belong in the Code of Federal Regulations.

---

[85] Gulf of Mexico Energy Security Act of 2006, 43 U.S.C. § 1331.
[86] 88 Fed. Reg. at 42138.

That conclusion is reinforced by the Major Questions Doctrine. "Congress … does not alter the fundamental details of a regulatory scheme in vague terms or ancillary provisions—it does not, one might say, hide elephants in mouseholes."[87] BOEM likely lacks statutory authority to require bonding generally, but to require bonding which will—by BOEM's own admission—have a "material" "adverse" effect on the energy sector, is unthinkable in a post-West Virginia world. Because no statute cited by BOEM gives it a clear charge to rewrite the playbook on offshore oil leasing and disrupt our nation's energy supply in the process, the Proposed Rule must fail.

V. The Proposed Rule Arbitrarily Departs from the 2020 Proposed Rule, Persisting with All of the Bad, and Leaving out All of the Good

The existing joint-and-solidary chain-of-title arrangement sufficiently distributes risk of loss, even with regard to the taxpayer. Indeed, it is Louisiana's view that supplemental bonding is unnecessary. So long as a creditworthy party exists in the chain of title—an able party that freely chose to transfer its interest to a particular party—there exists no meaningful decommissioning burden on the federal government. The 2020 Proposed Rule understood this reality and carved out from the bonding scheme properties that were "already covered" with sufficiently solvent predecessors-in-title.[88] With no such carve-out in this new Proposed Rule, BOEM guarantees that every OCS property, present and future, will be doubly-insured (or at least doubly assured).

Eliminating the phenomenon of double insurance, wherein the same risk is insured coextensively, is the focus of many in the maritime industry given its status as a significant and highly inefficient externality.[89] Yet here, BOEM is trying to codify a needless stacking of already- sufficient financial devices and schema upon each other to protect against a straw man that has, in its

---

[87] Whitman v. Am. Trucking Ass'ns, 531 U.S. 457, 468 (2001); see also West Virginia v. EPA, 142 S. Ct. 2587, 2608 (2022).

[88] 85 Fed. Reg. 65904 (Oct. 16, 2020).

[89] Cf. Michigan v. EPA, 576 U.S. 743, 753 (2015) (noting "the reality that so much wasteful expenditure devoted toone problem may well mean considerably fewer resources to deal effectively with other (perhaps more serious) problems").

entire history, cost the same to the American taxpayer as the 2022 "Visit Tunisia" initiative with USAID.[90]

VI. The Proposed Rule Operates as Mere Pretext for Further Handicapping Domestic Petroleum Production, in Yet Another Transparent Attempt by the Administration to Render Offshore Oil and Gas Exploration Economically Unfeasible

The Proposed Rule is fundamentally flawed in multiple ways, but what makes it uniquely troubling is its obvious insincerity. While the Proposed Rule reflects a valiant attempt at facial neutrality, it cannot be read without considering the Biden Administration's pervasive animosity toward domestic oil production.[91] In the absence of context, the Proposed Rule is more puzzling than concerning given the apparent lack of need for the supplemental bonding. But in the context of the Biden Administration's overt hostility toward fossil fuels, the haze dissipates and it becomes clear that the Proposed Rule is not intended to solve the problem of unfunded well decommissioning. Rather, the Proposed Rule is intended to solve what the Biden Administration views as Interior's ultimate problem—domestic petroleum production itself.[92] And if one is not yet convinced, look no further than BOEM's disparate treatment of renewables in the Proposed Rule. BOEM has provided no data that a wind lessee is more financially sound—or less likely to abandon an OCS property—than a petroleum lessee. Yet the Proposed Rule arbitrarily excludes renewables with no real explanation. Whether through favoritism or not, no matter how much it desires to do so, BOEM is forbidden from exterminating the petroleum industry through rulemaking.

---

[90] Launch of the "Visit Tunisia" Project, UNITED STATES AGENCY FOR INTERNATIONAL DEVELOPMENT, https://www.usaid.gov/tunisia/press-releases/feb-2-2022-launch-visit-tunisia-project (Feb. 2, 2022) ($50 million forTunisian tourism versus $58 million in taxpayer-funded decommissioning of OCS wells).

[91] E.O. 14,008 of Jan 27, 2021.

[92] S.O. 3395 of Jan. 20, 2021, U.S. DEPT. OF THE INTERIOR.

VII. The Proposed Rule Is a Clumsy Attempt to Circumvent the Mandatory Leasing Requirements Of OCSLA—As Recognized by an Injunction Favoring Louisiana—by Arbitrarily Reducing the Profitability of Oil and Gas Leasing While Excluding Favored Renewables

From inauguration day onward, the Administration has unashamedly thumbed the scale in favor of renewables, while it concurrently sandbagged oil production. In the intervening years, the federal courts have admonished such thumbing and sandbagging—and BOEM has been the subject of that admonishment on multiple occasions—yet it marches on. See, e.g., Louisiana v. Biden, 543 F. Supp. 3d 388 (W.D. La. 2021), vacated for lack of specificity and remanded, 45 F. 4th 841 (5th Cir. 2022), on remand, 622 F. Supp. 3d 267 (W.D. La. 2022). While no longer taking the forthright approach, BOEM appears as adamant and resolved as ever to switch the tap from our Gulf to the Persian Gulf. It is unnecessary to make an in-depth attempt to reconcile BOEM's supposed view that a barrel of oil in Jeddah is better for the nation than a barrel of oil in Calcasieu Parish, yet, the illogic persists.

Ideology notwithstanding, whether BOEM would like to facilitate lease-sales on the OCS is irrelevant, because it must. BOEM exists for one reason: to broker the capture of energy on the OCS. As noted above, the judiciary has rejected BOEM's attempt to halt leasing outright. "Congress has already decided that the OCS should be used to meet the nation's need for energy."[93] BOEM can't circumvent Congress's command by killing the industry through a thousand regulatory cuts.[94] Indeed, Congress—attuned to BOEM and Interior's strategy—recently reminded BOEM of its mandate.[95] BOEM's requirement of excessive bonding will do far more to reduce acreage sold on the OCS than its previous actions with Lease Sales 257 and 258.

* * * * *

The Proposed Rule is deeply flawed and inconsistent with BOEM's statutory mandates.

BOEM should accordingly withdraw the Proposed Rule.

---

[93] Ctr. for Biological Diversity v. U.S. Dep't of Interior, 563 F.3d 466, 485 (D.C. Cir. 2009).

[94] 43 U.S.C. § 1344(a).

[95] E.g., "The Secretary shall conduct Lease Sale 258." Inflation Reduction Act of 2022, Pub. L. No. 117-169 §50264(b).

For Louisiana,

Jeff Landry
Louisiana Attorney General

Exhibit A. A Cost Benefit Analysis of Increased OCS Bonding

July 13, 2023

I. Executive Summary

The offshore oil and gas industry remains conflicted between investors looking for a return on capital, a push by certain groups to eliminate oil and gas drilling, and a global need for sustained affordable energy prices. A decrease in oil and gas investments and related drilling over recent years has resulted in an unprecedented rise in commodity prices and inflation not seen in 40 years. Recent and potential bankruptcies by independent operators in the Outer Continental Shelf ("OCS") of the Gulf of Mexico ("GOM") have sparked potential action by the Bureau of Ocean Energy Management ("BOEM") and the current Presidential Administration (the "Administration") to revisit previously proposed regulations to increase surety bonding requirements - all in an effort to protect the U.S. taxpayer. Regardless of the genuine intent of all interested parties, further constraints on oil and gas capital are the greatest threat to the U.S. taxpayer.

A. Background

Opportune LLP ("Opportune"), a leading global business advisory firm, previously released a July 2016 independent study (the "2016 Study") assessing the economic effect that BOEM NTL 2016-N01 (the "2016 NTL")[96] would have had on the offshore oil and gas industry (the "Industry"), Gulf Coast and United States.

[96] Notice to Lessees ("NTL") No. 2016-N01 issued July 14, 2016.

Federal regulations have always required previous owners to remain responsible for the decommissioning costs of wells, pipelines, and other facilities (commonly referred to as plugging and abandonment, or "P&A") that existed at the time of the sale of such properties.[97] As a result, there had never been a single instance in which the U.S. taxpayer was required to bear the cost to P&A an OCS property. Regardless of those facts, the 2016 NTL would have changed the way many independent oil and gas operators fund P&A costs in the GOM's OCS by requiring them to post supplemental surety bonds or other collateral to guarantee coverage for 100% of the future P&A costs attributable to the oil and gas properties in which they own a working interest.

The stated purpose of the 2016 NTL was to protect the U.S. taxpayer from ever having to pay offshore decommissioning costs for an oil and gas company that falls into bankruptcy; however, Opportune's 2016 Study concluded that the proposed changes were wholly disproportionate to any potential risk.

After publication of the 2016 Study, liquidity across the Industry deteriorated under the weight of bloated balance sheets resulting from the combination of overheated commodity prices from 2011 to 2015 and capital markets that fueled excessive acquisition and development spending during the shale boom. The result was a volumetric supply overhang from late 2014 to 2016 that prevented commodity prices from recovering enough to offset the Industry's constrained capital. Commodity prices stabilized through 2019 as global demand eroded excess supply and the Industry began to recover; however, the COVID-19 pandemic shut down the global economy and "threw the supply chain and demand completely out of whack".[98] Within the first three months of 2020, U.S. production dropped approximately 20% as private and public companies shut-in wells and slashed drilling budgets. The front-month May 2020 WTI crude contract settled at an unprecedented low of negative $37.73 per barrel in April 2020.

Since July 2016, a total of 12 U.S. independent offshore oil and gas companies filed bankruptcy with respect to approximately $18.8 billion in total capital, including approximately $4.0 billion in future P&A liabilities[99]. All related P&A liabilities were either assumed by the reorganized Debtor, subsequent buyers of the underlying assets, co-lessees and/or predecessors in

[97] 30 CFR §556.62(d).

[98] President Biden, White House Press Briefing, May 10, 2022.

[99] See Exhibit A – Summary of Offshore Operator Bankruptcies: 2016 – 2023. Amounts include two Ch. 11 filings by Fieldwood Energy. $18.8 billion represents the combination of $14.8 billion in Pre-petition Capital plus $4.0 billion inP&A Obligations.

title, or funded through a liquidation trust – at a miniscule cost to the American taxpayer.

B. 2023 Update

BOEM published a new proposed rule on June 29, 2023[100] (the "Proposed Rule") to increase financial assurance requirements on the Industry. BOEM states[101] that the proposed changes "advance the Biden-Harris Administration's federal oil and gas reform agenda, which was outlined in a report from the Department of the Interior developed in response to Executive Order 14008."[102]

The Proposed Rule requires OCS companies to provide additional financial assurance (bonds) if i) their S&P/Moody's credit ratings and that of their current co-owners are lower than BBB-/Baa3; and ii) the current value of their proved oil and gas reserves is less than three times than the associated undiscounted P&A liability. As such, Opportune is updating the 2016 Study to independently review the following:

- Current BOEM estimated P&A liabilities for OCS leases and the underlying historical data aggregated by The Bureau of Safety and Environmental Enforcement ("BSEE") (the "BOEM Database")[103]
- Actual historical costs of OCS P&A liabilities borne by U.S. taxpayers
- Current distribution of OCS P&A liabilities by major integrated ("Majors"), large independent ("Large Independents") and small independent ("Small Independents") lessees[104]

---

100 Federal Register, Vol. 88, No. 124, June 29, 2023.

101 www.boem.gov/newsroom/press-releases/boem-proposes-stronger-financial-assurance-requirements-offshore-oil-and.

102 Exec. Order No. 14008, *Tackling the Climate Crisis at Home and Abroad*. 86 Fed. Reg. 7619, January 27, 2021.

103 Estimated P&A and Collateral obtained from https://www.boem.gov/oil-gas-energy/risk-management/property-lists-0as of March 15, 2023.

104 Majors and Large Independents are defined as having a tangible net worth, or market capitalization, of over $10.0 billion as of March 15, 2023. Small Independents are defined as all other OCS lessees who have a tangible net worth that is either less than $10.0 billion or unavailable from public sources. See Appendix C of the Opportune 2022 Studyfor a listing of lessees considered to be Majors and Large Independents (including subsidiaries and successors); SmallIndependents represent all other lessees.

- Estimated effect to U.S. taxpayers and the Gulf Coast as a result of implementing the Proposed Rule

Opportune shows within its revised study (the "Opportune 2023 Study") how the perceived benefits of additional bonding requirements remain wholly disproportionate to any potential risk. This Opportune 2023 Study reiterates how the current rule has adequately protected U.S. taxpayers for decades. Implementing the Proposed Rule would force independent lessees out of business, achieving the opposite public policy objective than purportedly intended by BOEM.

The Opportune 2023 Study has been conducted to independently calculate the OCS P&A liability, assess the risk such liability poses to the U.S. taxpayer, and perform a cost-benefit analysis of how the Proposed Rule would economically affect the Industry, Gulf Coast and United States. This study was performed by Opportune's valuation, petroleum engineering[105] and financial reporting[106] professionals through a series of interviews with Industry representatives, commercial banks and surety brokers, and includes an analysis of independently obtained P&A cost data and market research.

The results of the Opportune 2023 Study and recommended solutions are presented withinthe report and appendices.

C. Key Findings

The current system works: U.S. taxpayer funding of decommissioning liabilities is infinitesimal.

Historically, federal regulations have required previous owners (i.e., the "previous chain of title") to remain responsible for the P&A costs of wells and facilities that existed at the time of the sale of such properties to previous owners.[107] Current regulations operate to greatly reduce or maintain the current minimal risk to taxpayers. For example, a current lessee would have to default, file for bankruptcy, unsuccessfully auction the properties due to no interested buyers, and have no solvent co-owner(s) or previous owner in the chain of title for the taxpayer to be liable. In the limited cases where funding for the P&A liabilities of a bankrupt owner is uncertain, a decommissioning

[105] Ralph E. Davis Associates is a wholly-owned subsidiary of Opportune LLP.
[106] Opportune LLP is not a CPA firm.
[107] 30 CFR §556.62(d).

trust or cash escrow has been established to fund the future decommissioning. Unlike the $2.1 trillion paid to other industries since 1970[108], the potential Industry P&A cost to the U.S. taxpayer is infinitesimal.

For the first time in Industry history, BSEE announced a Draft for Proposal seeking bids to decommission 15 orphan wells located in the Matagorda Island, High Island, and West Delta areas of the OCS[109]. The decommissioning costs will be funded from the 2021 Infrastructure and Jobs Act appropriations. Based on BOEM's database, the P&A liability associated with the orphan wells and their related infrastructure is approximately $37.7 million[110] – an infinitesimal amount (~0.8%) compared to the $4.5 billion in royalties and revenue the U.S. government (i.e., taxpayer) received in 2021 alone from offshore Gulf of Mexico operations.

BOEM estimated P&L liabilities are overstated and the underlying assumptions remain opaque and flawed.

The ongoing and future decommissioning of oil and gas infrastructure in the OCS is no small task and the process is costly. Opportune believes, however, that BOEM's $84.8 billion[111],[112] estimate of the existing P&A liability is not transparent and remains dramatically overstated.

Since 2016, BSEE has subsequently improved its cost database by collecting actual decommissioning cost data from OCS lessees; however, it is not clear whether the costs captured within the BSEE database agree with the same assumptions underlying each lessee's audited financial statements. Not utilizing audited financial information unnecessarily adds costly government bureaucracy and begs the question of whether a black box is overstating OCS P&A liabilities.

BSEE has also begun applying a combination of a probability-weighting[113] of historical cost data and a deterministic value approach ("Deterministic") to better assess the underlying risk of related properties when calculating individual lease P&A liabilities. According to the law firm Vinson & Elkins LLP, "The deterministic value approach is calculated by

---

[108] See Exhibit B – Summary of Historical U.S. Government Bailouts, by Industry.

[109] "Orphaned Wells on the U.S. Outer Continental Shelf", BSEE Virtual Industry Day Presentation; February 2, 2022.

[110] See Exhibit C.

[111] Note that the collective tangible net worth of the Majors and Large Independents exceeds $1,274 billion and $552billion, respectively, per Capital IQ as of March 15, 2023.

[112] Note the $84.8 billion represents the sum of the undiscounted P70 and the deterministic costs of all leases in the OCS,excluding ROW and RUE properties, per the BOEM Database.

[113] BSEE applies P50, P70 or P90 cases to a log-normal distribution of historical cost data to estimate a 50%, 70% or90% probability that the actual abandonment costs will be less than the underlying BSEE costs.

BSEE based on its evaluation of decommissioning cost data but the model does not include elements of randomness; given a particular input, the same output will be produced every time."[114] BOEM proposes to use a "P70" probability-weighting, plus Deterministic, value to set the amount of any required supplemental financial assurance; however, Opportune believes a P50 probability-weighting is more representative of a log-normal distribution's statistical average.

Finally, the Proposed Rule places bonding requirements on lessees that are based on the unrealistic undiscounted assumption that all active OCS leases might be decommissioned at once (tomorrow). Opportune estimates that the net present value of the P50 (plus Deterministic) liability to P&A all OCS leases is approximately $16.4 billion ("OCS P&A Liability").

The underlying risk to the U.S. Taxpayer is overstated.

The implied risk to the U.S. taxpayer underpinning the Administration's desire for the Proposed Rule is overstated. BOEM holds each lessee responsible for 100% of the lease P&A liability, regardless of the amount of working interest actually owned by the individual lessee. Surety brokers state that the BOEM database cites the full P&A amount for each lease in order to maximize coverage, particularly when one or more of the lease's owners are exempt; however, the individual lessee's pro rata obligation is cited by BOEM when requesting additional bonding via a demand letter. BOEM's intent is unclear as the public information available on its website overstates the total OCS P&A Liability and the underlying risk to the U.S. taxpayer.

The OCS P&A Liability associated with properties in which the Majors and Large Independents are not part of the current ownership or previous chain-of-title (the "Uncovered Properties") is only $1.2 billion, which is ~7% of the total OCS P&A Liability calculated by Opportune. Approximately $761 million in bonding has already been posted (to the benefit of BOEM) with respect to the Uncovered Properties, leaving an estimated uncovered risk to the taxpayer of $391 million (the "Uncovered Risk"), or ~2% of the total OCS P&A Liability calculated by Opportune.

---

114 "BSEE Considerations in Establishing Decommissioning Liability Estimates for Offshore Facilities", Vinson & ElkinsLLP, June 2021.

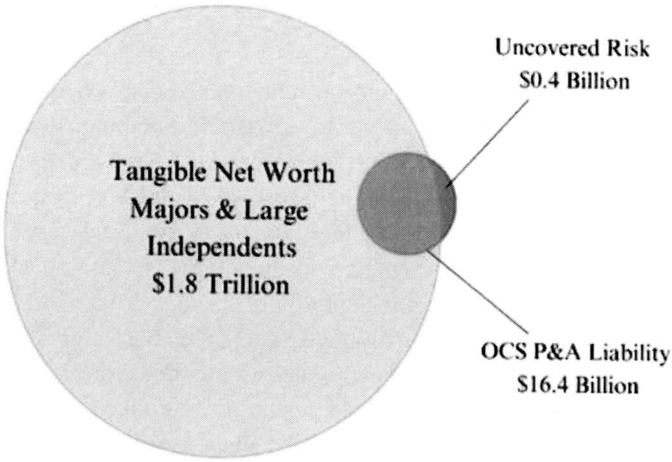

Small Independents will be unable to obtain the required bonding; the Proposed Rule will become a catalyst to spur the bankruptcy risk from which it was intended to protect the U.S. taxpayer.

The Proposed Rule cites an estimated $9.2 billion[115] in supplemental bonding to effectively insure the $391 million in Uncovered Risk noted above.

Beyond the cost[116] of such additional supplemental bonds, the lack of capacity in the surety bond market guarantees that the additional bonding requirements are untenable, as most Small Independents will simply not be able to obtain the supplemental bonding required for their existing properties, let alone the supplemental bonding required to drill the future wells necessary to support their capital structure. Under increasing pressure by ESG activists, surety markets have threatened to exit the offshore sector since publication of the 2016 NTL, thereby drastically reducing available bonding capacity.

To secure any available bonding, Small Independents would have to provide cash collateral that their balance sheets cannot currently support due to recent asset impairments that permanently reduced their net worth, as U.S. GAAP[117] prevents asset values from being written back up as commodity prices recover in the future. The overall opportunity cost of tying up this amount of the Small Independents' liquidity is enormous, resulting in a dollar-for-dollar reduction in their ability to develop properties and pay down debt, inevitably increasing the risk of additional bankruptcies.

---

[115] Federal Register, Vol. 88, No. 124, June 29, 2023.
[116] Approximately 1-4% of the total surety bond's face amount.
[117] Generally Accepted Accounting Principles.

The Proposed Rule will result in dramatically reduced future production, economic activity, and U.S. royalty revenue.

In our updated analysis, Opportune relies on BOEM's contention that their Proposed Rule will result in increased bonding premiums for "small" companies of approximately $257 million per year. Using our development model, we estimate that these higher costs will lead to less offshore development activity, lower economic growth, reduced hydrocarbon production, and lower federal royalties. In our model we assume a received price of $61 per barrel of oil equivalent ("boe", using a 6:1 gas:oil ratio and other assumptions), an assumed finding and development cost of $25/boe, along with other economic and operational assumptions, and find that the increased bonding requirements could result in a production decrease of approximately 55 million barrels of oil equivalent (boe) and approximately 36,000 fewer jobs over a 10-year time horizon. Additionally, the Opportune 2023 Study indicates that the US economy (particularly in the Gulf Coast states) will experience a decline in Gross Domestic Product (GDP) of as much as $9.9 billion over the same period. Beyond the opportunity cost of wells not drilled and the resulting overall impact to the Gulf Coast, the reduction in development may result in estimated lost royalties to the U.S. Department of the Interior of approximately $573 million over the 10- year period. All told, total direct and indirect costs of Additional Bonding Requirements to the taxpayer may total over $10.5 billion.

Cost-Benefit Analysis - U.S. Taxpayer

There is a better way than the Proposed Rule: increased communication with operators and a focus on Sole-Liability Properties.

The Opportune 2023 Study concludes that the estimated costs to U.S. taxpayers and the Industry will drastically exceed the benefits generated by the Proposed Rule. The Opportune 2023 Study shows the current system adequately addresses BOEM's concerns about the potential risk to taxpayers.

BOEM began improving the process during the prior Administration by collecting actual decommissioning data and considering feedback from stakeholders about the potential impact of Sole-Liability Properties. Under the current Administration, BOEM should abandon the Proposed Rule in favor of continued active discussions with OCS lessees to avoid any perception of a politicized process aimed to diminish drilling and development of the OCS.

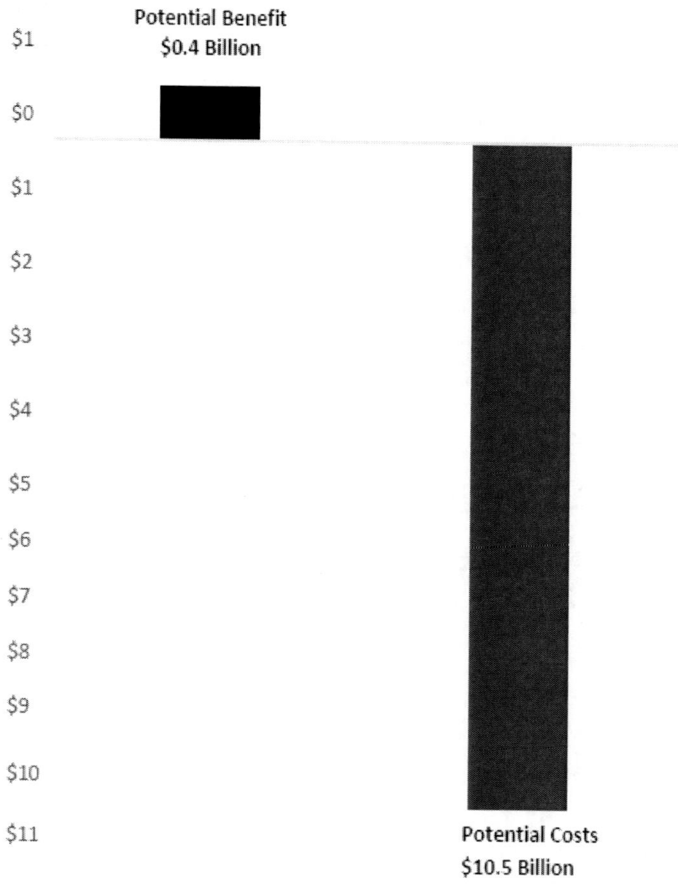

The Opportune 2023 Study proposes additional solutions to support OCS lessees and protect taxpayers.

The Opportune 2023 Study also puts forward proposals, which include using basic and widely accepted valuation techniques to estimate the P&A liability and financial health of OCS lessees; improving the process by which BOEM gathers P&A cost and collateral information; and introducing a new cash reserve to be funded by current OCS lessees to supplement current co-owners or previous owners in the chain of title and further protect the U.S. taxpayer. The following table provides a high-level summary of the solutions proposed, the details of which are further described within the Opportune 2023 Study:

Opportune Study – Recommended Solutions for BOEM Consideration

Valuation	Data Gathering	Financial Assurance
• Value P&A liabilities using the discounted ARO[118] balance audited andincluded in the financial statements of every lessee • When assessinglessees' financial strength and credit rating: • Useforward strip pricing to reflectthe current market valueof proved reserves. • Exclude the audited ARO liability from thevalue of proved reserves to avoida double jeopardy effect when comparing to the related P&A liability	• Avoid duplication of costs and efforts by requiringboth private and public OCS lessees to report the future P&A costs and dates, by property, underlying the ARO amount included in their audited financial statements. • Upload amounts noted above to a national BSEE database, similar to Federal requirements for U.S. refineries[119].	• Replace any additional bonding requirements with a cash reserveself-funded by 1% of each OCS lessee's pro rata share of future production until such reserve equals the Uncovered Risk ($0.4B). • Absent capacity in the bond market, create a federal agency to issuebonds on Sole-LiabilityProperties. If the U.S. government believes the taxpayer is really atrisk, allow taxpayers toinherently participateinthe upside of the premium float.

II. Background Information

The purpose of the Opportune 2023 Study is to perform a cost-benefit analysis of proposed financial assurance rules developed by BOEM for the plugging and abandonment of the energy infrastructure in the Gulf of Mexico's outer continental shelf. The Opportune 2023 Study independently calculates the OCS P&A liability, assesses the risk such liabilities pose to the U.S. taxpayer, and perform a cost-benefit analysis of the Proposed Rule's economic effect on the Industry, Gulf Coast and United States.

Opportune LLP, a leading global business advisory firm, performed this study using its valuation, petroleum engineering and financial reporting[120] professionals through conducting interviews with Industry representatives,

---

[118] Asset Retirement Obligation, which is the term used for P&A in Accounting Standards Codification 410-20.

[119] 40 CFR § 80.75, 80.125 and 80.150.

[120] Opportune LLP is not a CPA firm.

commercial banks and surety brokers, along with analyzing independently-obtained P&A cost data and market research.

This section provides information on the i) offshore drilling and infrastructure, ii) decommissioning requirements, iii) decommissioning cost estimates, and iv) historical financial assurance requirements. Note that certain background information previously included in the 2016 Study has been excluded from this update, or otherwise summarized, to limit duplication and remove details that may no longer be relevant.

A. Offshore Drilling and Decommissioning Requirements

In 1938, a predecessor to ExxonMobil completed the first producing offshore well in a water depth of 14 feet off the Louisiana coast. The first offshore well out of the sight of dry land was drilled in 1947 by a predecessor to Anadarko, as operator for predecessors of ConocoPhillips and BP. As the U.S. population grew post-World War II and became more dependent on oil and gas production, offshore drilling moved further offshore and in to depths exceeding 1,000 feet of water by 1970. As development of the shallower water (depths of 500 feet or less; the "Shelf") increased, the Majors moved further offshore in search of discoveries large enough in size to meet their internal reserve replacement and potential profit requirements.

Title 30, Section 250 of the Code of Federal Regulations ("CFR") requires that all wells be plugged, and platforms removed within one year after a lease terminates. In addition, all other lease structures, decommissioned pipelines and obstructions should be removed and cleared from the seafloor within five years of such infrastructure no longer useful for operations.

Section 250.1701 further states that (emphasis added):

> "Lessees and owners of operating rights are ***jointly and severally responsible*** for meeting decommissioning obligations for facilities on leases, including the obligations related to lease-term pipelines, ***as the obligations accrue***[121] and until eachobligation is met."

[121] Meriam-Webster defines accrue as follows: "to increase in value or amount gradually as time passes: to grow or buildup slowly".

Although estimates vary, the latest published independent research[122] notes that approximately 6,900 oil and gas structures have been installed approximately 5,300 structures havebeen removed in the Gulf of Mexico since offshore drilling first began, leaving approximately 1,600 active structures as of August 2022. Additionally, over 46,000 wells have been drilled, of which over 38,000 wells have been plugged, leaving an active inventory of approximately 8,000 wells.

B. Decommissioning Cost Estimates

BSEE is responsible for approving all decommissioning activities, as well as aggregating and estimating the amount of historical P&A cost and future liabilities. BSEE monitors potential idle wells and infrastructure based on information from the Technical Information Management System ("TIMS"), which is the main data system used by the Department of the Interior ("Interior"). BSEE notifies lessees of their related decommissioning liabilities and requests a plan and schedule for decommissioning the lessee's applicable assets.

Lessees must report the outcome of all activities to plug wells, remove platforms or other facilities, decommission pipelines, and clear sites around such infrastructure. BSEE estimates the costs of decommissioning activities based on well depth and the number and types of wells, pipeline segments, and structures on a lease.

BSEE may also review and periodically update its costs estimates based on the occurrence of certain events, which may include, but are not limited to, determining a lease may be sold or the lessee is under financial distress, or upon a request by the lessee to review lease obligations and estimates. BSEE is required to input all cost estimates and activities into TIMS.

C. Historical Financialassurance Rules and Proposals

Title 30 CFR Section 556 requires that all operators provide a non-cancellable bond, payable upon demand to BOEM Regional Director, in the amount of $50,000 per lease or $300,000 for an area-wide bond when there is no

---

122 Mark J. Kaiser, Shallow-water Gulf of Mexico Decommissioning Market Valued at $6.3 Billion, www.offshore-mag.com, August 2022.

operation or activity on the lease. Lease-specific bonds ramp up to $200,000 or $500,000 per lease, depending on whether there is existing exploration or development, respectively. Area-wide bonds increase for the same reasons up to $500,000 or $3 million, respectively.

The Regional Director of BOEM may require operators to provide supplemental bonds based on BOEM's assessment of an operator's ability to meet current or future decommissioning obligations. The Regional Director's assessment may be based on the operator's financial capacity in excess of existing and anticipated lease and other obligations, in addition to the operator's historical operating record, current production and estimated proven reserves of future production. Historically, most companies have been exempt from supplemental bonding, provided: i) net worth exceeds $65 million and is at least two-times the amount of their estimated P&A liabilities; and ii) total company liabilities of no more than 2 to 3 times the value of the adjusted net worth.

GAO Report to Congressional Requesters– December 2015

Due to the bankruptcy of ATP Oil & Gas ("ATP") in 2012 and the subsequent financial distress of a number of offshore oil and gas producers in subsequent years, certain members of the U.S. Congress requested GAO to review Interior's management of liabilities from offshore oil and gas production. In its December 2015 Report to Congressional Requesters, GAO issued a report (the "GAO Report") that: (1) examined the Interior's (a) procedures for overseeing decommissioning and estimating its costs, (b) procedures for obtaining financial assurances for these liabilities; and (2) challenges Interior's managing of these liabilities.

To better ensure that the government obtains sufficient financial assurances to cover decommissioning liabilities in the event of lessee default, the GAO Report called for the following actions to be taken by the Interior:

Ensure that BSEE collects all relevant data associated with decommissioning from lessees.

- Direct BSEE to establish documented procedures for estimating decommissioning liabilities.
- Develop a plan and set a time frame to ensure that Interior's data system for managing offshore oil and gas activities includes processes to accurately and completely record estimated decommissioning liabilities.

- Develop a plan and set a time frame to ensure that Interior's data system for managing offshore oil and gas activities will be able to identify, capture, and distribute data on decommissioning liabilities and financial assurances in a timely manner.
- Ensure that BOEM completes its plan to revise its financial assurance procedures, including the use of alternative measures of financial strength.
- Revise BOEM's regulations to establish a clear deadline for the reporting of transfers to require that lessees report the transfer of rights to lease production revenue.

Summary of Boem Notice To Lessees– July 2016

In a November 2015 letter to GAO, Interior stated that it generally agreed with GAO's draft findings and committed to implementing GAO's recommendations to document procedures, improve the data system, and revise financial assurance procedures and regulations. In its letter, Interior referred to a proposed Notice to Lessees to be issued by BOEM that would provide updated criteria for determining a lessee's ability to meet future decommissioning and other liabilities and the potential need for additional security.

On July 14, 2016, BOEM issued NTL No. 2016-N01, Requiring Additional Security (the "2016 NTL"), which clarifies the procedures and criteria that BOEM Regional Directors are to use in determining if and when additional security may be required for OCS leases, rights-of-way ("ROW"), and rights-of-use and easement ("RUE"). NTL No. 2016-N01 was scheduled to be effective September 12, 2016, and contained the following provisions.

Evaluation of Financial Ability to Carry Out Obligations

The Regional Director would evaluate the ability of lessees to carry out present and future obligations annually to determine whether additional security must be provided. Periodic reviews to evaluate financial ability could be done at any time at the discretion of the Regional Director, but would likely be done when BOEM became aware of any:

- Material or adverse change in a lessee's financial strength or OCS obligations;
- Performance deficiencies, such as incidents of noncompliance, civil penalties or failure to adhere to any term or condition of, or obligation

imposed by, a lease, exploration or development and production plan, development operations coordination document, or permit;

- Change in operator or ownership; or
- Violation of Interior or other applicable regulations.

Factors Considered in Evaluation of Financial Ability and Determination of Self-Insurance

The Regional Director's evaluation of financial ability would be based on information submitted by lessees and other information in BOEM's possession demonstrating financial capacity, financial strength, stability, reliability, and record of compliance. The result of such evaluation would determine whether, and how much, additional security would be required. In making this determination, the Regional Director would consider 100 percent of decommissioning and other liability for every lease, ROW, and RUE in which in which a lessee holds an ownership interest, or provides a guarantee.

Financial Capacity as evidenced by recent (not more than 12 months old) audited financial statements, would be demonstrated in part by the following financial criteria:

- Cash Flow from Operations/Total Debt
- Current Ratio
- Earnings Before Interest and Taxes (EBIT)/Interest Expense
- Quick Ratio
- Return on Assets
- Return on Equity
- Total Debt/Capital
- Total Debt/Earnings Before Interest, Taxes, Depreciation and Amortization (EBITDA)
- Total Debt/Equity

BOEM established minimum thresholds for each of the nine ratios above, as well as the number of such thresholds that BOEM would require lessees to exceed, to determine that the lessee had adequate Financial Capacity.

- Projected Financial Strength would be based upon the estimated value of existing OCS lease production and proven reserves of future production.

- Business Stability would be based upon five years of continuous operation and production on the OCS or onshore.
- Reliability would be based upon lessee's credit rating from Moody's or Standard and Poor's, or from tradereferences.
- Record of Compliance would be based upon whether the lessee or any of its affiliates or subsidiaries had been:
  - Assessed civil penalties by either BOEM or BSEE;
  - Found by BOEM and/or BSEE to be non-compliant with any lease, plan, or permit term or condition;
  - Cited by any other agency(ies) with jurisdiction on the OCS, for non- compliance with any regulation; and/or
  - Cited for non-payment or under-payment of rentals, royalties, interest bills, civil penalties, or inspection fees, and such non-payment or under-paymenthas been referred to the U.S. Treasury for collection within the past five years.

Based on the above criteria, the Regional Director could determine that a lessee had the ability to self-insure some portion of, or all of, any additional security obligations for a lease, ROW, or RUE; however, the Regional Director would not permit the use of self-insurance in an amount in excess of 10 percent (10%) of the lessee's tangible net worth. The Regional Director could determine, based on a lessee's credit rating that it cannot apply self-insurance to sole liability properties. Sole liability properties are leases, ROWs, or RUEs for which a lessee is the only liable party and no prior interest holders who would be liable to BOEM to meet the obligations arising from such properties.

Determination, Notification, and Timing of Additional Security

If the Regional Director determined that a lessee's financial ability along with existing financial assurance was not sufficient to cover its obligations, the Regional Director would notify the lessee in writing. The Regional Director's notification will either: (1) propose an amount of additional security required and give the lessee opportunity to meet with BOEM to discuss this amount; or (2) order that the provide the required additional security or present to BOEM a tailored plan to phase in the additional security requirement.

Tailored Plans to Meet Additional Security Requirements

If a lessee used any type of financial assurance other than surety bonds or Treasury securities to meet its additional security requirement, or if a lessee requested the ability to phase-in its additional security requirement, the lessee would be required to develop a tailored plan to meet that requirement; however, additional security for sole liability properties could not be phased-in. If requested, BOEM would provide guidance in formulating and developing a tailored plan.

If BOEM cannot approve a tailored plan within 180 days of its submission, BOEM could require the lessee to provide the full amount of the required additional security within 30 days of the date on which the 180-day period ends. Once approved, a lessee could request a modification of its approved plan. Pending the decision, the lessee would be required to adhere to the approved plan and timetable for compliance.

If a lessee failed to provide additional security in a timely manner, BOEM could assess penalties under 30 CFR part 550, subpart N; request BSEE to suspend production or other operations in accordance with 30 C.F.R. § 250.173; or initiate action to cancel the lease, pursuant to 30 C.F.R. § 556.1102.

III. Post-2016 Regulatory and Economicactivity

A. Executive Orders – 2017 to 2021

In December 2016, BOEM began implementing the 2016 NTL and started issuing orders to lessees and grant holders to provide additional security; however, BOEM subsequently issued a Notice to Stakeholders extending implementation of the 2016 NTL to allow additional time for the BOEM and Trump Administration to review the financial assurance program with all parties.

On March 28, 2017, President Trump issued Executive Order 13783—Promoting Energy Independence and Economic Growth, which directed Federal agencies to "review all existing regulations and other agency actions and, ultimately, to suspend, revise, or rescind any such regulations or actions that unnecessarily burden the development of domestic energy resources beyond the degree necessary to protect the public interest or otherwise comply with law."

On April 28, 2017, the Trump Administration issued Executive Order 13795 mandating BOEM to further review the 2016 NTL to determine whether modifications (reductions or additional regulations) to the financial assurance program are necessary, while minimizing unnecessary regulatory burdens.

On June 22, 2017, BOEM issued a third Note to Stakeholders announcing that it was in the final stages of its review of the 2016 NTL and had determined that "more time was necessary to work with industry and other interested parties." As such, BOEM extended the 2016 NTL implementation timeline beyond June 30, 2017, "except in circumstances where there would be a substantial risk of nonperformance of the interest holder's decommissioning liabilities."

In October 2019, the President issued E.O. 13891, Promoting the Rule of Law Through Improved Agency Guidance Documents that aimed to prohibit federal administrative agencies from issuing binding rules through guidance documents.

In response to Executive Orders 13783, 13795, and 13891, the Department of the Interior ("Interior") published a new proposed rule in October 2020[123] to: (1) Modify the evaluation process for requiring additional security; (2) streamline the evaluation criteria; and (3) remove restrictive provisions for third-party guarantees and decommissioning accounts. Interior stated that the October 2020 proposal would "provide greater protection as the financial assurance would be focused on the riskiest properties."

On January 20, 2021, President Biden signed E.O. 13990, "Protecting Public Health and the Environment and Restoring Science to Tackle the Climate Crisis." This order, among other things, instructed agencies to review the actions taken between January 20, 2017, and January 20, 2021, and consider publishing a notice of proposed rulemaking suspending, revising, or rescinding that action.

B. Interior Proposed Rule – June 2023

On June 29, 2023, the Department of the Interior ("Interior") published a new proposed rule (the "Proposed Rule").[124] The Proposed Rule (1) modifies the criteria used for determining whether oil, gas, and sulfur lessees, right-of-use

---

[123] Federal Register, Vol. 85, No. 201, October 16, 2020.

[124] Federal Register, Vol. 88, No. 124, June 29, 2023.

and easement (RUE) grant holders, and pipeline right-of-way (ROW) grant holders may be required to provide bonds or other financial assurance above the current prescribed base bonds; (2) removes existing restrictive provisions for third-party guarantees and decommissioning accounts and adds new criteria under which a bond or third- party guarantee that was provided as supplemental financial assurance may be canceled; and (3) clarifies bonding requirements for RUEs serving Federal leases.

Bonding and Other Security Requirements

The Proposed Rule allows the Regional Director to require additional security when: (1) a lessee or grant holder does not meet the proposed financial strength threshold; or (2) when the lease does not have sufficiently valuable proved oil and gas reserves to be sold to another company that could assume the abandonment obligations.

Leases

BOEM's existing evaluation process for leases is based on the current lessee's ability to carry out current and future obligations. Under the existing regulations, the Regional Director uses five criteria to evaluate the need for additional security: (1) financial capacity; (2) projected financial strength; (3) business stability; (4) reliability in meeting obligations based upon credit rating or trade references; and (5) record of compliance with laws, regulations, and lease terms.

BOEM's Proposed Rule streamlines the evaluation process to use only two criteria to determine if any supplemental financial assurance on a lease may be required: (1) An investment grade issuer credit rating[125], either from an NRSRO, as identified by the United States Securities and Exchange Commission (SEC), or a proxy credit rating determined by BOEM based on the company's audited financial statements; or (2) the 3-to-1 ration of the value of proved oil and gas reserves on a lease to the decommissioning liability associated with these reserves.

BOEM proposes to eliminate the "business stability" criterion currently used. The existing regulation bases business stability on five years of continuous operation and production of oil and gas; BOEM has determined that there is little correlation between being in business for five or more years and a company's ability to carry out its present and future obligations. BOEM

[125] Investment grade credit rating is defined as greater than or equal to either BBB- from S&P or Baa3 from Moody's.

also conducted a historical analysis of historical offshore company bankruptcies and determined that "whether a company was in business for five or more years had no relationship to its likelihood to declare bankruptcy".

BOEM also proposes to eliminate the existing "record of compliance" criterion. BOEM reviewed BSEE's Incidents of Noncompliance ("INCs") and Increased Oversight List and determined that the number of OCS properties owned by a company, not its financial health, is the best predictor of the number of INCs. Companies with a larger number of offshore assets inspected by BSEE accumulated a far greater number of INCs than offshore companies with fewer assets, irrespective of the company's overall financial health. The "record of compliance" criterion was also difficult to fairly apply because not all incidents are considered equal.

Additionally, BOEM proposes to replace the existing "financial capacity" and "reliability" criteria with issuer credit rating or proxy credit rating. BOEM has found credit rating to be the most reliable indicator of its overall credit risk or its ability to meet its financial commitments. If a lessee does not have a credit rating, they could submit audited financial statements for BOEM to determine a proxy credit rating. BEOM stated that "audited financial statements, prepared in accordance with Generally Accepted Accounting Principles (GAAP) and accompanied by an auditor's certificate, provide a level of certainty that the financial statements accurately represent the company's economic position and operational performance and would allow BOEM to accurately determine if additional security is needed." If a lessee does not meet the credit rating or proxy rating criteria, the lessee could avoid additional security if BOEM determined that the properties contained proved reserves exceeding three times the decommissioning costs (using the BSEE P70 estimated value) associated with the production of such properties.

Right-of-Use and Easement Grants

BOEM's existing regulations concerning right of-use and easement ("RUE") grants donot prescribe a base bond amount. The Proposed Rule replaces this vague requirement with a *cross-reference to the specific criteria governing bond demands with respect to leases,including the consideration of credit ratings of co-grant holders and lease holders*. The value of proved reserves will not be considered because the grant holder is not entitled to any interest in the reserves.

The Proposed Rule clarifies that any RUE grant holder must provide base financial assurance in a specific amount, regardless of whether the RUE serves a State or Federal OCS lease. BOEM is proposing to establish a Federal RUE base financial assurance requirement that matches the existing $500,000 base financial assurance requirement for State RUEs. The Proposed Rule clarifies that a grant holder may be required to provide supplemental assurance for the RUU, above the $500,000 base assurance, if the grant holder does not meet the criteria rating or proxy credit rating proposed to be used for lessees.

BOEM also proposes a requirement to establish a $500,000 area-wide financial assurance, that would satisfy the base requirement for any RUE holder that owns one or more RUEs within the same OCS area, regardless of whether the lease is a State or Federal lease. BOEM is also proposing to allow any lessee that has posted area-wide lease financial assurance, to modify that lease surety bond to also cover any RUE(s) in the area owned by the same lessee. The ability to use areawide lease financial assurance to cover the RUE base financial assurance obligation would be subject to the requirement that the area-wide lease financial assurance would be in an amount equal to or greater than the RUE base financial assurance requirement.

BOEM also proposes a new regulation to establish conditions under which the assignment of RUE interests may be disapproved. If an assignee has not satisfied all regulatory obligations, any BOEM or BSEE order, or has not satisfied the financial assurance requirements, BOEM may disapprove the assignment.

Pipeline Right-of-Way Grants

Existing BOEM bonding requirements for pipeline right-of-way ("ROW") grants require a $300,000 area-wide base bond, although the Regional Director may require additional security if the $300,000 area-wide bond is determined to be insufficient. *BOEM is proposing to revise the financial assurance regulations to provide that the Regional Director will demand that a pipeline ROW grant holder provide supplemental financial assurance when the grant holder does not meet the same credit rating or proxy credit rating criteria proposed to be used for lessees.* Proved reserves would not be considered because ROW grants do not authorize holders to produce hydrocarbon reserves.

Third-Party Guarantees

Existing guidance states that that a guarantor's total outstanding and proposed guarantees are not allowed to exceed 25 percent of its unencumbered net worth in the United States. This provision requires BOEM to consider the unencumbered net worth of the company in the United States, while another provision requires BOEM to consider the guarantor's unencumbered fixed assets in the United States. Interior stated that both criteria are difficult to apply when the companybeing evaluated has domestic and international assets, and BOEM has no ability to confirm whether the 25 percent net worth criteria has been exceeded.

BOEM proposes to (1) *apply the same credit rating or proxy credit rating criteria proposed for lessees and (2) remove the requirement for a third-party guarantee to ensure compliance* with the obligations of all lessees, operating rights owners, and operators on the lease,and would allow a guarantee limited to a specific amount, as agreed to by BOEM, or limited tothe liabilities of specific parties and (3) *allow third-party guarantees to be used as supplemental assurance for RUE or ROW grant, as well as a lease.* The Proposed Rule wouldalso allow BOEM to cancel a third-party guarantee under the same terms and conditions that apply to other types of financial assurance. The value of proved oil and gas reserves will not be considered because the guarantor would not have an interest and does not impact the guarantor's overall financial strength.

Lease-Specific Abandonment Accounts

The current guidance allows lessees to establish a lease-specific abandonment account in lieu of a bond. BOEM proposes to rename these accounts "Decommissioning Accounts" to remove any perceived limitation to a single lease, and to allow these accounts to be used to ensure compliance with additional security requirements for ROU and ROW grants, as well as a lease.

BOEM determined that the risk of loss through a bank failure is minimal, so it also proposes to *remove the requirements to pledge Treasury securities to fund the account* before the amount of funds in the account equals the $250,000 FDIC maximum insurable amount.

Transfer of Lease Interests

The Proposed Rule clarifies that BOEM will not approve the transfer of a lease interest, whether a record title interest or an operating interest, until the transferee complies with all applicable regulations and orders, including financial assurance.

BOEM Evaluation Methodology

Credit Ratings

BOEM is proposing to use an issuer credit rating or proxy credit rating to evaluate the financial health of lessees and grant holders. S&P's and Moody's rating methodologies analyses are wide-ranging and include factors beyond corporate financials, including the entity's ability to honor senior unsecured debt and debt-like obligations. If an entity does not have an issuer credit rating, BOEM would use S&P's Credit Analytics Credit Model to calculate proxy credit ratings based on audited financial information for the most recent fiscal year so that BOEM could compare the company with similar public companies in the same industry segment.

BOEM has determined that establishing an issuer credit rating threshold of BBB- (S&P) or Baa3 (Moody's), an equivalent credit rating provided by another SEC-recognized NRSRO, or an equivalent proxy credit rating, is the best means for accomplishing these objectives.

BOEM is proposing to add a new provision to the regulations that would authorize BOEM to require a company requesting a proxy credit rating to provide information on its ownership of other OCS facilities and leases. This new provision authorizes BOEM to take the contingent liabilities associated with the company's coownership of these assets into consideration in determining the appropriate proxy credit rating.

Valuing Proved Oil and Gas Reserves

For lessees requesting BOEM to consider the value of the underlying proved reserves in determining whether additional security is required, BOEM would require the lessee to submit a SEC price case reserve report[126] for the proved oil and gas reserves located on a given lease.

BOEM will use this proved oil and gas reserves per-lease value when determining whether the value of the reserves on any given lease exceeds three times the cost of the P70 decommissioning estimate associated with the production of those reserves. The valuationanalysis would be based on a reserve report for the proved oil and gas reserves that includes the following:

- The projected future production quantities of proved oil and gas reserves

---

[126] as defined by the SEC regulations at 17 CFR 210.4–10(a)(22).

- The production cost for those reserves
- The discounted future cash flows fromproduction
- The net present value of the proved oil and gasreserves

BOEM stated it believes that “a property with a high enough ‘reserves-to decommissioning cost’ ratio would likely be purchased by another lessee if a current lessee defaults on its obligations, thereby reducing the risk that decommissioning costs would be borne by the government, and consequently reducing the need for additional security.”

Phased Compliance

BOEM is proposing to phase in the new bonding requirements over a three-year period for existing leaseholders. BOEM will require that any company receiving a supplemental financial assurance demand post one-third of the total amount by the deadline listed on the demand letter. A second one-third would be required by the end of the second year (i.e., within 24 months of the receipt of the demand letter). The final one-third payment would be due within 36 months of receipt of the demand letter.

BOEM will discontinue collection of and return any supplemental financial assurance previously provided if a lessee’s credit rating improves above the required threshold.

Bond Appeals

BOEM is proposing a new requirement whereby any company seeking to stay a supplemental financial assurance demand pending appeal must, as a condition of obtaining a stay of the order, *post an appeals bond in the amount of supplemental financial assurance required.* If the appeal is successful, the amount of the appeals bond in excess of the amount of supplemental financial assurance determined to be required would be released. If the appeal is unsuccessful, the appeals bond could be replaced or converted into bonds to cover the supplemental financial assurance demand.

C. Update to BSEE P&A Calculations

Currently, BSEE displays the abandonment cost estimate as the sum of two different determinations for each lease, ROW or RUE: a distribution estimate and a deterministic estimate.

The distribution estimate applies a probabilistic approach using a P50, P70 and P90 distribution based on actual decommissioning costs reported to BSEE by OCS lessees since mid- 2016. A P50, P70 or P90 determination means that there is a 50%, 70% or 90% chance that the actual abandonment costs will be less than the BSEE estimated cost. As the certainty of the estimate increases, so does the cost estimate; therefore, P90 estimates will generally be higher than the P70 and P50 estimates.

The deterministic amount is calculated by BSEE based on its evaluation of abandonment cost data. Using this model, the output is determined by the input parameters and does not include elements of randomness.

Historically, BSEE relied upon their proprietary algorithms and methodologies to calculate abandonment estimates. The primary sources of data used in those estimates came from BSEE studies, lessee presentations to bankers, professional experience, industry publications, and other sources. The issue with this approach was that some data sources of information were decades old and did not reflect current technologies. In 2016, BSEE indicated that the algorithms to calculate abandonment estimates had not materially changed since 2011.

To address this issue, Interior published a rule obligating lessees and operating right owners to submit actual expenditures for abandonment activities. BSEE now relies on those reported expenditures to develop its cost estimates. Opportune noted the following about BSEE's updated cost estimates:

- Cost estimatesassume the U.S. Taxpayer would potentially pay for the abandonment; thus, the cost estimate uses a cost that represents what a third-party (as opposed to the lessee in question) would require to complete the abandonment work.
- Estimates include a third-party profitmargin.
- Estimates do not allow for economies of scale. The estimates assume all work will be performed on a one-off basis and not as a part of a larger project including multiple assets.
- Estimates do not consider any salvage values of components that could be recovered from the structures during abandonment.

D. Summary of 2016– 2022 Industryand OCS Economicactivity

Subsequent to publication of the 2016 Study, liquidity across the Industry deteriorated under the weight of bloated balance sheets resulting from the combination of overheated commodity prices from 2011 to 2015 and capital markets that fueled excessive acquisition and development spending during the shale boom. The result was a volumetric supply overhang from late 2014 to 2016 that prevented commodity prices from recovering enough to offset the Industry's constrained capital.

Since July 2016, a total of 12 U.S. small independent oil and gas companies filed bankruptcy with respect to approximately $18.8 billion in total capital, including approximately $4.0 billion in future P&A obligations.[127] All related P&A liabilities were either assumed by the reorganized Debtor, subsequent buyers of the underlying assets, co-lessees and/or predecessors in title, or funded through a liquidation trust – at no cost to the American taxpayer.

Commodity prices stabilized through 2019 as global demand eroded excess supply and the Industry began to recover. Oil and gas bankruptcy fears waned and the BOEM delayed its previous proposals to streamline the financial assurance regulations. However, a once-in-a-century COVID-19 pandemic that shut down the global economy and "threw the supply chain and demand completely out of whack."[128] Within three months of the start of 2020, U.S. production fell from a record 13.1 million barrels a day to 10.5 million and stayed there for nearly a year.[129] Consequently, public and private companies across the Industry shut-in production and drastically cut drilling plans. Companies that couldn't survive on their own were acquired, merged with peers to create enough scale, or were forced into bankruptcy.

---

[127] See Exhibit A – Summary of Offshore Operator Bankruptcies: 2016 – 2023. Amounts include two Ch. 11 filings by Fieldwood Energy. $18.8 billion represents the combination of $14.8 billion in Pre-petition Capital plus $4.0 billion in P&A Obligation.

[128] President Biden, White House Press Briefing, May 10, 2022.

[129] Bloomberg, February 16, 2022.

World Liquid Fuels Production and Consumption Balance
Million Barrels per Day

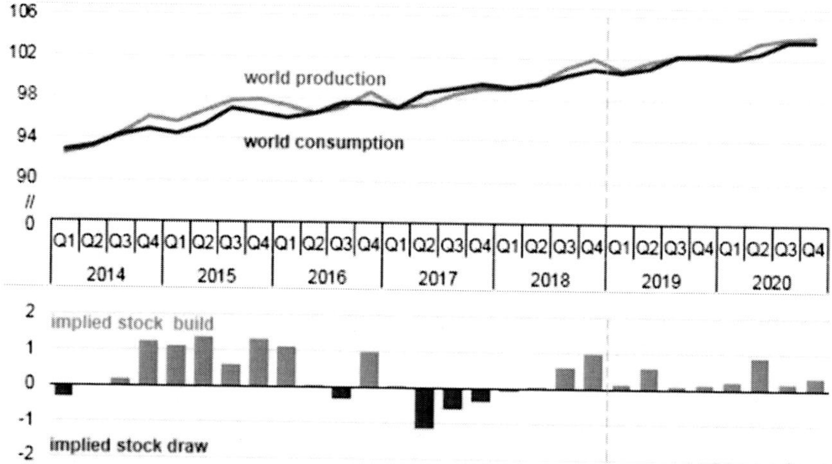

Source: U.S. Energy Information Administration, Short-Term Eergy Outlook, January 2019.

Monthly Crude Oil Spot Prices (Jan 2014 – Dec 2019)

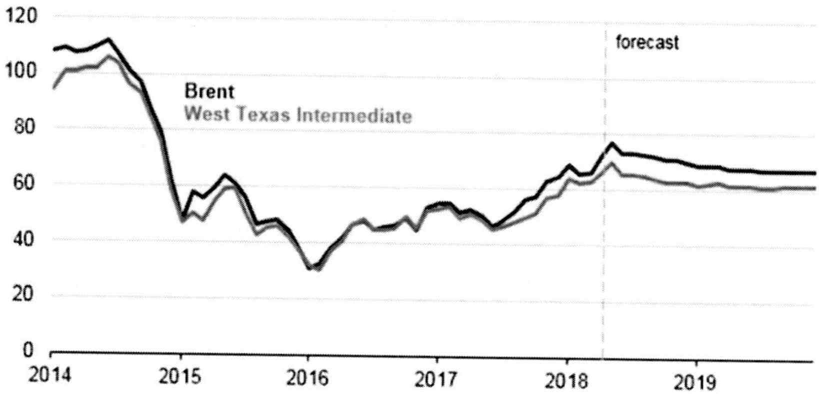

Source: U.S. Eergy Information Administration, Short-term Energy Outlook, June 2018.

IV. Opportune 2023 Study

As a result of the Proposed Rule, Opportune is updating its previous study to independently review the following:

- Current BOEM estimated P&A liabilities for OCS leases and the underlying BSEE historical data
- Actual historical costs of OCS P&A liabilities borne by U.S. taxpayers
- Current distribution of OCS P&A liabilities by major integrated ("Majors"), large independent ("Large Independents") and small independent ("Small Independents") operators[130]
- Estimated effect to U.S. Taxpayers and the Gulf Coast

A. Independentcost-Benefitanalysis for the U.S. Taxpayer

Cost-benefit analysis is a widely used technique for evaluating a possible course of action. It is a systematic way to look at the expected benefits compared to the expected costs for the course of action. To be rational and justifiable, the expected benefits must, at least, outweigh the costs. The cost-benefit analysis presented in the Opportune 2023 Study examines benefits of the proposed bonding requirements compared to the costs to Industry and the regional and national economy. The stakeholder is taken to be the American public, either as taxpayers funding the federal government or as citizens being impacted by a government policy. *Details of Opportune's cost-benefit analysis are given in Appendix A.*

The benefit of the Proposed Rule is to protect the public from paying for decommissioning of OCS oil and gas infrastructure using tax dollars because lessees default on their obligations. Under the Proposed Rule, the risk to the taxpayer is a function of the number of jointly and severally liable lessees in the chain of title and the financial strength of the individual lessees. If a lease includes a Major or a Large Independent, with high financial strength in the chain of title, then the risk of default for these leases is near zero. If there is no Major or Large Independent, the risk is tied to the financial condition of the Small Independents in the chain of title. Current regulations operate to greatly reduce or maintain the current minimal risk to taxpayers. For example, a current lessee would have to default, file for bankruptcy, unsuccessfully

130 Majors and Large Independents are defined as having a tangible net worth, or market capitalization, of over $10.0 billion as of March 15, 2023. Small Independents are defined as all other OCS operators who have a tangible net worththat is either less than $10.0 billion or unavailable from public sources. See Appendix C of the Opportune 2022 Study for a listing of lessees considered to be Majors and Large Independents (including subsidiaries and successors); Small Independents represent all other lessees.

auction the properties due to no interested buyers, and have no solvent co-owner(s) of previous owner in the chain of title for the taxpayer to be liable. The greater the number of companies in the chain of title for the lease, the less likely it is that they would all default and the less risky the lease. The riskiest leases would be those with only one lessee in the chain of title. In such case, the risk of default is directly tied only to the individual lessee's financial strength.

Opportune analyzed BOEM lease data to understand the benefit of reducing the risk. The results of Opportune's analysis indicate that 1,706 leases have decommissioning liabilities considering all water depths with a total estimated OCS P&A liability of over $16.4 billion. Of those leases with liabilities, 378 have no Major or Large Independent in the chain of title. These leases have a total estimated P&A liability of about $1.2 billion. The vast majority of leases, over 93 percent of the total P&A liabilities, or $15.3 billion, have a Major or a Large Independent in thechain of title and have a minimal likelihood of default. Only 7 percent of total P&A liabilities in the OCS are associated with Small Independents.

Risk Category	Number ofLeases	Total Cost	ARO	Percent of Total	Collateral	Percent of Total	Un-covered	Percent Liability Covered
Leases - Active								
Major Included	1,063	65,100	13,670	83.1%	1,571	47.9%	12,100	11.5%
Large Independent Included	265	8,391	1,620	9.9%	520	15.9%	1,100	32.1%
Small Independents								
>= 5 Lessees	184	7,560	410	2.5%	334	10.2%	77	81.3%
4 Lessees	18	280	21	0.1%	44	1.3%	-22	205.1%
3 Lessees	35	519	77	0.5%	83	2.5%	-6	107.9%
2 Lessees	63	2,152	455	2.8%	271	8.3%	183	59.7%
1 Lessee	78	843	189	1.1%	29	0.9%	160	15.5%
Subtotal Independents	*378*	*11,354*	*1,152*	*7.0%*	*761*	*23.2%*	*391*	*66.1%*
Subtotal All Leases	*1,706*	*84,845*	*16,442*	*100.0%*	*2,852*	*87.0%*	*13,590*	*17.3%*
General Allocations	-			-	425	13.0%	-	-
Total	*1,706*	*84,845*	*16,442*	*100.0%*	*3,276*	*100.0%*	*13,166*	*19.9%*

Opportune also analyzed the collateral associated with these leases. The results indicate that collateral requirements increase with increasing risk. For the lowest risk category, leases with at least one Major in the chain of title, collateral covers only about 11.5% of the total P&A liability. Coverage for Small Independents increases to about 66% on average. This level of coverage for the higher risk categories suggests that the previous bonding requirements did act to place extra security where it was needed.

Opportune concludes that the benefit in the reduction of risk from the Proposed Rule is relatively small because the riskiest categories do not comprise a large part of the total liability. If the increase in collateral from the new bonding requirement provides 100 percent coverage for all leases without a Major or Large Independent, then the benefit is a reduction in risk of $391 million(the "Uncovered Risk"). This amount is determined as the total P&A liability for leases with no Major or Large Independent in the chain of title, $1.2 billion, less collateral already in place of $761 million.

BOEM's goal of protecting taxpayers from paying decommissioning liabilities comes at asignificant cost. This cost results from changes in the way producers will choose to operate as a result of increased bonding requirements. Using more of producers' available financial resources to protect the public reduces the funds available for exploration and production activities in the OCS. In turn, these reductions in spending will have a negative impact on economic activity of theGulf Coast states as well as in other parts of the country.

The potential impact of the Proposed Rule is difficult to predict but is likely to be very significant. Majors are expected to be largely unaffected; however, Small Independents are expected to significantly reduce their development spending.

In our analysis, we assume that the funds used to pay the premiums for the proposed supplemental bonding are funds that would have otherwise been spent on exploration or development drilling or recompletions in the Gulf OCS. Our model assumes a $25/boe finding anddevelopment cost and other parameters that are described in the Appendix. We also assumed that the lost positive cash flows that would have been generated by developing reserves would have been re-invested in additional future development activities in the year after they are received. As a result, the estimated reduction in future activity is greater than what would be suggested merely by considering the bonding premiums.

These development activities would not only result in additional capital spending, but also lead to additional operating cost expenditures. We have assumed that both classes of expenditure would be spent in the Gulf Coast states (Alabama, Florida, Louisiana, Mississippi, and Texas) since this is where most of the industries that support the offshore oil industry, and their employees, reside.

The broader economic effects of this decreased spending are estimated using the parameters from the Regional Input-Output Modeling System ("RIMS II") that is provided by the Bureau of Economic Analysis, US Department of Commerce. RIMS II is a tool for economic planners that provides a systematic analysis of the economic impacts of projects or programs on regional economies. The economic impact is estimated for the Gulf Coast states, namely, Alabama, Florida, Louisiana, Mississippi, and Texas, based on RIMS II multipliers that were generated in June 2023.

The decrease in spending over the ten-year study period is estimated under our Base Case assumptions to be approximately $4.7 billion. The result in a reduction of approximately 36,000 jobs. Similarly, output, earnings, and value added are all reduced by billions of dollars as a result of the decrease in spending. *In our Base Case, the decrease in spending results in a $9.9 billion reduction in GDP.*

The results of this cost-benefit analysis indicate the reduction in the risk for taxpayers is not justified given the impact on Industry and the regional/national economy. The benefits of the Proposed Rule are relatively small and the cost is very high in terms of the decreased developmentof OCS resources and associated losses in production, Industry revenue, taxes, economic activity, and jobs. The benefit of reduced risk of approximately $391 million is far outweighed by the decline OCS development. Based on the Base Case estimate, producers will see 55 million boe less production, over $2.8 billion less revenue, as well *as nearly $573 million in lost royalties to the federal government.*

Total direct and indirect costs of the Proposed Rule to the taxpayer are estimated to be approximately $10.5 billion.

B. Independent Analysis of Additional Bonding Requirements and Recommendations for Improvement

GAO suggestions and the resulting NTL were predicated on the false presumption that the U.S. taxpayer is at risk vis-à-vis decommissioning costs posing potential financial liabilities to thefederal government. Existing federal regulations[131] require previous owners to remain responsible for the future decommissioning costs of assets sold if current owners fail to meet their obligations.

---

[131] Title 30 CFR Part 556, Subpart I § 556.62 and § 556.64.

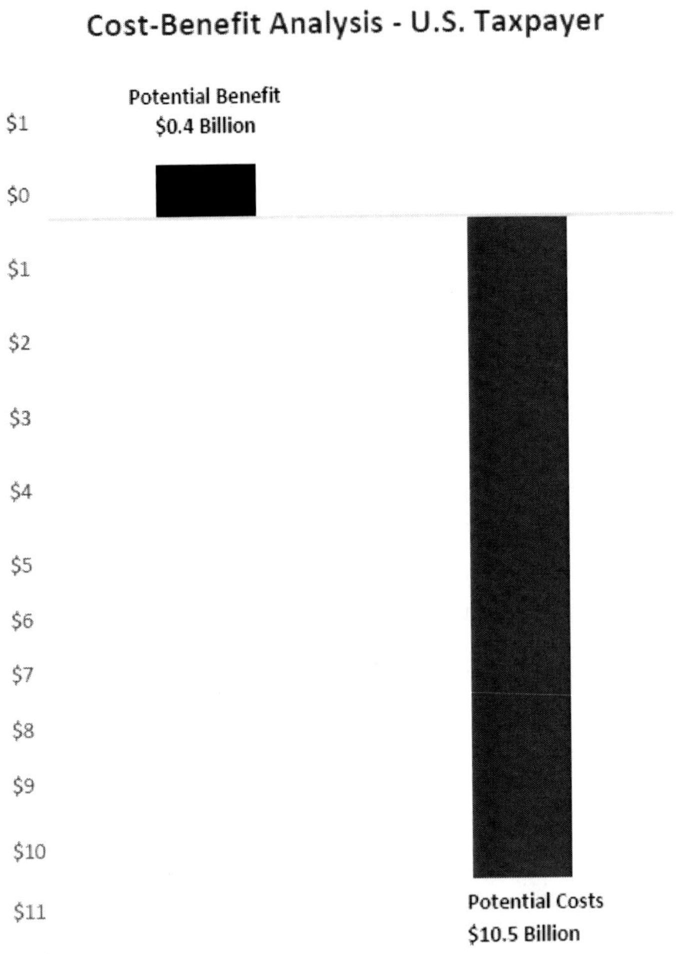

Additionally, sellers acknowledge that they may be held responsible for the future P&A of the properties being sold by typically requiring buyers to post private bonds or pay additional cash to the seller upon closing of the sale. Unlike the $2.1 trillion paid to other industries since 1970, the potential Industry P&A cost to the U.S. taxpayer is infinitesimal.[132,133]

---

[132] See Exhibit B – Summary of Historical U.S. Government Bailouts, by Industry.

[133] Per the 2015 GAO Report, Interior officials cite an event in 1989 whereby a company in Ch. 11 proceedings was permitted to divert $13 million in royalties toward decommissioning costs as part of the company's agreement to fundtwo decommissioning trusts.

For the first time in Industry history, BSEE recently announced a Draft for Proposalseeking bids to decommission 15 orphan wells located in the Matagorda Island, High Island, and West Delta areas of the OCS.[134] The decommissioning costs will be funded from the 2021Infrastructure and Jobs Act appropriations. Based on BOEM's database, the P&A liability associated with these wells and their related infrastructure is approximately $37.7 million[135] – an infinitesimal amount (~0.8%) compared to the $4.5 billion in royalties and revenue the U.S. government (i.e., taxpayer) received in 2021 from offshore Gulf of Mexico production.

To put in perspective how minimal the U.S. taxpayers' risk is in relation to the chain oftitle's wherewithal to pay, in a worst-case scenario, BSEE's $84 billion undiscounted estimate ofthe total P&A liability represents only 4.6% of the $1.8 trillion in collective tangible net worth ofthe Majors and Large Independents.

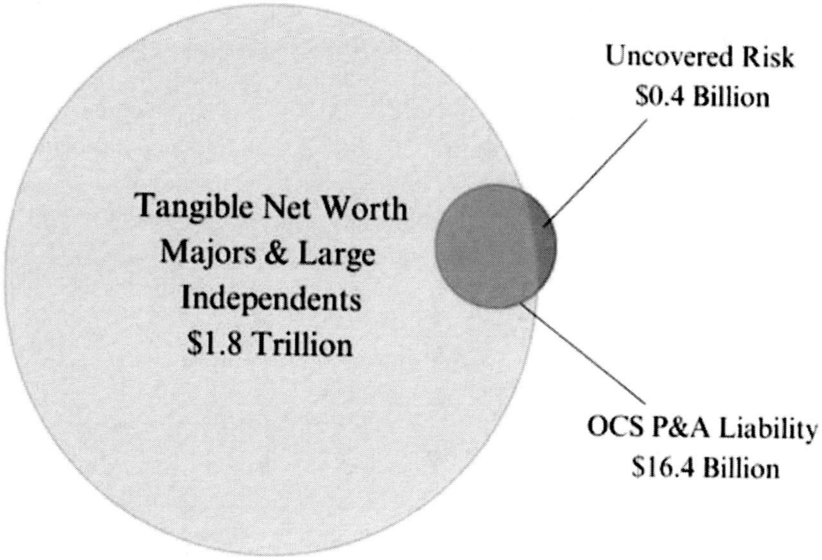

Out of BSEE's $11.3 billion undiscounted liability related to Small Independents, $2.9 billion (3.5% off the total undiscounted liability) relates to properties with less than three lessees. Given the Industry's history for always

134 "Orphaned Wells on the U.S. Outer Continental Shelf", BSEE Virtual Industry Day Presentation; February 2, 2022.

135 See Exhibit C.

covering P&A obligations and the relatively small P&A amount that relates to uncovered leases, there appears to be very little risk to the U.S. taxpayer.

Leases by Risk Category	BOEM Estimated Total Cost[136]
Major Included	65,100,035,911
Large Independent	8,390,974,983
No Major or Large Ind -1	842,588,808
No Major or Large Ind -2	2,151,525,433
No Major or Large Ind -3	519,034,023
No Major or Large Ind -4	280,293,184
No Major or Large Ind -5	26,215,872
No Major or Large Ind >5	7,533,964,549
Grand Total	$ 84,844,632,763

Estimate of the P&A Liability is Overstated

BSEE's $84 billion estimate of the P&A liability falsely assumes that all related decommissioning costs occur tomorrow, not over the next 20 or 30 years in which such costs will actually be incurred – that is, BSEE fails to discount the estimated future costs back to today's present value.

The Financial Accounting Standards Board ("FASB"), the authoritative accounting standard-setting organization for the U.S., issued Accounting Standards Codification 410-20, Asset Retirement Obligations ("ASC 410-20) to address the accounting for environmental remediation liabilities (also referred to as asset retirement obligations, or "ARO") that results from the normal operation of a long-lived asset and that is associated with the retirement of that asset.[137]

Paragraph 25-4 of ASC 410-20 address the timing of when an ARO should be recorded, as follows (emphasis added):

> "An entity shall recognize the fair value of a liability for an asset retirement obligation in the period in which it is incurred if a reasonable estimate of fair value can be made. If a reasonable estimate of fair value cannot be made in the period the asset retirement obligation is incurred, the liability shall be recognized when a reasonable estimate of fair value can be made. If a tangible long-lived asset with an existing asset retirement obligation is acquired, a liability for that obligation shall be

[136] Note the $84.8 billion represents the sum of the undiscounted P70 and the deterministic costs of all leases in the OCS,excluding ROW and RUE properties, per the BOEM Database.

[137] ASC 410-20, paragraph 15-2.

recognized at the asset's acquisition date as if that obligation were incurred on that date."

Paragraph 25-7 of ASC 410-20 addresses the required valuation of an ARO, as follows (emphasis added):

> "The obligation to perform the asset retirement activity is unconditional even though uncertainty exists about the timing and (or) method of settlement. *Thus, the timing and (or) method of settlement may be conditional on a future event. Accordingly, an entity shall recognize a liability for the fair value of a conditional asset retirement obligation if the fair value of the liability can be reasonably estimated.* In some cases, sufficient information about the timing and (or) method of settlement may not be available to reasonably estimate fair value. *An expected present value technique incorporates uncertainty about the timing and method of settlement into the fair value measurement.* Uncertainty is factored into the measurement of the fair value of the liability through assignmentof probabilities to cash flows."

Every oil and gas company with operations in the OCS has measured and recorded the present value of its ARO, in accordance with ASC 410-20. These valuations and the assumptions used therein are then audited annually by each company's respective independent external certified public accountant ("CPA") firm as a part of the annual audit of their historical financial statements.

Opportune's mathematical calculation uses four variables: 1) estimated cost to decommission; 2) estimated time until the decommissioning occurs; 3) an inflation rate factor to arrive at an estimated future cash out flow ("FCOF"); and 4) an estimated credit adjusted risk free rate ("CARF") to calculate the present value ("PV") of the estimated FCOF as of the date of evaluation. *See Opportune's independent calculation of the OCS P&A Liability within Appendix B.*

In determining the estimated P&A dates for the fields in BOEM's database, we utilized an analysis conducted by Opportune's reserve engineering firm Ralph E. Davis Associates ("RED"). Because most of the reservoirs in the Gulf of Mexico have water drives, decline curve analysis is not suitable to estimate their remaining life. RED's analysis used the operators' reported estimatesof reserves originally in place (at the field level effective 2019), combined with the reported historical field-level production, to generate a forecast of percentage or original reserves remaining. These forecasts were extrapolated to 99% or more of the original reserves to determine each field'sremaining life.

The inflation rate should approximate the life of the assets in calculating a P&A obligation.In this case, we used 2.5% based on the consumer price index annual expect growth over the next 10 years as published in June 2022 the Livingston Survey.

In calculating the Credit Adjusted Risk Free Rate ("CARF") typically we relied on a corporate cost of debt as the discount rate. The cost of debt is a reasonable proxy of an entity's ability to meet its debt obligations. We've concluded on the CARF for each lease, as follows, based on the mix of lessees and their respective credit ratings:

- Majors only: 6.5%
- Large Independents and other lessees: 8.0%
- Small Independents only: 9.5%.

Based on the analysis and procedures described above we have concluded that the *OCS P&A Liability is approximately $16.4 billion.* See *Appendix B* for the report and exhibits detailing the ARO calculations.

Financial Strength of OCS Companies is Understated

Although the financial strength of many companies operating in the OCS is currently diminished, certain errors in BOEM's estimates thereof should be eliminated so that financial strength can be fairly assessed in all economic environments.

First, the estimated P&A liability (or audited ARO) should be added back to the calculation of tangible net worth. By not doing so and then using the estimated tangible net worth as a measuring stick of whether a company has the financial wherewithal to pay its future P&A obligations, BOEM is effectively counting the P&A liability twice.

When an ARO liability is initially recorded, ASC 410-20 requires that an equal, but offsetting ARO asset be recorded at the same time so that there is no immediate impact to a company's earnings.

Over time, the ARO liability is "accreted" (to expense) up to the estimated future cost that will be spent at the P&A date; however, the related ARO asset is "depleted" (also to expense) down to zero as the related reserves are produced and/or "impaired" due to reductions in economic reserves and prices.

Offsetting ARO Asset and Liability at Inception

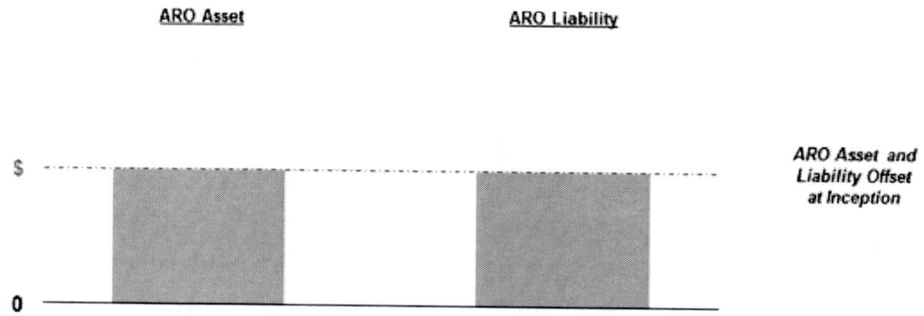

In the current economic environment, most E&P companies have experienced significant impairments of their reserves and related assets, which likely means that little to none of the initially recorded ARO assets remain on their balance sheets. As conditions improve, the ARO liability continues to increase until the P&A date, while the ARO asset decreases or no longer exists at all. As such, the company's tangible net worth is being reduced by the full amount of its ARO liability.

Changes in ARO Asset and Liability Over Time

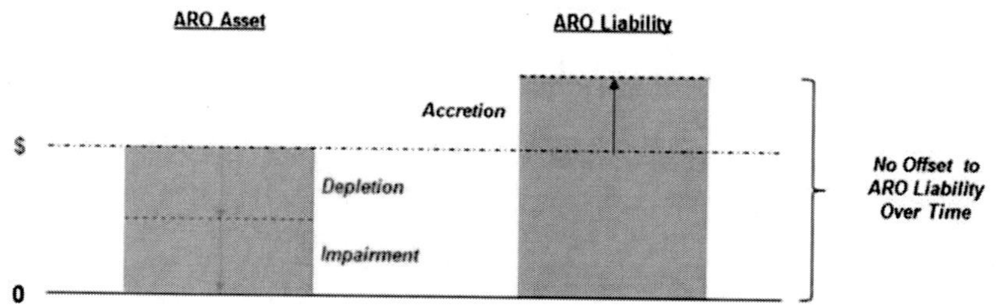

Second, the impaired historical book value of the company's proved reserves may permanently underestimate its financial strength in future periods. When assessing financial strength, BOEM should incorporate the value of each company's proved oil and gas reserves using forward strip pricing, discounted at 10% ("PV10"), rather than using the historical book value of proved reserves.

GAAP requires companies to record impairments, immediately, when economic conditions dictate under the accounting rules; however, GAAP never allows impaired asset values to be written back up as the underlying economic conditions improve and/or reserve volumes increase. As commodity prices rise over time, debt and liabilities (including the PV of the ARO liability) on a company's financial statements reflect current reality, but the offsetting assets are forever stuck at their historical impaired (and further depleted) cost basis. As a result, assessing a company's financial strength may be artificially underestimated without considering the market value of its proved reserves.

Consideration was given to valuing proved reserves using the Standardize Measure.[138] Standardized Measure is a required disclosure by the Securities and Exchange Commission to conservatively compare reserve values across all public companies (not required for public companies), regardless of whether they use the Full Cost and Successful Efforts method of accounting for oil and gas operations. Because the Standardized Measure utilizes historical average pricing, the results are conservatively high in periods of declining prices and conservatively low in periods of rising prices; therefore, it's not the best measure for measuring current financial strength.

Incorporating the PV10 value of proved reserves is forward-looking with respect to both price and a company's expected drilling plans.

Use Standard Industry Forms and Independent CPA Firms to Improve Data Gathering

The Proposed Rule states that, absent a credit rating from a SEC-approved Nationally Recognized Statistical Rating Agency, a proxy credit rating should be determined by BOEM usingthe entity's audited financial statements. To this point, the Proposed Rule states, "BOEM has concluded that audited financial statements, prepared in accordance with Generally Accepted Accounting Principles (GAAP) and accompanied by an auditor's certificate, provide a level of certainty that the financial statements accurately represent the company's economic position and operational performance."

As such, BOEM should avoid a duplication of costs and efforts by utilizing the audited financial statements of every public and private OCS lessee when determining each entity's P&A liabilities and the value of future production underlying such liabilities.

---

[138] Valuation based on SEC pricing, which is the arithmetic average of the first of month pricing for the preceding 12months.

BOEM would also benefit from following practices already in place within the refining industry where CFR[139] requires that independent CPA firms attest to the accuracy and completeness of certain biofuel information submitted to the Environment Protection Agency ("EPA"). Under the Energy Policy Act of 2005, the EPA sets annual quotas dictating what percentage of the motorfuels consumed in the United States must be represented by biofuel blended into fossil fuels. Individual quotas are then assigned to refineries based on the volume of fuel they introduce into the U.S. market. Refineries track the percentage of blended biofuel in each batch of production through an assigned Renewable Information Number ("RIN"), which is uploaded to an EPA database through a "unified form report". To ensure compliance, CFR requires that independent CPA firms being engaged to perform Agreed-Upon Procedures.

By analogy, legislators should consider amending CFR to require that *OCS lesseesconfidentially file their ARO balances by the type of structure on each property (e.g., wells platforms and pipelines), quarterly, on forms specified by BOEM. In addition, OCS lessees should be required, annually, to engage the same independent CPA firm that already audits the ARO balances within their financial statements (on an annual basis) to perform agreed- upon procedures attesting that the ARO information submitted to BOEM is complete and accurate.*

Supplemental Bonding Is Not the Only Mechanism Available

In February 2016, a presentation[140] by BOEM's Gulf Coast Regional Director stated (emphasis added):

- "BOEM's role is to *encourage oil and gas development activities on the OCS* to increase the nation's energy independence and promote U.S. taxpayers' interests while simultaneously protecting natural resources and the environment."
- "The current general lease surety bond is too low to effectively cover any substantive offshore decommissioning costs, as necessary. Therefore, *BOEM must ultimately rely on the sufficiency of its supplemental bond program* for the purpose of assuring decommissioning performance."

---

[139] 40 CFR Title 80 Subpart F § 80.75, 80.125 and 80.150.
[140] BEOM Feb 2016 presentation.

- "BOEM strives to ensure appropriate procedural and operational safeguards *without unduly discouraging exploration and development.*"

Chain of title regulations in the CFR have always encouraged development of the OCS by allowing Majors and Large Independents to shift their budgets away from the re-development and P&A of older assets in the shallow water toward more material development projects in the deepwater. Meanwhile, Smaller Independents used the older infrastructure acquired to improve economics on new drilling projects that are more material to them than the larger companies that initially sold the assets.

P&A liabilities in the North Sea are also a concern; however, governmental authorities under the Decommissioning Security Agreement of October 2015 (the "DSA") do not require supplemental bonding until the P&A liability exceeds the estimated discounted present value of the underlying reserves. Simply, the DSA rule encourages future offshore oil and gas exploration and development.

BOEM should consider a proposal that all OCS producers self-fund the uncovered P&Aliability of the Uncovered Properties ("Uncovered Risk")[141] through an additional 1% overriding royalty, payable to BOEM as beneficiary, which would supplement (but not eliminate) the chain of title in the event current OCS lessees fail to fulfill their P&A obligation (the "BOEM Reserve").

Based on current commodity prices and approximately 627 million BOE in OCS production during 2022, OCS producers would self-fund the present value of the Uncovered Risk within approximately one year. Once the BOEM Reserve equaled the present value of the Uncovered Risk, the additional 1% override would stop until i) additional net liabilities were incurred or accreted; or ii) the BOEM Reserve was reduced by funds used to pay the P&A obligations of bankrupt lessees.

Absent capacity in the bond market, the Industry is adapting by developing cutting-edge financial insurance products to eliminate the uncertainty of future P&A funding and help defray the residual risk that remains with the previous chain of title. *If the U.S. government believes the taxpayer is really at risk, the Administration should allow a federal agency to issue bonds onSole-Liability Properties, allowing the taxpayers to participate in the upside of the associatedpremium float.*

---

[141] $0.4 billion, as discounted by Opportune.

V. Summary Conclusion

Historically low commodity prices and the exit of Majors and Large Independents from the shallow water, or from the OCS entirely, raises concerns of whether the U.S. taxpayer is at risk for the cost of existing P&A obligations if current lessees file bankruptcy. However, existing CFR regulations and BOEM boding rules have historically protected the U.S. taxpayer through previous economic and commodity price down-turns, and there is little-to-no evidence to suggest additional risks to the taxpayer exist today.

The BOEM-estimated $84 billion in undiscounted existing P&A liabilities is miniscule compared to the tangible net worth of the Industry, particularly the Majors and Large Independents who remain potentially responsible for the decommissioning costs of the assets they sold. Only $11.3 billion of the BOEM's estimate relates solely to Small Independents as current or chain of title owners; in today's dollars, the discounted un-bonded liability related solely to Small Independents is $0.4 billion.

Under the Proposed Rule, the balance sheets of Small Independents will be unable to fund collateral requirements to obtain surety bonds in a market that is too thin and for P&A liability estimates that are overstated. Although tailored plans may alleviate Industry concerns that existing wells will be shut-in, the uncertainty over future bonding that remains will prevent Small Independents from drilling new wells or selling their assets, both of which will lead to additional bankruptcies.

Increased bonding requirements is not the only solution. The Opportune 2023 Study proposes solutions that encourages the exploration and development of the OCS, while also protecting the environment and the chain of title from the uncertainty of unmet P&A obligations.

Exhibits

Exhibit A– Summary of Offshore Operator Bankruptcies: 2016– 2023

Company Name	Filing Date	*Pre-petition Capital	*Post-petition Capital	*Difference	*P&A Obligation		Notes
Whistler Energy II, LLC	3/24/2016	$156.5	$75.0	$(81.5)	$78.8		Argonaut Insurance retained its collateral rights post-bankruptcy related to P&A bonds. $75mm was held in escrow for Argo, the remaining amount Argo asserted (~$3mm was treated as a GUC).
Energy XXI LTD	4/14/2016	$2,858.0	$301.7	$(2,556.3)	$613.0		All decommissioning obligations were assumed by the Reorganized Debtors
Bennu Titan LLC	8/11/2016	$180.4	Liquidated	$(180.4)	$3.6		Same as above. cases jointly administered
Bennu Oil & Gas, LLC	11/30/2016	$180.4	Liquidated	$(180.4)	$28.0-41.0		BOG's trustee sold the company's Clipper assets in the Gulf of Mexico in July 2017 to Murphy Exploration & Production Company for USD 3.3m in cash plus the assumption of USD 28m to USD 41m in plugging and abandonment (P&A) liabilities.
Stone Energy Corp.	12/14/2016	$1,507.9	$386.0	$(1,121.9)	$139.0		$75m escrowed for P&A obligations. A condition precedent to the Plan going effective was the resolution of issues related to the provision of additional collateral required by BOEM.
Castex Energy Partners, L.P.	10/16/2017	$390.0	$90.0	$(300.0)	$17.0		Plan of Reorganization states the debtors will continue with their decommissioning obligations
Cobalt International Energy, Inc.	12/14/2017	$2,840.8	Liquidated	$(2,840.8)	$205.0	1	Assets sold to 5 parties. Each sale contemplates the assumption of P&A liabilities with the purchase of the asset.
Fieldwood Energy LLC	2/15/2018	$3,287.0	$2,186.0	$(1,101.0)	$844.0	2	Reorganized Fieldwood Energy assumed liability for future environmental remediation and obligations related to plugging and abandoning wells and decommissioning facilities.
Fieldwood Energy LLC	8/3/2020	$1,800.0	$1,030.0	$(770.0)	$1,179.0	3	For assets purchased during case, P&A obligations were assumed as part of the sale. For abandoned properties, Fieldwood anticipated BSEE seeking payment from predecessor operators.
Arena Energy, LP	8/20/2020	$1,068.7	Liquidated	$(1,068.7)	$530.0		Sold all assets to San Juan Offshore VIII LP (owned by Lime Rock) and P&A obligations were assumed by Buyer.
Castex Energy 2005	2/26/2021	$251.7	$160.0	$(91.7)	$36.0		"Castex's Chapter 11 plan called for the creation of a liquidation trust to fund the company's plugging and abandonment obligations for certain oil and gas wells. Aside from those funded P&A obligations, Castex proposed to abandon all of its other interests in oil and gas wells." - Debtwire April 20, 2022 article. In April 2022, DOI agreed to withdraw a $245m decommissioning claim.
MLCJR LLC (Cox Operating)	5/14/2023	$270.0	Pending Case	Pending	$329.0		Companies reserve report forecasts PV10 of P&A obligations at $329mm, however, the company also has P&A obligations related to asset purchases for an undisclosed amount. At this time, no Plan of Reorganization has been filed proposing treatment of P&A obligations.
		$14,791.4	**$4,228.7**	**$(10,292.6)**	**$3,974.4**		

* Capital, difference, and P&A amounts in millions of dollars

1 205mm as per BSEE claim for future obligations; Annual Report shows 7.4mm in ARO

2 844mm figure includes Apache, P&A obligation excluding Apache is 346

3 Includes 177mm to BOEM, 504mm to third parties (excluding Apache), and 498mm to Apache.

I. Whistler Energy II, LLC

Whistler Energy II, LLC	
Prepetition Capital Structure	
(in millions)	
Note Purchase Agreement	125.00
Unsecured Notes	31.50
Total Debt*	156.50

*does not include accrued and unpaid interest.

Postpetition Capital Structure	
(in millions)	
Revolving Credit Facility	75.00
Series A Units	
Series B Units	
Total	75.00
Notes	

Argonaut Insurance retained its collateral rights post-bankruptcy related to P&A bonds. $75mm was held in escrow for Argo, the remaining amount Argo asserted(~$3mm was treated as a GUC).

II. Energy XXI LTD

Energy XXI Ltd.	
Prepetition Capital Structure	
(in millions)	
Revolving Credit Facility	99.00
Second lien notes	1,450.00
Total Debt	946.00
Convertible notes	363.00
Total Debt*	2,858.00

*does not include 227.7 million in undrawn letters of creditunder RCF.

Postpetition Capital Structure	
(in millions)	
Revolving Credit Facility	227.74
Term Loan	74.00
Total	301.74
Notes	
Dkt. 1340 "The Reorganized Debtors shall assume and succeed to all decommissioning obligations in connection with the Federal Lease Interests assumed pursuant to the Plan and BOEM's consent. The assumed and succeeded obligations shall include all obligations due under the applicable statutes, regulations and any idle iron plan accepted and approved by the Department of the Interior."	

III. Bennu Titan LLC

Bennu Titan LLC	
Prepetition Capital Structure	
(in millions)	
Total Debt	180.40
Postpetition Capital Structure	
(in millions)	
Liquidated	
Notes	
Case jointly administered with Bennu Oil & Gas, LLC	

IV. Bennu Oil & Gas, LLC

Bennu Oil & Gas,LLC	
Prepetition Capital Structure	
(in millions)	
Total Debt	180.40
Postpetition Capital Structure	
(in millions)	
Liquidated	
Notes	
Dkt. 171 "The purchase price for the Assets shall include (i) Three Million Three Hundred Thousand and 00/100 Dollars (USD $3,300,000.00) or such higher amount as the (Purchaser bids at the Auction and (ii) Purchaser shall agree toindemnify the Estate for all decommissioning or plugging and abandonment or other environmental Labilities related to the Assets (the "Purchase Price"), subject to adjustment as set forth in Section 3.2. The Purchase Price shall be paid by the Purchaser to the Seller at the Closing in cash, by wire transfer in immediately available funds)." "Acknowledgment of Obligations. Purchaser acknowledges that as the initiallessee under Lease Serial Number OCS-G 22939, dated July 1, 2001, between Purchaser and the United States of America, Purchaser may have certain obligations under applicable Law to plug, dismantle or abandon the Wells, and nothing contained herein shall either (a) increase the aforementioned obligations; or (b) relieve any Person from the rights of Purchaser to seek contribution, subrogation, and/or indemnity from, or assert any other related rights against, such Person for plugging and abandonment, decommissioning, or other environmental-related Liabilities associated with the Assets, all of which rights Purchaser hereby expresslyreserves unto itself"	

V. Stone Energy Corp.

Stone Energy Corporation	
Prepetition Capital Structure	
(in millions)	
Bank Borrowings	447.00
Term Loan	11.70
Total Debt	458.70
Senior Convertible Notes	279.20
Senior Notes	770.00
Total Debt	1,507.90
Postpetition Capital Structure	
(in millions)	
Revolving Credit Facility	150.00
Second Lien Notes	225.00
Building Loan	11.00
Total	386.00
Notes	
Dkt. 216-2 "$75 million escrow on the Effective Date related to projected plugging andabandonment expenditures which shall be reduced dollar for dollar for any payments made by the Reorganized Debtors related to any plugging and abandonment related liabilities" A condition precedent to the Plan going effective is the resolution of issuesrelated to the provision of additional collateral required by BOEM.	

VI. Castex Energy Partners, L.P.

Castex Energy Partners, L.P.	
Prepetition Capital Structure	
(in millions)	
Total Debt*	390.00

*does not include accrued unpaid interest, fees, and expenses.

Postpetition Capital Structure	
(in millions)	
Revolving Credit Facility	90.00
Notes	
Plan of Reorganization states the debtors will continue withtheir decommissioning obligations.	

VII. Cobalt International Energy, Inc.

Cobalt International Energy	
Prepetition Capital Structure	
(in millions)	
First lien notes	500.00
Second lien notes	934.70
Total Debt	1,406.10
Total Debt	2,840.80
Postpetition Capital Structure	
(in millions)	
Liquidated	
Notes	
Dkt. 562 details out the sale of all assets to 5 parties. Each sale contemplates the assumption of P&A liabilities with the purchase of the asset.	

VIII. Fieldwood Energy LLC

Fieldwood Energy LLC	
Prepetition Capital Structure	
(in millions)	
First Lien Debt Obligations	1,282.00
Second Lien Debt Obligations	518.00
Total Debt	1,800.00
Postpetition Capital Structure	
(in millions)	
Credit Bid Sale	1,030.00
Total	1,030.00
Notes	
Following the consummation of the credit bid deal, Fieldwood's Chapter 11 plan calls for the company to undergo a divisive merger into Fieldwood Energy I LLC (FWE I) and Fieldwood Energy III LLC (FWE III). FWE I will vest with shelf assets that Fieldwood previously acquired from Apache Corp, as well as related liabilities tied to the plugging, abandonment and decommissioning of those oil and gaswells. FWE III, meanwhile, will own, operate, plug and abandon any assets that are neither legacy Apache assets nor assets transferring to the purchaser under the credit bid transaction. Related to theAbandoned Properties, Dkt. 1285 states " Abandoned Properties: Immediately upon the occurrence ofthe Effective Date, certain of the Debtors' assets (the "Abandoned Properties") will be abandoned pursuant to sections 105(a) and 554(a) of the Bankruptcy Code. The Debtors anticipate that BSEE will issue orders compelling either all or certain entities who are in the chain of title (collectively, the "Predecessors") and/or current co- working interest owners	

(collectively, the "CIOs") for each of the Abandoned Properties to perform the P&A Obligations for each respective property. A schedule of the oil and gas leases, rights-of-way, and rights-of-use and easement related to the Abandoned Properties is annexed hereto as Exhibit F12 (which may be amended, modified, or supplemented from time to time).13 As further detailed below, the Debtors have taken several steps to facilitate the safe and orderly operational transfer of the Abandoned Properties currently operated by the Debtors and are working to reach long-term commercial agreements similar to the FWE I and FWE IV structures with interested Predecessors for assuming operational control for Abandoned Properties operated by the Debtors. The Debtors (i) have dedicated approximately $6 million, in addition to amounts spent in the ordinary course, on safety related repairs and improvements on the AbandonedProperties and (ii) have provided Predecessors detailed operational information on the Abandoned Properties. Additionally, for any Predecessor with whom a consensual arrangement has not yet beenagreed to, Credit Bid Purchaser will offer a 90-day transition period post-Effective Date during which Credit Bid Purchaser will offer to manage at the requesting Predecessor's cost and on its behalf any of the Abandoned Properties."

IX. Fieldwood Energy LLC

Fieldwood Energy LLC	
Prepetition Capital Structure	
(in millions)	
First Lien Term Loans & Reserve BasedTerm Loans	1,143.00
Prepetition FLLO Loans	518.00
Total Debt	*1,626.00*
Total Funded Debt	3,287.00
Prepetition RBL Facility (Undrawn)	148.00
Postpetition Capital Structure	
(in millions)	
First Lien Term Loan	1,143.00
Second Lien Term Loan	518.00
Rights Offering	*525.00*
Total	2,186.00
Notes	
Dkt. 34 "Additionally, Reorganized Fieldwood Energy will assume liability for future environmental remediation and obligations related to plugging and abandoning wells and decommissioning facilities. In connection with the assumption of this plugging and abandonment and decommissioning liability and to secure its obligations with respect thereto,Reorganized Fieldwood Energy will be required to post an aggregate of approximately $240 million 3 of surety bonds for the benefit of Seller and an additional prior owner of certain of thePurchased Assets."	

X. Arena Energy, LP

Arena Energy, LP	
Prepetition Capital Structure	
(in millions)	
RBL Facility	
Principal Outstanding	599.64
Total Debt	29.94
RBL Obligation	629.58
Term Loan	
Principal Outstanding	410.01
Unpaid Interest	29.06
Term Loan Obligation	*439.07*
Total Debt Obligation	1,068.65
Postpetition Capital Structure	
(in millions)	
Liquidated	
Notes	
Sold all assets to San Juan Offshore VIII LP (owned by Lime Rock). Dkt. 13 "The InitialPlan Sponsor Proposal was attractive for a variety of reasons, including that it: (a) contemplated the acquisition of all of the Debtors' assets and equity as a going concern on an "as is, where is" basis, without potential title or environmental contingencies; (b) provided for the assumption of all the Debtors' plugging, abandonment, and decommissioning liability and the replacement of the Debtors' existing surety bonds, and included a bonding support letter demonstrating that existing surety providers of the Debtors were willing to support the proposal and continue to provide surety bonding to thePlan Sponsor"	

XI. Castex Energy 2005

Castex Energy 2005	
Prepetition Capital Structure	
(in millions)	
Total Secured Debt	199.60
P&A Liabilities	36.10
Total Debt	*16.00*
Total Funded Debt	251.70
Postpetition Capital Structure	
(in millions)	
Revolving Credit Facility	105.00
Term Loan	*55.00*

Total Debt	160.00
Notes	
"Castex's Chapter 11 plan called for the creation of a liquidation trust to fund the company's plugging and abandonment obligations for certain oil and gas wells. Aside from those funded P&A obligations, Castex proposed to abandon all of its other interests in oil and gas wells." Debtwire April 20, 2022 article. In April 2022, DOI agreed to withdraw a $245m decommissioning claim.	

XII. MLCJR LLC (Cox)

MLCJR Capital Structure		
$ in millions	Maturity	Prepetition Outstanding Amount
Prepetition Revolver Facility	8/31/2024	80.0
ISDA Agreement	n/a	190.0
Prepetition Funded Obligations		270.0

Exhibit B – Summary Of Historical U.S. Government Bailouts, by Industry[142]

Industry	*Size in 2008 U.S. Dollars ($ Billions)*
Aerospace / Airlines	$ 20.0
Auto	$ 29.0
Banking	$ 1,862.8
Insurance	$180.0
Municipality	$ 9.4
Railroad	$ 3.2
	$ 2,104.4

[142] Information and, in some cases, direct excerpts in this section were based on the ProPublica article History of U.S.Gov't Bailouts, Updated April 15, 2009.

Industry / Corporation	Year	What Happened	Size in 2008 U.S. Dollars ($ Billions)
Penn Central Railroad	1970	In May 1970, Penn Central Railroad, then on the verge of bankruptcy, appealed to the Federal Reserve for aid on the grounds that it provided crucial national defense transportation services. The Nixon administration and the Federal Reserve supported providing financial assistance to Penn Central, but Congress refused to adopt the measure. Penn Central declared bankruptcy on June 21, 1970, which freed the corporation from its commercial paper obligations. To counteract the devastating ripple effects tothe money market, the Federal Reserve Board told commercial banks it would provide the reserves needed to allow them to meet the credit needs oftheir customers.	$3.2
Lockheed	1971	In August 1971, Congress passed the Emergency Loan Guarantee Act, which could provide funds to any major business enterprise in crisis. Lockheed was the first recipient. Its failure would have meant significant jobloss in California, a loss to the GNP and an impact on national defense.	$ 1.4
Franklin National Bank	1974	In the first five months of 1974 the bank lost $63.6 million. The FederalReserve stepped in with a loan of $1.75 billion.	$7.8
New York City	1975	During the 1970s, New York City became over-extended and entered a period of financial crisis. In 1975 President Ford signed the New YorkCity Seasonal Financing Act, which released $2.3 billion in loans to the city.	$9.4
Chrysler	1980	In 1979 Chrysler suffered a loss of $1.1 billion. That year the corporation requested aid from the government. In 1980 the Chrysler Loan GuaranteeAct was passed, which provided $1.5 billion in loans to rescue Chrysler from insolvency. In addition, the government's aid was to be matched by U.S. and foreign banks.	$4.0
Continental IllinoisNational Bank andTrust Company	1984	Then the nation's eighth largest bank, Continental Illinois had suffered significant losses after purchasing $1 billion in energy loans from the failed	$9.5

Industry / Corporation	Year	What Happened	Size in 2008 U.S. Dollars ($ Billions)
		Penn Square Bank of Oklahoma. The FDIC and Federal Reserve devised a plan to rescue the bank that included replacing the bank's top executives.	
Savings & Loan	1989	After the widespread failure of savings and loan institutions, President George H. W. Bush signed, and Congress enacted the Financial Institutions Reform Recovery and Enforcement Act in 1989.	$293.3
Airline Industry	2001	The terrorist attacks of September 11 crippled an already financially troubled industry. To bail out the airlines, President Bush signed into law the Air Transportation Safety and Stabilization Act, which compensated airlines for the mandatory grounding of aircraft after the attacks. The act released $5 billion in compensation and an additional $10 billion in loan guarantees or other federal credit instruments.	$18.6
Bear Stearns	2008	JP Morgan Chase and the federal government bailed out Bear Stearns when the financial giant neared collapse. JP Morgan purchased Bear Stearns for $236 million; the Federal Reserve provided a $30 billion credit line to ensure the sale could move forward.	$30.0
Fannie Mae/ Freddie Mac	2008	On Sep. 7, 2008, Fannie and Freddie were essentially nationalized: placed under the conservatorship of the Federal Housing Finance Agency. Under the terms of the rescue, the Treasury has invested billions to cover the companies' losses. Initially, Treasury Secretary Hank Paulson put a ceiling of $100 billion for investments in each company. In February, Tim Geithner raised it to $200 billion. The money was authorized by the Housing and Economic Recovery Act of 2008.	$400.0
American International Group(A.I.G.)	2008	On four separate occasions, the government has offered aid to AIG to keep it from collapsing, rising from an initial $85 billion credit line from the Federal Reserve to a combined $180 billion effort between the Treasury ($70 billion) and Fed ($110 billion). ($40	$180.0

Industry / Corporation	Year	What Happened	Size in 2008 U.S. Dollars ($ Billions)
		billion of the Treasury's commitment is also included in the TARP total.)	
Auto Industry	2008	In late September 2008, Congress approved a more than $630 billion spending bill, which included a measure for $25 billion in loans to the auto industry. These low-interest loans are intended to aid the industry in its push to build more fuel-efficient, environmentally-friendly vehicles. The Detroit 3 -- General Motors, Ford and Chrysler -- will be the primary beneficiaries.	$25.0
Troubled Asset Relief Program	2008	In October 2008, Congress passed the Emergency Economic Stabilization Act, which authorized the Treasury Department to spend $700 billion to combat the financial crisis. Treasury has been doling out the money via an alphabet soup of different programs. Here's our running tally of companies getting TARP funds.	$700.0
Citigroup	2008	Citigroup received a $25 billion investment through the TARP in October and another $20 billion in November. (That $45 billion is also included in the TARP total.) Additional aid has come in the form of government guarantees to limit losses from a $301 billion pool of toxic assets. In addition to the Treasury's $5 billion commitment, the FDIC has committed $10 billion and the Federal Reserve up to about $220 billion.	$280.0
Bank of America	2009	Bank of America has received $45 billion through the TARP, which includes $10 billion originally meant for Merrill Lynch. (That $45 billion is also included in the TARP total.) In addition, the government has made guarantees to limit losses from a $118 billion pool of troubled assets. In addition to the Treasury's $7.5 billion commitment, the FDIC has committed $2.5 billion and the Federal Reserve up to $87.2 billion.	$142.2
			$ 2,104.4

Exhibit C– Summary of Current Gulf of Mexico Orphaned Assets

On February 2, 2022, BSEE hosted a virtual workshop to discuss and provide information in anticipation of issuing a Request for Proposal (RFP) for the decommissioning of orphaned wells and infrastructure located in the Gulf of Mexico Outer Continental Shelf. Below is a summary of those assets and the estimated decommissioning costs.

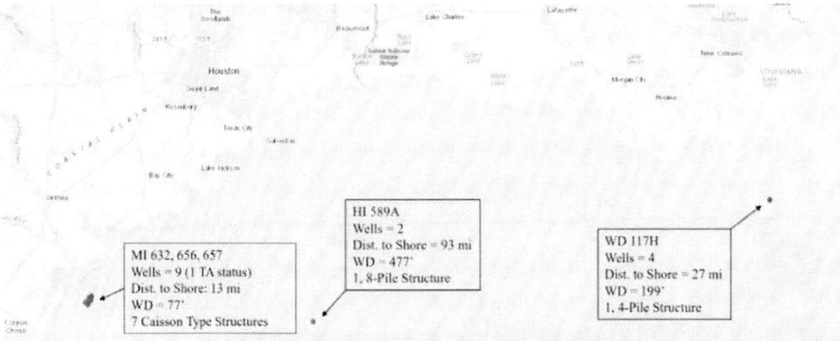

Wells identified for decommissioning

Lease	Area Block	Well Name	API Number	Well Status	Water Depth	MD TVD	Spud Date	Comp Date	Last Produced
G03096	MI 656	002	427034056702	COM	77'	5850' 5847'	23-Oct-06	02-Dec-06	Aug-2010
G03096	MI 656	003	427034056800	COM	75'	5831' 5820'	05-Dec-06	17-Feb-07	Sep-2013
G03096	MI 656	004	427034056900	COM	77'	5850' 5849'	13-Jan-07	25-Mar-07	Jan-2012
G03091	MI 632	007	427034057100	COM	74'	5850' 5849'	16-Feb-07	30-Apr-07	Sep-2009
G03091	MI 632	008	427034057500	COM	70'	2380' 2380'	17-Apr-08	29-Apr-08	Jul-2011
G03091	MI 632	009	427034057600	COM	75'	2359' 2359'	10-May-08	02-Jun-08	Sep-2013
G04139	MI 657	D001	427034057200	COM	71'	3560' 2350'	28-Jan-08	01-Apr-08	Sep-2013
G04139	MI 657	D002	427034057300	TA	71'	3485' 2315'	18-Feb-08	N/A	N/A
G04139	MI 657	D003	427034057400	COM	71'	3032' 2047'	08-Mar-08	13-Apr-08	Sep-2013
G01101	WD 117	H001	177204017301	COM	200'	9796' 9040'	13-May-06	26-Jul-06	Oct-2015
G01101	WD 117	H002	177204017500	COM	205'	11418' 10294'	25-Mar-06	19-Aug-06	Aug-2015
G01101	WD 117	H003	177204017200	COM	200'	11646' 10118'	27-Dec-05	18-Sep-06	Oct-2015
G01101	WD 117	H004	177204017400	COM	200'	9570' 8706'	23-Feb-06	28-Sep-06	Feb-2015
G27532	HI A589	A001	427094116502	COM	477'	11934' 11451'	05-Mar-08	07-Jul-08	Jul-2016
G27532	HI A589	A002	427094116600	COM	477'	10702' 10401'	08-Apr-08	23-Jun-08	Apr-2014

Well Locations

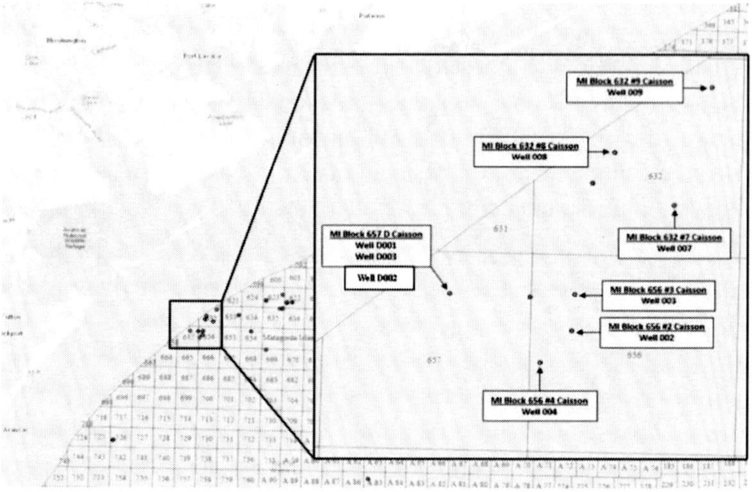

HI Wells (Block A 589)

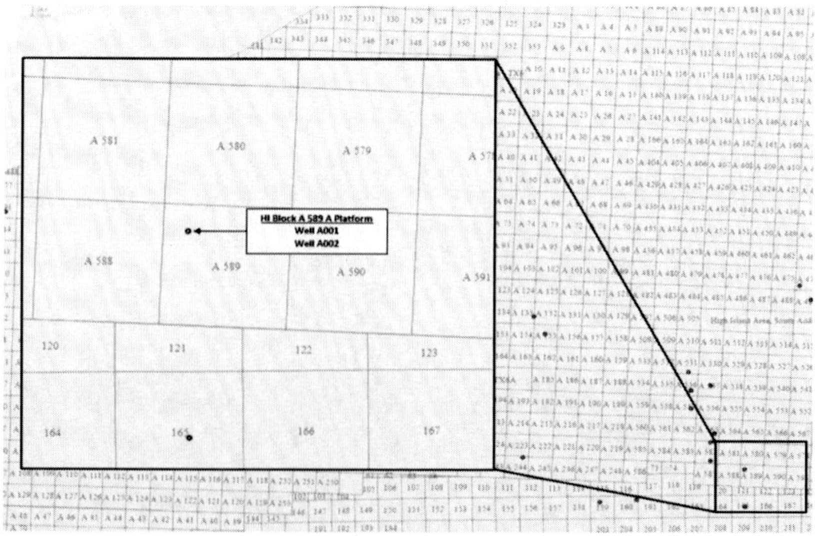

WD Wells (Block 117)

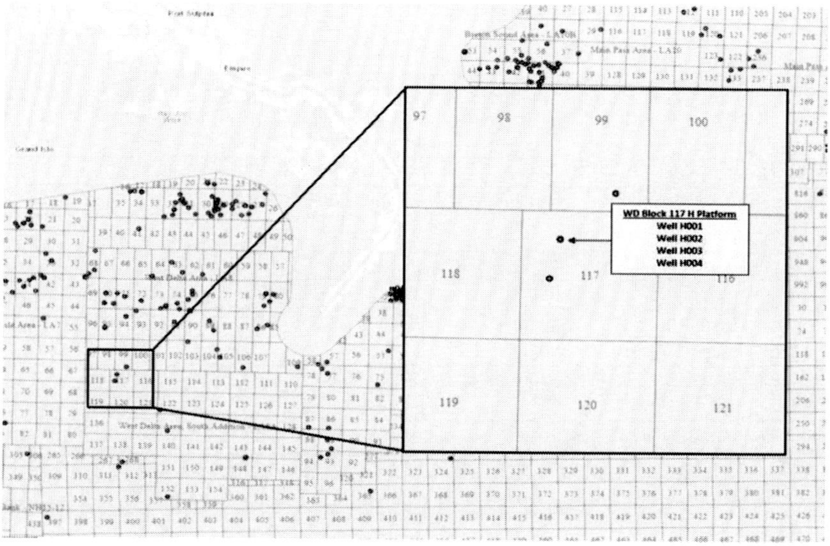

Lease	API Number	Well Name	Protraction Area	Block Number	Surface Location Latitude	Surface Location Longitude
OCS-G 03096	427034056702	002	Matagorda Island (MI)	656	28.033860	-96.576398
OCS-G 03096	427034056800	003	Matagorda Island (MI)	656	28.041805	-96.575704
OCS-G 03096	427034056900	004	Matagorda Island (MI)	656	28.026901	-96.583688
OCS-G 03091	427034057100	007	Matagorda Island (MI)	632	28.061401	-96.552252
OCS-G 03091	427034057500	008	Matagorda Island (MI)	632	28.073045	-96.566436
OCS-G 03091	427034057600	009	Matagorda Island (MI)	632	28.087336	-96.543297
OCS-G 04139	427034057200	D001	Matagorda Island (MI)	657	28.042068	-96.604989
OCS-G 04139	427034057300	D002	Matagorda Island (MI)	657	28.042090	-96.605001
OCS-G 04139	427034057400	D003	Matagorda Island (MI)	657	28.042090	-96.604976
OCS-G 01101	177204017200	H003	West Delta (WD)	117	28.817060	-89.798527
OCS-G 01101	177204017400	H004	West Delta (WD)	117	28.817065	-89.798550
OCS-G 01101	177204017500	H002	West Delta (WD)	117	28.817080	-89.798521
OCS-G 01101	177204017301	H001	West Delta (WD)	117	28.817085	-89.798543
OCS-G 27532	427094116502	A001	High Island (HI)	A 589	27.893154	-94.323778
OCS-G 27532	427094116600	A002	High Island (HI)	A 589	27.893135	-94.323787

Abandonment Cost Estimates

Matagorda Island (MI 632, 656, 657)					
Type	Assets	P50	P70	P90	Deterministic
Pipelines Decom Cost	8	$0.00	$0.00	$0.00	$2,425,082.00
Platforms Decom Cost	9	$7,400,283.00	$9,679,223.00	$13,022,378.00	$0.00
Platforms Site Clear Cost	9	$743,256.00	$1,300,158.00	$2,529,096.00	$0.00
Wells Decom Cost	9	$2,626,056.00	$4,461,845.00	$7,119,552.00	$0.00
Totals		$10,769,595.00	$15,441,226.00	$22,671,026.00	$2,425,082.00

High Island (589 A)					
Type	Assets	P50	P70	P90	Deterministic
Pipelines Decom Cost	1	$0.00	$0.00	$0.00	$1,482,300.00
Platforms Decom Cost	1	$0.00	$0.00	$0.00	$8,433,619.00
Platforms Site Clear Cost	1	$260,731.00	$321,433.00	$424,275.00	$0.00
Wells Decom Cost	2	$0.00	$0.00	$0.00	$1,162,902.00
Totals		$260,731.00	$321,433.00	$424,275.00	$11,078,821.00

West Delta (117 H)					
Pipelines Decom Cost	3	$0.00	$0.00	$0.00	$966,222.00
Platforms Decom Cost	1	$2,418,075.00	$2,868,687.00	$3,525,366.00	$0.00
Platforms Site Clear Cost	4	$6,503,860.00	$10,571,159.00	$20,832,103.00	$0.00
Wells Decom Cost	4	$3,299,686.00	$4,193,375.00	$5,487,095.00	$0.00
Totals		$12,221,621.00	$17,633,221.00	$29,844,564.00	$966,222.00

Assignment History

Matagorda Island (Blocks 632, 656 and 657)

Lease Number	Area	Block	Date Received	Date Returned	Date Approved	Assignment Type	Assignor Name	Assignee Name	Returned Remark
G03091	MI	632	8/4/2004		8/25/2004	OR	Exxon Mobil Corporation	LLOG Exploration Offshore, Inc.	
G03091	MI	632	9/30/2004		11/9/2004	OR	LLOG Exploration Offshore, Inc.	Hunt Petroleum (AEC), Inc.	
G03091	MI	632	9/30/2004		11/9/2004	OR	LLOG Exploration Offshore, Inc.	Hassie Hunt Exploration Company	
G03091	MI	632	9/30/2004		11/9/2004	OR	LLOG Exploration Offshore, Inc.	Haroldson L. Hunt, Jr. Trust Estate	
G03091	MI	632	9/30/2004		11/9/2004	OR	LLOG Exploration Offshore, Inc.	Margaret Hunt Trust Estate	
G03091	MI	632	9/30/2004		11/9/2004	OR	LLOG Exploration Offshore, Inc.	Lyda Hunt-Margaret Trusts-Lyda Hill	
G03091	MI	632	9/30/2004		11/9/2004	OR	LLOG Exploration Offshore, Inc.	Bushhill, L.P.	
G03091	MI	632	9/30/2004		11/9/2004	OR	LLOG Exploration Offshore, Inc.	AGH Energy, LLC	
G03091	MI	632	3/7/2005		5/25/2005	RT	Exxon Mobil Corporation	Paloma Offshore, LLC	
G03091	MI	632	3/7/2005		5/25/2005	OR	Exxon Mobil Corporation	Paloma Offshore, LLC	
G03091	MI	632	3/7/2005	5/25/2005		OR	Paloma Offshore, LLC	Exxon Mobil Corporation	Attempts more than 2 levels of OR.
G03091	MI	632	8/28/2008		9/12/2008	OR	Haroldson L. Hunt, Jr. Trust Estate	Hunt Petroleum (AEC), Inc.	
G03091	MI	632	8/28/2008		9/12/2008	OR	Margaret Hunt Trust Estate	Hunt Petroleum (AEC), Inc.	

Lease Number	Area	Block	Date Received	Date Returned	Date Approved	Assignment Type	Assignor Name	Assignee Name	Returned Remark
G03091	MI	632	8/28/2008		9/12/2008	OR	Lyda Hunt-Margaret Trusts-Lyda Hill	Hunt Petroleum (AEC), Inc.	
G03091	MI	632	8/28/2008		9/12/2008	OR	Bushhill, L.P.	Hunt Petroleum (AEC), Inc.	
G03091	MI	632	8/28/2008		9/12/2008	OR	AGH Energy, LLC	Hunt Petroleum (AEC), Inc.	
G03091	MI	632	5/25/2011	6/9/2011		OR	XTO Offshore Inc.	Matagorda Island Gas Operations, LLC	DOO description incorrect
G03091	MI	632	5/25/2011	6/9/2011		OR	HHE Energy Company	Matagorda Island Gas Operations, LLC	DOO description incorrect
G03091	MI	632	7/6/2011		8/3/2011	OR	XTO Offshore Inc.	Matagorda Island Gas Operations, LLC	
G03091	MI	632	7/6/2011		8/3/2011	OR	HHE Energy Company	Matagorda Island Gas Operations, LLC	
G03096	MI	656	8/4/2004		8/25/2004	OR	Exxon Mobil Corporation	LLOG Exploration Offshore, Inc.	
G03096	MI	656	9/30/2004		11/9/2004	OR	LLOG Exploration Offshore, Inc.	Hunt Petroleum (AEC), Inc.	
G03096	MI	656	9/30/2004		11/9/2004	OR	LLOG Exploration Offshore, Inc.	Hassie Hunt Exploration Company	
G03096	MI	656	9/30/2004		11/9/2004	OR	LLOG Exploration Offshore, Inc.	Haroldson L. Hunt, Jr. Trust Estate	
G03096	MI	656	9/30/2004		11/9/2004	OR	LLOG Exploration Offshore, Inc.	Margaret Hunt Trust Estate	
G03096	MI	656	9/30/2004		11/9/2004	OR	LLOG Exploration Offshore, Inc.	Lyda Hunt-Margaret Trusts-Lyda Hill	
G03096	MI	656	9/30/2004		11/9/2004	OR	LLOG Exploration Offshore, Inc.	Bushhill, L.P.	

Lease Number	Area	Block	Date Received	Date Returned	Date Approved	Assignment Type	Assignor Name	Assignee Name	Returned Remark
G03096	MI	656	9/30/2004		11/9/2004	OR	LLOG Exploration Offshore, Inc.	AGH Energy, LLC	
G03096	MI	656	3/7/2005		5/25/2005	RT	Exxon Mobil Corporation	Paloma Offshore, LLC	
G03096	MI	656	3/7/2005		5/25/2005	OR	Exxon Mobil Corporation	Paloma Offshore, LLC	
G03096	MI	656	3/7/2005	5/25/2005		OR	Paloma Offshore, LLC	Exxon Mobil Corporation	Attempts more than 2 levels of OR.
G03096	MI	656	8/28/2008		9/12/2008	OR	Haroldson L. Hunt, Jr. Trust Estate	Hunt Petroleum (AEC), Inc.	
G03096	MI	656	8/28/2008		9/12/2008	OR	Margaret Hunt Trust Estate	Hunt Petroleum (AEC), Inc.	
G03096	MI	656	8/28/2008		9/12/2008	OR	Lyda Hunt-Margaret Trusts-Lyda Hill	Hunt Petroleum (AEC), Inc.	
G03096	MI	656	8/28/2008		9/12/2008	OR	Bushhill, L.P.	Hunt Petroleum (AEC), Inc.	
G03096	MI	656	8/28/2008		9/12/2008	OR	AGH Energy, LLC	Hunt Petroleum (AEC), Inc.	
G03096	MI	656	5/25/2011	6/9/2011		OR	XTO Offshore Inc.	Matagorda Island Gas Operations, LLC	DOO description incorrect
G03096	MI	656	5/25/2011	6/9/2011		OR	HHE Energy Company	Matagorda Island Gas Operations, LLC	DOO description incorrect
G03096	MI	656	7/6/2011		7/19/2011	OR	XTO Offshore Inc.	Matagorda Island Gas Operations, LLC	
G03096	MI	656	7/6/2011		7/19/2011	OR	HHE Energy Company	Matagorda Island Gas Operations, LLC	
G04139	MI	657	8/4/2004		8/25/2004	OR	Exxon Mobil Corporation	LLOG Exploration Offshore, Inc.	
G04139	MI	657	9/30/2004		11/9/2004	OR	LLOG Exploration Offshore, Inc.	Hunt Petroleum (AEC), Inc.	
G04139	MI	657	9/30/2004		11/9/2004	OR	LLOG Exploration Offshore, Inc.	Hassie Hunt Exploration Company	
G04139	MI	657	9/30/2004		11/9/2004	OR	LLOG Exploration Offshore, Inc.	Haroldson L. Hunt, Jr. Trust Estate	

Lease Number	Area	Block	Date Received	Date Returned	Date Approved	Assignment Type	Assignor Name	Assignee Name	Returned Remark
G04139	MI	657	9/30/2004		11/9/2004	OR	LLOG Exploration Offshore, Inc.	Margaret Hunt Trust Estate	
G04139	MI	657	9/30/2004		11/9/2004	OR	LLOG Exploration Offshore, Inc.	Lyda Hunt-Margaret Trusts-Lyda Hill	
G04139	MI	657	9/30/2004		11/9/2004	OR	LLOG Exploration Offshore, Inc.	Bushhill, L.P.	
G04139	MI	657	9/30/2004		11/9/2004	OR	LLOG Exploration Offshore, Inc.	AGH Energy, LLC	
G04139	MI	657	3/7/2005		5/25/2005	RT	Exxon Mobil Corporation	Paloma Offshore, LLC	
G04139	MI	657	3/7/2005		5/25/2005	OR	Exxon Mobil Corporation	Paloma Offshore, LLC	
G04139	MI	657	3/7/2005	5/25/2005		OR	Paloma Offshore, LLC	Exxon Mobil Corporation	Attempts more than 2 levels of OR.
G04139	MI	657	8/28/2008		9/12/2008	OR	Haroldson L. Hunt, Jr. Trust Estate	Hunt Petroleum (AEC), Inc.	
G04139	MI	657	8/28/2008		9/12/2008	OR	Margaret Hunt Trust Estate	Hunt Petroleum (AEC), Inc.	
G04139	MI	657	8/28/2008		9/12/2008	OR	Lyda Hunt-Margaret Trusts-Lyda Hill	Hunt Petroleum (AEC), Inc.	
G04139	MI	657	8/28/2008		9/12/2008	OR	Bushhill, L.P.	Hunt Petroleum (AEC), Inc.	
G04139	MI	657	8/28/2008		9/12/2008	OR	AGH Energy, LLC	Hunt Petroleum (AEC), Inc.	
G04139	MI	657	5/25/2011	6/9/2011		OR	XTO Offshore Inc.	Matagorda Island Gas Operations, LLC	DOO description incorrect
G04139	MI	657	5/25/2011	6/9/2011		OR	HHE Energy Company	Matagorda Island Gas Operations, LLC	DOO description incorrect
G04139	MI	657	7/6/2011		7/19/2011	OR	XTO Offshore Inc.	Matagorda Island Gas Operations, LLC	
G04139	MI	657	7/6/2011		7/19/2011	OR	HHE Energy Company	Matagorda Island Gas Operations, LLC	

High Island (Block A 589)

Lease Number	Area	Block	Date Received	Date Returned	Date Approved	Assignment Type	Assignor Name	Assignee Name	Returned Remark
G27532	HI	A 589	4/15/2014		4/16/2014	RT	ATP Oil & Gas Corporation	Bennu Oil & Gas, LLC	

West Delta (Block 117)

Lease Number	Area	Block	Date Received	Date Returned	Date Approved	Assignment Type	Assignor Name	Assignee Name	Returned Remark
G01101	WD	117	4/3/2000	4/18/2000		OR	Chevron U.S.A. Inc.	W & T Offshore, Inc.	Assignor does not own operating rights.
G01101	WD	117	6/23/2000		7/13/2000	OR	Kewanee Industries, Inc.	W & T Offshore, Inc.	
G01101	WD	117	7/30/2004		8/24/2004	OR	Kewanee Industries, Inc.	Chevron U.S.A. Inc.	
G01101	WD	117	9/9/2004		10/12/2004	OR	Chevron U.S.A. Inc.	Anglo-Suisse Offshore Partners, LLC	
G01101	WD	117	11/24/2004		2/28/2005	OR	Chevron U.S.A. Inc.	Anglo-Suisse Offshore Partners, LLC	
G01101	WD	117	11/24/2004		2/28/2005	OR	Kewanee Industries, Inc.	Chevron U.S.A. Inc.	
G01101	WD	117	1/10/2005		3/8/2005	RT	Chevron U.S.A. Inc.	Anglo-Suisse Offshore Partners, LLC	
G01101	WD	117	5/4/2005	12/21/2005		P	Anglo-Suisse Offshore Pipeline Partners, LLC	Anglo-Suisse Offshore Partners, LLC	Cannot assign a lease term pipeline

Appendices

Appendix A– Cost-Benefit Analysis of Additional Bonding Requirements

I. Cost-Benefit Analysis

Cost-benefit analysis is a widely used technique for evaluating a possible course of action. It is a systematic way to look at all of the expected benefits of the course of action compared to all of the expected costs. For the course of action to be rational and justifiable, the expected benefits must, at least, outweigh the costs. The technique is often used by public policymakers in determining if a proposed policy is sound as well as by businesses in making decisions.

The cost-benefit analysis presented in this Study examines the proposed bonding requirements expected to be implemented by BOEM. In conducting this cost-benefit analysis, the stakeholder is taken to be the American public, either as taxpayers funding the federal government or as citizens being impacted by a government policy. Here, BOEM is merely an intermediary managing the public's resources. Similarly, oil & gas companies affected by the new bonding requirements as lessees are also intermediaries making investment/spending decisions based on the new bonding requirements. Benefits and costs are ultimately measured as impacts to the public.

II. Determination of Benefits

The goal of BOEM in implementing the Proposed Rule is to protect the public from paying for decommissioning of offshore oil and gas infrastructure in the OCS using tax dollars because lessees default on their obligations. To address these concerns, BOEM has proposed to tighten requirements for supplemental bonding for individual lessees seeking to increase the level of financial assurances.

BOEM's approach includes certain factors that operate to mitigate some of the risk of default. Key among these factors is BOEM requirement that the lessees have "joint and several" liability for the decommissioning costs. In particular, BOEM attributes all decommissioning liabilities for a lease to any waived lessee even if there are other lessees present on the lease. The waived lessee is, with all other lessees, jointly and severally liable for decommissioning and relies on its financial strength to secure the costs of decommissioning, on behalf of all jointly and severally liable parties.

Consider how a default leading to taxpayer funding might arise. Typically, leases have more than one lessee in the chain of title. For the taxpayer to have to pay decommissioning costs on the lease, all of the lessees in the chain of title would have to default on their obligation simultaneously. Thus, the risk to the taxpayer of default on an individual lease becomes a function of the number of jointly and severally liable lessees and the financial condition of the individual lessees.

Using these factors, leases can be divided into a several risk profiles. First is a lease with a waived lessee in the chain of title. Joint and several liability ensures that any decommissioning costs are placed on the waived lessee. Typically, the waived party is a Major or a Large Independent with high financial strength, so the risk of default for these leases is near zero. Next, if there is no Major or Large Independent in the chain of title, the risk of default falls to Small Independents. The risk is tied to the financial condition of these individual companies. However, all of the companies in the chain would have to default simultaneously. The greater the number of companies in the chain of title for the lease, the less likely it is that they would all default and the less risky the lease. Obviously, the riskiest leases would be those with only one lessee in the chain of title. In such cases, the risk of default is directly tied only to the lessee's financial condition.

It is important to acknowledge the presence of systematic and unsystematic risk in the above profiles. Systematic risk refers to exogenous changes that affect all lessees at the same time. The recent downturn in the price of oil is a classic example. While all lessees are impacted by the downturn at the same time, they are not all impacted the same. Majors and Large Independents are less affected than Small Independents due to their greater financial strength. Unsystematic risk refers to changes affecting individual lessees that are independent of other lessees. Even in the current market, the more lessees in the lease chain of title, the less likely all will default simultaneously.

The following risk categories are used in this study for the purpose of evaluating the magnitude of the decommissioning liabilities that fall into different risk profiles in order to assess the benefit of the new bonding requirements. Note that these categories are presented from least risky to most risky.

1) *Major Included in Lease* – This category includes leases that have at least one Major in the chain of title (see Appendix C). These leases are viewed as having nearly zero risk of default since the Majors are

very large with robust balance sheets and are very unlikely todefault on these obligations.

2) *Large Independent Included in Lease* – This category includes leases that have no majorin the chain of title, but have at least one Large Independent company (see Appendix C). These companies are large enough to be viewed as having very low risk of default although slightly above the majors.
3) *No Major or Large Independent, 5 or More Small Independents Included in Lease* – This category includes no Majors or Large Independents, but has five or more Small Independents in the chain of title. Individually, the Small Independent companies are smaller, so each has a higher possibility of defaulting than a Major or Large independent. However, because there are five or more companies in the chain of title for the lease and all would have to default simultaneously, the likelihood of default is small, but viewed as higher than for categories 1 and 2.
4) *No Major or Large Independent, 4 Small Independents Included in Lease* – Same as category 3 above, except four lessees are in the chain of title for the lease. The risk is viewed as higher than for category 3 because there are fewer lessees on the lease.
5) *No Major or Large Independent, 3 Small Independents Included in Lease* – Same as category 4 above, except three lessees are in the chain of title for the lease. The risk is viewed as higher than for category 4 because there are fewer lessees on the lease.
6) *No Major or Large Independent, 2 Small Independents Included in Lease* – Same as category 5 above, except two lessees are in the chain of title for the lease. The risk is viewed as higher than for category 5 because there are fewer lessees on the lease.
7) *No Major or Large Independent; 1 Small Independent Included in Lease* – This category includes only one lessee in the chain of title for the lease. This company is not a Major or a Large Independent. This category is viewed as having the highest risk of the seven categories because the possibility of default is tied only to one Small Independent.

To understand the distribution of lease decommissioning liabilities among these risk categories, Opportune analyzed selected lease data for all water depths from the BOEM website using the P&A calculation. Most companies active in deep water regions of the OCS are larger due to the high capital requirements to explore and produce these areas. However, a few independents operate in depths greater than 500 ft. These companies are captured in the data for all depths.

The results of Opportune's analysis of the BOEM lease data are shown in Table 1 and graphically in Figure 2. In the data, 1,706 leases have decommissioning liabilities considering all water depths with a total estimated P&A liability of over $16.4 billion. Of those leases with liabilities, 378 have no Major or Large Independent in the chain of title. These leases have a total estimated ARO liability of about $1.2 billion. The vast majority of leases, over 93 percent of the total ARO liabilities, or $15.3 billion, have a Major or a Large Independent in the chain of titleand have a minimal likelihood of default. Only 7 percent of total P&A liabilities in the OCS are associated with only Small Independents.

Table 1. Summary of ARO by Risk Category – All Depths

Risk Category	Number of Leases	Total Cost	ARO	Percent of Total	Collate-ral	Percent of Total	Un covered	Percent Liability Covered
Leases - Active								
Major Included	1,063	65,100	13,670	83.1%	1,571	47.9%	12,100	11.5%
Large Independent Included	265	8,391	1,620	9.9%	520	15.9%	1,100	32.1%
Small Independents								
>= 5 Lessees	184	7,560	410	2.5%	334	10.2%	77	81.3%
4 Lessees	18	280	21	0.1%	44	1.3%	-22	205.1%
3 Lessees	35	519	77	0.5%	83	2.5%	-6	107.9%
2 Lessees	63	2,152	455	2.8%	271	8.3%	183	59.7%
1 Lessee	78	843	189	1.1%	29	0.9%	160	15.5%
Subtotal Independents	*378*	*11,354*	*1,152*	*7.0%*	*761*	*23.2%*	*391*	*66.1%*
Subtotal All Leases	*1,706*	*84,845*	*16,442*	*100.0%*	*2,852*	*87.0%*	*13,590*	*17.3%*
General Allocations	-			-	425	13.0%	-	-
Total	*1,706*	*84,845*	*16,442*	*100.0%*	*3,276*	*100.0%*	*13,166*	*19.9%*

Breaking the data for Small Independents down by risk category provides further insight into the risk exposure. First, $410 million of the total $1.2 billion in ARO liabilities for Small Independents, or 36 percent, are associated with leases having five or more lessees. Whilesomewhat risker than leases that have a Major or Large Independent, these leases still have a low likelihood of default since all of the lessees would have to default simultaneously. Second, the liability associated with the highest risk category is only $189 million or

16 percent of the total ARO liability for Small Independents and a mere 1 percent of all ARO liabilities for the OCS.

Source: Opportune Analysis.

Figure 2. Breakdown of ARO by Risk Category – All Depths.

In addition to the decommissioning liability data, the BOEM website provides a measure of the level of financial assurance associated with each lease under the previous bonding requirements. The website refers to this measure as "collateral." Collateral is used as measure of how much of the decommissioning liability are covered with the balance being uncovered liabilities. Collateral listed on the website total approximately $3.4 billion for all 1,706 leases with decommissioning liabilities or about 20 percent of total ARO liabilities.

Table 1 also summarizes the collateral associated with the leases by risk category. Breaking up the collateral for the leases among the various risk categories indicates that collateral requirements increase with increasing risk. For the lowest risk category, leases with at least one Major in the chain of title, collateral covers only about 11.5 percent of the total ARO liability. Coverage for Small Independents increases to about 66.1 percent on average. This level of coverage for the higher risk categories suggests that the previous bonding requirements did act to place extra security where it was needed.

The Proposed Rule will likely result in increased collateral coverage across all risk categories. While the impact on collateral will not be fully known for some time as implementation proceeds, it appears that it will provide only limited benefit to the public. For the low-risk categories, increased collateral represents an unnecessary burden since these categories are well secured through large companies in the chain of title. However, even if the new bonding requirements provide 100 percent coverage for the riskiest categories, the benefit is small because these categories do not comprise a large part of the total liability. Specifically, if the increase in collateral from the new bonding requirement provides 100 percent coverage for all Small Independents, this decreases uncovered liability by $391 million. This reduction in uncovered liabilities, then, is the measure of the benefit of the Proposed Rule to the public.

III. Determination of Costs

Achieving BOEM's goal of protecting the public from paying for decommissioning of offshore oil and gas infrastructure in the OCS using tax dollars because lessees default on their obligations comes at a significant cost to the public. This cost results from changes in the way producers will choose to operate as a result of increased bonding requirements. Using more of producer's available financial resources to protect the public reduces the funds available to them for exploration and production activities in the OCS. In turn, these reductions in spending will have a negative impact on the economic activity of the Gulf Coast states as well as in other parts of the country.

The potential impacts of the Proposed Rule are difficult to predict. Increased bonding requirements will have a negative impact on the creditworthiness of each producer. Some producers could become insolvent and be reorganized or auctioned off. Others will experience a decrease in the amount of capital each can raise for exploration and production activities. As a minimum, producers will reduce spending on activities in the OCS. Producers' reduced spending is the major driver for the purposes of analyzing the costs of the increased bonding requirements.

The decrease in spending will have direct effects on the OCS producers as well as on federal tax revenues. Reduced spending will result in decreased production and less development of OCS by producers. Current production may decrease from reduced spending on operating expenses, although producers will first attempt to maintain current production in the face of funding constraints. Future production will be less as a result of less investment in exploration and development. This will directly reduce producer's current and future revenue and curtail many expansion plans.

Finally, the reduction in current and future production will directly impact royalty revenue to the federal government.

The decrease in spending will have indirect effects on economic activity and growth as well. These indirect effects will have the greatest impact in states along the Gulf Coast where most of the spending takes place. The rest of the country can also be expected to see some negative impact. Major indirect effects will include a decrease in economic output in the affected regions and a decrease in employment by producers and suppliers. In this analysis, Opportune looks at these effects over a 10-year timehorizon.

A. Decrease in Spending

The expected decrease in operator spending is difficult to predict. Major operator spending in the deepwater Gulf is expected to remain unchanged by changes in bonding requirements; however, independent producers on the shelf are expected reduce their spending due to the loss of available capital that accompanies bonding requirement expenditures. The distribution of spending reductions will not be uniform among producers.

The magnitude of the decrease depends on a number of factors such as the availability and cost of surety bonds and the outlook for the oil and gas industry. Effects of the new bonding requirement are assumed to be independent of other regulations coming into effect, such as the recent Well Control Regulation, which affect OCS producers.

Economic Impact Model

We developed an economic model to determine the impact of bonding requirements on development activity in the Gulf, and the related impact on the regional economies of the bordering coastal states. Our model is driven by the following high-level assumptions:

1) The premium expenses associated with the increased bonding requirements will be $257 million per year, based on the BOEM estimates for small companies. The requirements are phased in uniformly over the first three years.
2) Bonding requirements have no direct impact on existing production rates.
3) Funds spent on bonding requirements are funds that would otherwise be invested in Gulf of Mexico exploration and development activities.

4) The positive cash flows resulting from these exploration and development activities would likewise be re-invested in additional exploration and development activities in the Gulf of Mexico.
5) The resulting cash flows (revenues and costs) can be modelled using broad assumptions of finding and development costs, future commodity prices, operating costs, and average royalty rates, combined with an assumed production rate behavior of the developed hydrocarbon resources.
6) The production rate behavior of the future-developed resources can be modeled using a four-segment decline curve that describes the typical water drive nature of the reservoirs.
7) Capital and operating costs associated with these development activities would be spent in the Gulf Coast region
8) The economic impact of these expenditures can be estimated using the RIMS II multipliers from the United States Bureau of Economic Analysis.

Economic Impact Model Assumptions

Our base case economic assumptions are shown below. A key assumption is the finding and development (F&D) cost. From this value, we determine the amount of resources developed and placed on production for each $1 million spent. In the section that follows are the parameters that describe the four-segment production curve.

Exploration & Development Assumptions	
Finding and Development Cost, $/boe	25.00
EUR/$MM spent, boe	40,000
Initial production rate, boe/mo	750
Flat rate period, years	0.5
Initial period duration, years	2.0
Initial period decline rate, %pa	30%
Water breakthrough period duration, years	0.5
Water breakthrough period final rate, fraction	50%
Post-breakthrough decline rate, %	10%
Final production rate, fraction of initial	5%
Royalty rate, %	17%
Revenue, $/boe (net)	61.00
Opex, $/boe (gross)	20.00
Reinvestment rate, %	100%

Our model relies on several assumptions that are difficult to validate. Nevertheless, we think our straightforward approach should provide a reasonable estimate of the magnitude of economic impact.

In the case shown above, we have assumed an average finding and development cost of $25/barrel of oil equivalent (boe). Therefore, each $1 million of exploration and development spending would place 40,000 boe on production.

This production would be recovered according the production decline assumptions. In our model, there are four segments to the production rate behavior: 1) a flat period of no decline, 2) an initial decline period of gradual decline, 3) a water breakthrough period during which hydrocarbon production rapidly declines, and 4) a post-breakthrough period of gradual decline to the economic limit. We express the economic limit as a percentage of the initial rate. By providing the parameters that describe these four segments, we can calculate the initial production rate required to produce the developed reserves.

Using the assumptions above, the following monthly production function is generated per$1 million spent. This production is scaled up by the amount of exploration and development assumed to occur.

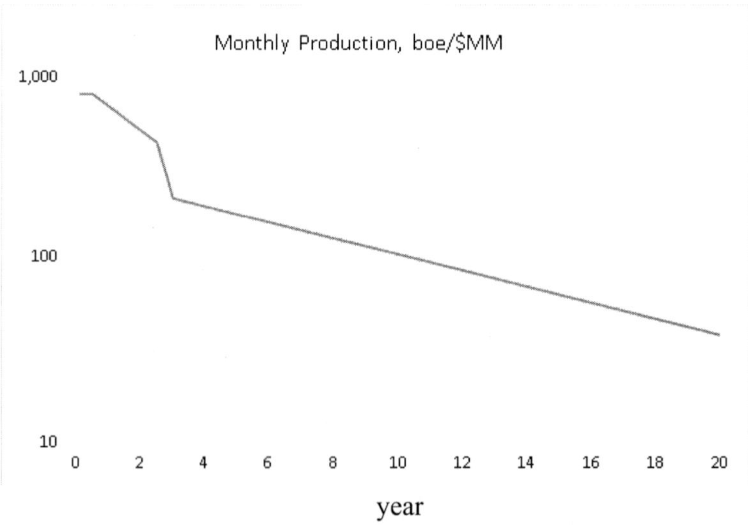

In our model, the positive free cash flow generated from these investments is re-invested in similar projects. We assume that positive free cash flow from one year, whether from reductionsin bonding requirements or newly developed production, is spent evenly across the following year.

Annual Cash Flows	2024	2025	2026	2027	2028	2029	2030	2031	2032	2033
Reduced bonding expenses, $MM	85.7	171.3	257.0	257.0	257.0	257.0	257.0	257.0	257.0	257.0
Incremental Investment, $MM	85.7	171.3	269.6	302.0	350.7	400.5	445.3	489.4	535.4	582.3
Gross production, mmboe		0.4	1.5	3.1	4.7	6.1	7.6	9.1	10.6	12.2
Royalty, mmboe		0.1	0.2	0.5	0.8	1.0	1.3	1.5	1.8	2.1
Net production, mmboe		0.3	1.2	2.5	3.9	5.1	6.3	7.5	8.8	10.1
Revenue, $MM		20.8	74.4	154.9	237.2	311.3	384.1	460.1	537.8	617.1
Opex, $MM		8.2	29.4	61.2	93.7	123.0	151.7	181.8	212.4	243.8
Net cash flow		12.6	45.0	93.7	143.5	188.3	232.4	278.4	325.3	373.3

Our model generates a forecast of incremental capital investment and operating expense expenditures over the ten-year forecast period. We use the sum of these expenditures to estimate their impact on the economies of the states that border the Gulf (Texas, Louisiana, Mississippi, Alabama, and Florida) using the RIMS II multipliers. The multipliers we used were retrieved from the Bureau of Economic Analysis website in June 2023 and are applied over the entire region rather than state-by-state.

In the RIMS II system, there are several types of multipliers that are used to estimate various economic impacts on the region resulting from increased expenditure. "Final Demand Change" is the term that defines the amount of incremental spending in the region. In our model, this is the sum of the capital invest and operating expense expenditures resulting from the change in exploration and development activity.

1) Final Demand Output – The output multipliers represent the total change in local sales per dollar of final demand change.
2) Final Demand Earnings – The earnings multipliers measure the total change in local household earnings per dollar of final demand change. Earnings consist of wages and salaries and of proprietors' income, which is the net earnings of sole-proprietors and partnerships. Employer contributions for health insurance are also included.
3) Final Demand Employment – The employment multipliers measure the total change in the number of local jobs per million dollars of final demand change. Employment consists of full- and part-timejobs.

4) Final Demand Value Added – The value-added multipliers measure the total change in local value added per dollar of final demand change. Value added is comparable to regional measures of GDP.

Using our assumptions and the RIMS II multipliers, *we estimate a final demand change of $9.9 billion* if the increased bonding premiums estimated by BOEM for small companies in the GOM OCS were instead spent on exploration and development activities. The table below shows the estimated economic impact of this level of final demand change.

Metric	Multiplier	Impact
Final demand output	2.0945	$9,922 million
Final demand earnings	0.5908	$2,799 million
Final demand employment	7.6391	36,189 jobs
Final demand value added	1.2310	$5,832 million

IV. Conclusion - Comparison of Benefits and Costs

The results of this cost-benefit analysis indicate the reduction in the risk that taxpayers will have to pay for decommission costs provided by Additional Bonding Requirements is not justified given the impact on Industry and the regional/national economy. The benefits of Additional Bonding Requirements are relatively small, and the cost is very high in terms of the decreased development of OCS resources and associated losses in production, Industry revenue, taxes, economic activity, and jobs.

BOEM's approach is a very inefficient way to protect the taxpayer. It looks only at the absolute magnitude of the decommissioning costs and requires additional financial guarantees to cover a greater proportion of the total liabilities. It does not fully consider the actual risk of default at the lease level. For the low-risk leases, increased collateral represents an unnecessary burden since these categories are well secured through large companies in the chain of title. For risker leases, it does provide incremental protection. However, the incremental benefit is small, estimated in this analysis to be about $950 million, because these risky leases do not comprise a large part of the total liability.

By comparison, the costs of BOEM's plan are very high. Using our assumptions, Industry spending is expected to be over $4.7 billion less than it otherwise would over the 10-year time horizon because of bonding requirements. Future production from the OCS is estimated to be reduced by approximately 55 million boe. Associated with this reduction are approximately $2.8 billion in reduced revenue to the Industry, plus $573 million fewer royalties to the federal government. Gulf Coast states will forego significant increases in economic activity and jobs. Reducing the bonding expenses by the amount shown would generate over $9.9 billion of growth over the next 10 years and add 36,000 jobs.

Appendix B– Opportune Independent Calculation of the P&A Liability

I. Background Information

In order to assess the discounted P&A liability of the oil & gas infrastructure ("Subject Assets") located in the OCS of the Gulf of Mexico, Opportune has independently calculated the asset retirement obligation ("ARO") pertaining to the Subject Assets in accordance with Financial Accounting Standards Board ("FASB") Accounting Standards Codification ("ASC") Topic 410- 20: Asset Retirement Obligation ("ASC410-20").

ASC 410-20

ASC 410-20 is the accounting standard for the recognition and measurement of a liability for an ARO. The ARO for companies engaged in the production and transportation of hydrocarbons in the Gulf of Mexico stems from a legal obligation to decommission the infrastructure used for such activities when no longer used. The decommissioning includes plugging and abandonment of wells, decommissioning of pipelines, removal of production platforms, and site clearance.

The mathematical calculation uses three variables: 1) estimated cost to decommission, 2) estimated time until the decommissioning occurs, 3) an inflation rate factor to arrive at an estimated future cash out flow ("FCOF"), and an estimated credit adjusted risk free rate ("CARF") to calculate the present value ("PV") of the estimated FCOF as of the date of evaluation.

We discuss herein our independent estimate of the ARO. Our analyses have been developed in accordance with ASC 410-20.

Procedures

In general, the procedures used in our analysis consisted of the following:

- Determining the population of infrastructure assets in the Gulf of Mexico;
- Analysis of BOEM's estimated decommissioning costs by asset type;
- Analysis of general market data, including economic, financial, governmental, and environmental forces;
- Determination of market-based assumptions to support the inflation rates and CARF used in the calculation;
- Estimation of the expected life of the population of assets in the Gulf of Mexico

Estimated ARO

Based on the analysis discussed in this report, we believe that the ARO as of the evaluation date ("Evaluation Date") is $16.4 billion. The Evaluation Date is March 15, 2023. The population of infrastructure assets and the decommissioning costs for this population of assets are developed from a download of BOEM data on March 15, 2023.

In arriving at our conclusions, we calculated the obligation under generally accepted accounting standards based upon an investigation of economic and market factors as of the Evaluation Date. Opportune prepared the estimate on the basis of public information, Industry knowledge, third party financial and economic information, and other information.

In addition, Opportune has not independently verified the data obtained from certain public databases and other third-party sources of information utilized in our analysis. The results and conclusions presented in this Report may be materially affected to the extent that actual information differs from that which was provided to us.

II. Industry Overview[143]

Fluctuating commodity prices and unstable energy markets have caused the oil drilling and gas extraction industry to endure an extremely high level of revenue volatility over the five years to 2022. Revenue grew at the outset of

---

[143] Information and, in some cases, direct excerpts in this section were based on IBISWorld Industry Report: Oil Drilling& Gas Extraction in the US, February 2022.

the period as the world prices of crude oil and natural gas rebounded off of decade lows fueled by booming US production. Despite hydrocarbon prices having remained far below record highs achieved in 2012 and 2013, domestic production flourished as unconventional and highly efficient drilling techniques such as hydraulic fracturing and horizontal drilling have become upstream mainstays. More oil and gas companies have deployed these recovery methods to fulfill a greater portion of aggregate demand, leading revenue to grow an annualized 8.2 percent to $463.8 billion over the five years to 2022. Revenue is expected to increase 71.8 percent and 0.5 percent in 2021 and 2022, respectively, as global oil demand improves amid the retracement of macroeconomic restrictions instilled during the COVID-19 (coronavirus) pandemic.

Industry exports have been a catalyst for industry growth over the past five years. Since the introduction of the Energy Policy Conservation Act in 1975 in wake of the Arab oil embargo, exports of crude oil and natural gas from the United States have been banned. However, the Consolidated Appropriations Act of 2016 overturned this decision, enabling domestic producers to access foreign markets with strong energy appetite. This has given US producers a new avenue to offload burgeoning production which has stemmed from greater use of unconventional drilling methods. Since industry profit is largely a function of commodity prices, profit has exhibited significant fluctuations throughout the period.

Industry revenue is projected to decline an annualized 0.3 percent to $456.8 billion over the five years to 2027 as volatile global demand subsides and prices of industry-specific resources are pressured lower. Although IBISWorld forecasts world commodity prices to fall slightly, energy prices are expected to remain highly volatile. The future of the industry is expected to hinge on improvements in drilling technology and techniques. Improving technology is likely to assist operators in meeting environmental concerns and maximizing well efficiency. As industry operators deplete reserves, it will likely become necessary to improve efficiency and minimize waste. Operators may choose to replenish reserves either through acquisitions or exploration.

Recent Performance

Over the five years from 2017 through 2022, the oil drilling and gas extraction industry's production capacity has grown tremendously, benefiting from greater technological adoption and the push toward domestic energy independence. Greater use of hydraulic fracturing (fracking) and horizontal drilling techniques have enabled industry operators to restore production to fields historically viewed as marginal and uneconomic. Booming domestic production, along with 2015 legislation overturning a ban on US exports of crude, enabled the US to become a leading global supplier of energy during the period. However, burgeoning US capacity contributed to an amassing global energy oversupply which suppressed global hydrocarbon prices for most of the current period. Also, the ongoing economic turndown authored by the COVID-19 (coronavirus) pandemic has caused revenue to collapse as global travel and business prohibitions aimed at combatting the health crisis have weighed tremendously on oil demand. Overall, revenue is estimated to increase an annualized 8.2 percent to $463.8 billion over the five years to 2022, including an increase of 71.8 percent and 0.5 percent in 2021 and 2022, respectively, as economic activity and oil demand is restored.

Oil and natural gas are highly globalized commodities that serve a variety of purposes. Best known for gasoline, oil provides many downstream industries with component materials that are staples of everyday life. Additionally, natural gas is the fuel of choice for a sizeable share of electricity generation in the United States. Without these commodities, many facets of the global economy would not be able to function as they currently do. Consequently, this industry is highly dependent on global energy markets. When demand is strong, the industry generally benefits, and it becomes more economical to seek new deposits. However, the current period has also served as a reminder that supply plays a critical role in the global commodity market.

In late 2015, Congress overturned the 40-year-old ban on U.S. crude oil exports due to pressure from industry operators. Advocates of free trade of crude oil expect US producers will benefit from selling products abroad, which may command premiums. The United States exported record amounts of hydrocarbons during the period, aided by greater pipeline infrastructure from the United States to and within Mexico. Greater domestic production and reduced reliance on imports has led the US to post its first-ever petroleum trade surplus in September 2019, a feat that is anticipated to continue moving forward. However, the Energy Information Administration and IBISWorld research asserts that the United States will lose its positive balance of trade in 2021 and 2022 due to short-term disruptions in global energy markets and international trade flows. Over the five years to 2022, the value of exports is estimated to increase an annualized 22.2 percent to $139.8 billion, accounting for 30.1 percent of industry revenue in 2022. Conversely, due to increasing domestic production, diminishing world oil prices, and declining energy demand amid the pandemic, the value of imports has increased at a much more subdued annualized rate of 1.0 percent to $142.3 billion during the current period. Global commodity markets are expected to remain volatile; the strength of the global economic recovery will likely dictate near-term industry trends.

Industry Outlook

Over the five years to 2027, the oil drilling and gas extraction industry is expected to rebound alongside restored output and recovering energy prices. Global economic growth is anticipated to support commodity price gains at the outset of the outlook period as renewed economic activity jump-starts demand for oil and natural gas. Natural gas is also expected to remain a hotbed for investment as operators continue to extract from hydrocarbon-rich basins in North Dakota and the Appalachians. However, energy prices are likely to fall in the event energy producers supply the market with more hydrocarbons than what is required by downstream consumers. Overall, industry revenue is forecast to decline an annualized 0.3 percent to $456.8 billion over the next five years. Nevertheless, IBISWorld expects average industry profit to improve as recently volatile energy prices are anticipated to become more expectable. Gains in revenue will likely flow through to the bottom line as companies focus on containing costs and using technology to improve exploration and production efficiencies.

Global demand for oil recovered to pre-COVID-19 (coronavirus) levels in the second half of 2022 and continues to improve. Emerging economies will likely remain pivotal as they continue to build essential infrastructure and consume greater amounts of petroleum-based products. Accordingly, demand from emerging markets is anticipated to grow at a faster rate compared with more developed nations. As global demand increases, more companies will likely serve these emerging economies, slowing expansion opportunities in the US despite overall higher demand. Over the next five years, the number of industry enterprises is projected to increase an annualized 0.2 percent to 68,676 companies.

The volatile conditions that the industry has experienced in recent years are expected to moderately subside over the five years to 2027. Early during the period, revenue is expected to expand alongside increases in energy prices and resumed production. However, as large oil- producing countries are eager to increase output, new capacity amid prolonged economic disruptions could likely subdue long-term energy price growth. In addition, more crude oil controlled by the Organization of Petroleum Exporting Countries is expected to be available during this period, which is typically of lighter and sweeter grade than most refineries seek.

After the publication of the IBIS report, however, Russia invasion of Ukraine significantly impacted global energy markets. In particular, European countries began and are continuing to shift away from Russian supplies to alternative sources of hydrocarbons. In particular, we believe this will increase the demand for US natural gas as domestic LNG export and European LNG import facilities are expanded and new ones areconstructed.

Pent-up demand will likely drive industry revenue over the next five years, and US exports are expected to continue growing as a share of revenue. Though the US Energy Information Administration expects the United States to return to being net importer of crude oil in 2021 and 2022, recovering demand from global economies is expected to lift industry exports over the next five years. IBISWorld forecasts the value of exports to increase an annualized 0.1 percent to $140.3 billion over the five years to 2022. The value of industry imports is estimated to decline an annualized 1.0 percent to $135.3 billion during the outlook period.

Natural gas production is also forecast to expand. New producing fields, especially in the Marcellus Shale region of the Appalachian Basin and the Bakken Formation in North Dakota, will likely continue fueling supply growth over the next five years. Growing production is likely to spur employment growth, which is anticipated to grow an annualized 0.7 percent to 226,544 workers over the five years to 2027. Growing demand for natural gas will likely reflect its increasing use in electricity generation. Most new and nonrenewable electricity generation capacity planned for the United States during the period will likely be produced using natural gas. New technologies are expected to enhance natural gas-fired electricity generation, making it as or more affordable than coal-fired generation, which was formerly the lowest-cost fuel. Natural gas also has an environmental advantage over coal and crude oil in that its emission of most pollutants is lower. In addition, gas-fired generation has much lower capital costs than nuclear and coal-based generation, giving it a financial advantage in the uncertain environment surrounding electricity deregulation. Higher levels of natural gas output will likely meet most of the growth in demand, but imports are also expected to play an important role. For instance, Canada is forecast to continue expanding its exports of natural gas to the United States over the next five years.

III. Asset Retirement Obligation

FASB ASC Topic 410-20: Asset Retirement Obligation is the authoritative guidance for asset retirement obligations.

Per ASC 410-20-30-1:

> "An expected present value technique will usually be the only appropriate technique with which to estimate the fair value of a liability for an asset retirementobligation. An entity, when using that technique, shall discount the expected cash flows using a credit-adjusted risk-free rate. Thus, the effect of an entity's credit standing is reflected in the discount rate rather than in the expected cash flows. Proper application of a discount rate adjustment technique entails analysis of at least two liabilities—the liability that exists in the marketplace and has an observable interest rate and the liability being measured. The appropriate rate of interest for the cash flows being measured shall be inferred from the observable rate of interest of some other liability, and to draw that inference the characteristicsof the cash flows shall be similar to those of the liability being measured. Rarely, if ever, would there be an observable rate of interest for a liability that has cash flows similar to an asset retirement obligation being measured. In addition, an assetretirement obligation usually will have uncertainties in both timing and amount…"

The mathematical calculation under an expected present value technique uses three variables 1) estimated cost to decommission 2) estimated time until the decommissioning occurs and 3) an inflation rate factor to arrive at an estimated future cash out flow ("FCOF"). The estimated FCOF is then present valued back to a date of evaluation through an estimated credit adjusted risk free rate ("CARF") to arrive at a present value ("PV").

IV. ARO Methodology and Assumptions

ARO Methodology

Estimated Abandonment Costs

In conducting our analysis, we relied on BOEM's estimated abandonment costs. These costs are detailed on BOEM's website for the respective leases in the Gulf of Mexico. BOEM's data provide current cost estimates for the pipeline removal, platform removal, well bore plugging, and site clearance by lease for all federal leases in the Gulf of Mexico. The BOEM cost estimates are shown as a distribution (P50, P70, P90) based on actual expenditure data reported by OCS operators. We totaled the P70 estimate and the deterministic estimate by lease to arrive at an estimate of total decommissioning costs for the Gulf of Mexico of approximately $84.8 billion. We also totaled the P50 estimate and the deterministic estimate to arrive at our ARO estimate of approximately $16.4 billion.

Estimated Remaining Field Life

Estimating the remaining economic life of the producing wells and related facilities is a challenging problem, particularly under conditions of very limited technical data. In many hydrocarbon reservoirs around the world, the primary recovery mechanism is depletion drive, and the future production performance of the wells producing from such reservoirs can be forecast using decline curve analysis. Such production forecasts can then be extrapolated to an “economic limit” to determine the remaininglife.

Most offshore reservoirs produce under water drive conditions, however, and decline curve analysis is of limited utility. This is especially true in cases where the geologic structure results in rapid "watering-out" of the producers. Also, many offshore fields have multiple pay zones that are produced sequentially. Those "behind pipe" zones will extend the life of the field, but their future performance is not revealed in the historical performance of the actively producing zones. Predicting their performance requires detailed geologic and petrophysical data that was not available to us. Moreover, conducting such evaluations is prohibitively time-consuming for a study of this nature.

For the purpose of this study, we relied on the operators' estimate of each field's total recoverable reserves (including the cumulative production) that they disclose to BOEM on an annual basis. We combined this data with the reported cumulative production over time to calculate a time-series of the percent of recoverable reserves remaining to be produced. We observed that these series for each field often followed a decline trend that could be reasonably extrapolated on a semi-log graph. In most cases, we extrapolated these values to the point where 1% of theestimated recoverable reserves would remain, but in some cases (for instance, where they were already less than 1%) we extrapolated them to a lower value.

The estimated field life was matched to leases by area and block. Where a direct match could not be made the average for the area was used. Inactive fields, those awaiting decommissioning, were included with a zero life for calculation purposes which is equal to the estimated cost.

The estimated lives range from zero to fifty years with an average age of approximately 6.8 years. The actual decommissioning activities most likely will occur at a future date depending on individual circumstances and BOEM approval.

Inflation Rate

In performing our analysis, we have used an inflation rate of 2.5%. The 2.5% inflation factor was determined by the CPI growth of 2.5% over the next ten years from the June 2022 Livingston Survey.

Credit Adjusted Risk-Free Rate

In performing our analysis, we have used a CARF for each lease, as follows, based on the mixof lessees and their respective credit ratings:

- Majors only: 6.5%
- Large Independents and other lessees: 8.0%
- Small Independents only: 9.5%.

Typically, in valuing liabilities, we rely on a corporate cost of debt as the discount rate. The reason is the risk (or likelihood) the company will make "good" on the liability is comparable to the risk of the company paying debt liabilities. In our opinion this meets the requirements of a CAFR since essentially a company's cost of debt is a risk-free borrowing rate plus a premium for the credit risk of the company.

We have analyzed the credit ratings of companies with operations in the Gulf of Mexico and split the companies into the groups (Majors, Large Independents, and Small Independents) based on their tangible asset value. We looked at each company's S&P debt rating (when available) and the associated yields for long-term debt for the U.S. Energy Index per Bloomberg. Majors had credit ratings ranging from AA- to BBB+ with yields ranging from 4.81% to 5.72%, Large Independents had a range of credit ratings from A- to BB+ with yields ranging from 5.33% to 6.78%, and Small Independents had credit ratings ranging from BB to not rated with yields ranging from 7.67% to 8.36% (if the company's debt was not rated we assumed the lowest quality of debt quoted in the index of B+). We then estimated the rounded average energy yield for each group to arrive at our concluded discount rates of 6.5% for Majors, 8.0% for Large Independents, and 9.5%. for Small Independents.

V. Conclusion

Based on the analysis and procedures discussed herein, we have concluded that the Asset Retirement Obligation is $16.4 billion.

Appendix C – Lessees Considered to Be Majors and Large Independents

Companies Categorized as Majors

MMS Number	Business Name
00164	Aminoil Development, Incorporated
00368	Aminoil USA, Inc.
00735	Amoco Canyon Company
02244	Amoco Corporation
01679	Amoco Foundation, Inc.
00751	Amoco Pipeline Company
00114	Amoco Production Company
00635	ARCO Alaska, Inc.
00486	ARCO Pipe Line Company
00002	Atlantic Richfield Company
00967	Atlantic Richfield Company
00222	Aviara Energy Corporation
00368	BOAG Oil and Gas Company
00301	BP Alaska Exploration Inc.
00114	BP America Production Company
02367	BP Amoco Corporation
02367	BP Corporation North America Inc.
01680	BP Exploration & Oil Inc.
02481	BP Exploration & Production Inc.
00113	BP Exploration (Alaska) Inc.
00593	BP Exploration Inc.
00120	BP Exploration U.S.A., Inc.

MMS Number	Business Name
01680	BP Oil Company
00120	BP Oil Corporation
00751	BP Pipelines (North America) Inc.
00751	BP Pipelines (North America) Inc.
00751	BP Pipelines (North America) Inc.
00751	BP Pipelines (North America) Inc.
02350	BP Prod. Corp.
02442	Burlington Resources Inc.
01904	Burlington Resources Offshore Inc.
01904	Burlington Resources Offshore Inc.
02229	Burlington Resources Oil & Gas Company
02229	Burlington Resources Oil & Gas Company LP
00049	Burmah Oil & Gas Company
00368	Burmah Oil and Gas Company
00371	Burmah Oil Development 1974-1 Limited
00164	Burmah Oil Development, Inc.
00213	Burmah Oil Exploration, Inc.
00281	Burmah Oil Offshore Trading, Inc.
00095	Burmah Oil Western Exploration Company
00078	California Oil Company
02335	Chevron Corporation
02335	Chevron Corporation
02626	Chevron Natural Gas Pipe Line Company
00078	Chevron Oil Company
01443	Chevron Oil Company of the Netherlands
01750	Chevron PBC, Inc.
00400	Chevron Pipe Line Company
00078	Chevron U.S.A. Inc.
02544	Chevron U.S.A. LP
02335	ChevronTexaco Corporation
01866	CIECO Energy Ventures LLC
00222	Columbia Gas Development Corporation
00001	Conoco Inc.
00001	Conoco Inc.
01948	Conoco Offshore Inc.
00999	Conoco Pipe Line Company
02952	ConocoPhillips
00056	ConocoPhillips Company
01948	Continental Alaska Pipe Line Company
00001	Continental Oil Company
00999	Continental Pipe Line Company

Companies Categorized as Majors (Continued)

MMS Number	Business Name
02253	Coral Offshore Gathering, LLC
01934	Cross Timbers Oil Company
01240	Cross Timbers Oil Company, L.P.
01745	Cross Timbers Oil Company, L.P.
01240	Cross Timbers Partners
01216	Cross Timbers Production Company
01678	Domain Energy Production Corporation
01866	Domain Energy Ventures Corporation
02253	Enbridge Offshore (Gas Gathering) L.L.C.
02117	Enterprise Oil Gulf of Mexico Inc.
01840	EPEC Offshore Gathering Company
02289	Equilon Pipeline Company LLC
02356	Exxon Asset Holdings LLC
02295	Exxon Asset Management Company
00276	Exxon Corporation
00276	Exxon Mobil Corporation
00276	Exxon Mobil Corporation
00103	Exxon Pipeline Company
01227	Exxon San Joaquin Production Company
00039	ExxonMobil Oil Corporation
00103	ExxonMobil Pipeline Company
00005	Four Star Oil & Gas Company
00005	Four Star Oil and Gas Company
00005	Getty Oil Company
01107	Getty Pipeline, Inc.
00112	Gulf Oil Corporation
00037	Humble Oil & Refining Company
00103	Humble Pipe Line Company
00222	Hunt Petroleum (AEC), Inc.
00921	Jet Oil Company
00174	Kern County Land Company
01904	Meridian Offshore Company
01904	Meridian Oil Offshore Inc.
02128	MG Gas Services Inc.
02254	Mississippi Canyon Gas Pipeline, LLC
02254	Mississippi Canyon Gas Pipeline, LLC
02215	Mobil California Exploration & Producing Asset Company
02221	Mobil Corporation
02203	MOBIL E&P U.S. DEVELOPMENT CORPORATION

MMS Number	Business Name
02203	MOBIL E&P U.S. DEVELOPMENT CORPORATION
02209	Mobil E&P U.S. Development Fund, L.P.
00883	Mobil Eugene Island Pipeline Company
01055	Mobil Exploration and Producing North America Inc.
01933	Mobil Foundation, Inc.
00021	MOBIL NOC INC.
00039	Mobil Oil Corporation
00540	MOBIL OIL EXPLORATION & PRODUCING SOUTHEAST INC.
00565	Mobil Producing Texas & New Mexico Inc.
00565	Mobil-GC Corporation
00637	Mobil-TransOcean Company
02418	Mobile Mineral Corporation
00021	Newmont Oil Company
00114	Pan American Petroleum Corp.
01750	Pennzoil Exploration and Production Company
01750	Pennzoil Petroleum Company
00788	Phillips Oil Company
00056	Phillips Petroleum Company
01866	Range Energy Ventures Corporation
00092	Richfield Oil Corporation
00025	Seaboard Oil Company
00728	Shell California Production Inc.
01940	Shell Consolidated Energy Resources Inc.
02139	Shell Deepwater Development Inc.
02140	Shell Deepwater Production Inc.
00688	Shell Energy Resources Inc.
01728	Shell Frontier Oil & Gas Inc.
02168	Shell Gas Gathering Company
02253	Shell Gas Gathering Company, L.L.C.
02253	Shell Gas Gathering, LLC
01070	Shell Gas Pipeline Company
02254	Shell Gas Pipeline Company, L.L.C.
02621	Shell GOM Pipeline Company LLC
02117	Shell Gulf of Mexico Inc.
02117	Shell Gulf of Mexico Inc.
01967	Shell Land & Energy Company
00689	Shell Offshore Inc.
02128	Shell Offshore Properties and Capital II, Inc.
01839	Shell Oil & Gas Investment Limited Partnership
00117	Shell Oil Company

Companies Categorized as Majors (Continued)

MMS Number	Business Name
01845	Shell Onshore Ventures Inc.
00124	Shell Pipe Line Corporation
02289	Shell Pipeline Company LP
02289	Shell Pipeline Company LP
02147	Shell Seahorse Company
00832	Shell Western E&P Inc.
00049	Signal Oil and Gas Company
00044	Skelly Oil Company
00039	Socony Mobil Oil Co.
00113	Sohio Alaska Petroleum Company
00113	Sohio Natural Resources
00593	Sohio Petroleum Company
00113	Sohio Petroleum Company
00164	Southdown Burmah Oil Company
00113	Standard Alaska Production Company
00393	Standard Oil Company of California
00114	Stanolind Oil and Gas Company
00832	SWEPI LP
01866	Tammany Energy Ventures LLC
01866	Tammany Energy Ventures LLC
01866	Tammany Energy Ventures LLC
01866	Tammany Energy Ventures LLC
02253	Tejas Offshore Gathering, L.L.C.
00079	Tenneco Corporation
00121	Tenneco Exploration Company
00337	Tenneco Exploration II, Ltd.
00275	Tenneco Exploration, Ltd.
01840	Tenneco Gas Gathering Company
01678	Tenneco Gas Production Corporation
00014	Tenneco Inc.
00498	Tenneco OCS Company, Inc.
00500	Tenneco OCS Limited Partnership
00468	Tenneco Offshore Company, Inc.
01840	Tenneco Offshore Gathering Company
00081	Tenneco Oil Company
01866	TENNECO VENTURES CORPORATION
00174	Tenneco West, Inc.
00014	Tennessee Gas Pipeline Company
00014	Tennessee Gas Pipeline Company

MMS Number	Business Name
00080	Tennessee Gas Supply Co.
00014	Tennessee Gas Transmission Company
02142	Texaco California Inc.
00771	Texaco Exploration and Production Inc.
02758	Texaco Harvest LLC
00857	Texaco Oils Inc.
01107	Texaco Pipeline Inc.
00771	Texaco Producing Inc.
00025	Texaco Seaboard Inc.
02020	Texaco Trading and Transportation, Inc.
00002	The Atlantic Refining Company
00043	The Burmah Oil Western Company
02220	The Standard Oil Company
00111	The Standard Oil Company (Ohio)
00040	The Texas Company
00003	Union Oil Company of California
00003	Union Oil Company of California
00003	Union Oil Company of California
01550	Unocal Exploration Corporation
01113	Unocal Pipeline Company
01934	XTO Energy Inc.
00222	XTO Offshore Inc.
02079	Nexen Petroleum Offshore U.S.A. Inc.
00011	Transcontinental Gas Pipe Line Company, LLC
02227	Enterprise Field Services, LLC
01207	Petrobras America Inc.
01796	Manta Ray Gathering Company, L.L.C.
00207	Sea Robin Pipeline Company, LLC
02546	Trunkline Field Services LLC
00030	Trunkline Gas Company, LLC
02516	Agip Oil US L.L.C.
00162	Agip Petroleum Co., Inc.
02248	Agip Petroleum Exploration Co. Inc.
01221	Amax Oil & Gas Inc.
00059	Amerada Hess Corporation
00059	Amerada Petroleum Corporation
00320	American Petrofina Company of Texas
00148	Anadarko E&P Company LP
00981	Anadarko Petroleum Corporation
00170	Anadarko Production Company
02219	Anadarko US Offshore Corporation

Companies Categorized as Majors (Continued)

MMS Number	Business Name
02219	Anadarko US Offshore Corporation
02219	Anadarko US Offshore LLC
02219	Anadarko US Offshore LLC
02534	Apache Clearwater Operations, Inc.
00105	Apache Corporation
03165	Apache Deepwater LLC
01762	Apache Gathering Company
02820	Apache GOM Pipeline, Inc.
02767	Apache Offshore Holdings, LLC
00904	Apache Oil & Gas Transmission, Inc.
00106	Apache Oil Corporation
03296	Apache Shelf Exploration LLC
02851	Apache Shelf, Inc.
02851	Apache Shelf, Inc.
02851	Apache Shelf, Inc.
00128	Aquitaine Oil Corporation
01247	Aran Energy Corporation
00320	ATOFINA Petrochemicals, Inc.
02525	Belco Oil & Gas Corp.
00362	BHP Billiton Petroleum (Americas) Inc.
02277	BHP Billiton Petroleum (Deepwater) Inc.
02010	BHP Billiton Petroleum (GOM) Inc.
02552	BHP Holdings (Resources) Inc.
02245	BHP Minerals International Inc.
00362	BHP Petroleum (Americas) Inc.
02277	BHP Petroleum (Deepwater) Inc.
02010	BHP Petroleum (GOM) Inc.
00768	BHP Petroleum Company Inc.
02426	BHP Resources Inc.
01831	Bogert Oil Company
02361	British-Borneo Deepwater LLC
00075	Cabot Carbon Co.
00046	Cabot Carbon Company
00075	Cabot Corporation
01372	Cabot Exploration Corporation
01355	Cabot Oil & Gas Corporation
01241	Cabot Oil & Gas Production Corporation
00685	Cabot Petroleum Corporation
00148	Champlin Petroleum Company

MMS Number	Business Name
02716	Cimarex Energy Co.
00777	Cities Service Oil and Gas Corporation
00362	Clinton Oil Company
00282	CNG Producing Company
00886	CNG Transmission Corporation
00159	Coastal States Gas Producing Co.
00886	Consolidated Gas Transmission Corporation
00748	DelMar Operating, Inc.
00769	DelMar Operating, Inc. (P. M.)
01525	DelMar/JHM 1989 Exploration Limited Partnership
00919	DelMar/MM 1982 Exploration Limited Partnership
01016	DelMar/MM 1985 Exploration Limited Partnership
01526	DelMar/MM 1989 Exploration Limited Partnership
01527	DelMar/NEW 1989 Exploration Limited Partnership
01528	DelMar/NFC 1989 Exploration Limited Partnership
00918	DelMar/PM 1984 Exploration Limited Partnership
00441	Devon Corporation
02410	Devon Energy Corporation
02410	Devon Energy Corporation
01010	Devon Energy Corporation
01010	Devon Energy Corporation (Nevada)
02638	Devon Energy Operating Company, L.P.
01901	Devon Energy Operating Corporation
01853	Devon Energy Petroleum Pipeline Company
02421	Devon Energy Production Company, L.P.
02421	Devon Energy Production Company, L.P.
01777	Devon Louisiana Corporation
01551	Devon SFS Operating, Inc.
00282	Dominion Exploration & Production, Inc.
02576	Dominion Oklahoma Texas Exploration & Production, Inc.
02576	Dominion Oklahoma Texas Exploration & Production, Inc.
02044	Dominion Reserves Gulf Coast, Inc.
02023	Dominion Reserves, Inc.
00886	Dominion Transmission, Inc.
15049	Dominion Wind Development, LLC
01015	Elf Aquitaine Exploration, Inc. (M. M.)
00769	Elf Aquitaine Exploration, Inc. (P. M.)
01141	Elf Aquitaine Oil Programs, Inc.
00748	Elf Aquitaine Operating, Inc.
00128	Elf Aquitaine, Inc.
01525	Elf Aquitaine/JHM 1989 Exploration Limited Partnership

Companies Categorized as Majors (Continued)

MMS Number	Business Name
00919	Elf Aquitaine/MM 1982 Exploration Limited Partnership
01016	Elf Aquitaine/MM 1985 Exploration Limited Partnership
01526	Elf Aquitaine/MM 1989 Exploration Limited Partnership
01527	Elf Aquitaine/NEW 1989 Exploration Limited Partnership
01528	Elf Aquitaine/NFC 1989 Exploration Limited Partnership
00918	Elf Aquitaine/PM 1984 Exploration Limited Partnership
01500	Elf Exploration, Inc.
00898	Energen Resources MAQ, Inc.
00362	Energy Reserves Group, Inc.
02361	Eni Deepwater LLC
02516	Eni Oil US LLC
00162	Eni Petroleum Co. Inc.
00162	Eni Petroleum Co. Inc.
02248	Eni Petroleum Exploration Co. Inc.
02361	Eni Petroleum US LLC
02920	Eni Petroleum US LLC
02361	Eni Petroleum US LLC
02782	Eni US Operating Co. Inc.
02782	Eni US Operating Co. Inc.
01103	Enron Oil & Gas Company
02414	EOG Resources Omega LLC
01103	EOG Resources, Inc.
02401	Equitable Production (Gulf) Company
02483	Fina E&P, Inc.
00964	Fina Exploration, Inc.
00910	Fina Oil & Gas, Inc.
00320	Fina Oil and Chemical Company
00481	Finadel, Incorporated
01777	Flores & Rucks, Inc.
02077	Flores & Rucks, Inc.
02851	Forest Energy Resources, Inc.
01688	Freeport Interstate Pipeline Company
00012	Freeport Minerals Company
00428	Freeport Oil Company
00469	Freeport Oil Company
00428	Freeport Petroleum Company
00374	Freeport Pipeline Company
00012	Freeport Sulphur Company
02225	Freeport Sulphur Company

MMS Number	Business Name
01583	Freeport-McMoRan Acquisition Company
03303	Freeport-McMoRan Copper & Gold Inc.
02313	Freeport-McMoRan Energy LLC
03263	Freeport-McMoRan Exploration & Production LLC
03263	Freeport-McMoRan Exploration & Production LLC
01597	Freeport-McMoRan Inc.
01531	Freeport-McMoRan Oil & Gas Company
01583	Freeport-McMoRan Oil & Gas Company
03280	Freeport-McMoRan Oil & Gas LLC
03280	Freeport-McMoRan Oil & Gas LLC
03280	Freeport-McMoRan Oil & Gas LLC
01082	Freeport-McMoRan Resource Partners, Limited Partnership
02225	Freeport-McMoRan Sulphur Inc.
02313	Freeport-McMoRan Sulphur LLC
02820	FW GOM Pipeline, Inc.
02820	FW GOM Pipeline, Inc.
02250	GEL Offshore Pipeline, LLC
02250	GEL Offshore Pipeline, LLC
01570	Greenhill Petroleum Corporation
00819	Halliburton Company
00819	Halliburton Energy Services, Inc.
00818	Hardy Oil & Gas USA Inc.
03356	Hess Conger LLC
00059	Hess Corporation
00059	Hess Corporation
03359	Hess GOM Deepwater LLC
03360	Hess GOM Exploration LLC
03366	Hess Llano LLC
03365	Hess Shenzi LLC
03355	HESS STAMPEDE LLC
03358	HESS TUBULAR BELLS LLC
00919	Huffco 1982 Exploration Limited Partnership
00918	Huffco 1984 Exploration Limited Partnership
01015	Huffco Gas and Oil, Inc.
01141	Huffco Oil Programs, Inc.
00748	Huffco Petroleum Corporation
01016	Huffco/MM 1985 Exploration Limited Partnership
00769	Huffington Exploration Corporation
02169	Hydro Gulf of Mexico, L.L.C.
03280	IMONC LLC
03160	Kerr-McGee (Nevada) LLC

Companies Categorized as Majors (Continued)

MMS Number	Business Name
00035	Kerr-McGee Corporation
00707	Kerr-McGee Federal Limited Partnership I-1981
02859	Kerr-McGee Oil & Gas (Shelf) LLC
02219	Kerr-McGee Oil & Gas Corporation
00035	Kerr-McGee Oil Industries, Inc.
00647	KERR-McGEE PIPELINE CORP.
01831	Louis Dreyfus Natural Gas Corp.
01674	Louis Dreyfus Reserves Corp.
02249	Marathon Ashland Pipe Line LLC
02086	Marathon Energy Corporation
02250	Marathon Offshore Pipeline LLC
00724	Marathon Oil Company
00115	Marathon Oil Company
03123	Marathon Oil Corporation
00115	Marathon Petroleum Company
00310	Marathon Pipe Line Company
02250	Marathon Pipe Line LLC
02249	Marathon Pipe Line LLC
02851	Mariner Energy Resources, Inc.
02851	Mariner Energy Resources, Inc.
00818	Mariner Energy, Inc.
02169	Mariner Gulf of Mexico LLC
00270	McMoRan Exploration Co.
00270	McMoRan EXPLORATION CO.
02320	McMoRan Exploration Co.
00212	McMoRan Exploration Company
02320	McMoRan Exploration LLC
00648	McMoRan Offshore Exploration Co.
00477	McMoRan OFFSHORE EXPLORATION CO.
00428	McMoRan Offshore Exploration Co.
03178	McMoRan Offshore LLC
03179	McMoRan Offshore LLC
03178	McMoRan Offshore LLC
00477	McMoRan Offshore Production Co.
01888	McMoRan Oil & Gas Co.
00961	McMoRan Oil & Gas Co.
00428	McMoRan Oil & Gas Co.
02312	McMoRan Oil & Gas LLC
00634	McMoRan Pipeline Company

MMS Number	Business Name
00469	McMoRan-Freeport Oil Company
00428	McMoRan-Freeport Petroleum Company
02055	MESA Inc.
01025	Mesa Limited Partnership
00257	Mesa Offshore Co.
00753	Mesa Offshore Management Co.
01935	Mesa Operating Co.
01026	Mesa Operating Limited Partnership
00233	Mesa Petroleum Co.
01935	Mesa Sub 1, Inc.
02062	MidPar L.P.
00898	Minatome Corporation
00768	Monsanto Oil Company
02237	Noble Affiliates, Inc.
02425	Noble Drilling Exploration Company
02237	Noble Energy, Inc.
02237	Noble Energy, Inc.
02528	Norsk Hydro USA Oil & Gas, Inc.
15013	Occidental Development & Equities, LLC
01061	Occidental of Alaska, Inc.
01045	Occidental of California, Inc.
00157	Occidental Petroleum Corporation
01777	Ocean Energy, Inc.
02575	Ocean Energy, Inc.
02077	Ocean Energy, Inc.
02242	Ocean Energy, Inc.
02859	Offshore Shelf LLC
00325	Oil Development Company of Texas
02848	OXY Deepwater USA, Inc.
00346	Oxy Petroleum, Inc.
00777	OXY USA Inc.
01912	Parker & Parsley Acquisition Company
02062	Parker & Parsley Development L.P.
01375	Parker & Parsley Petroleum Company
02001	Parker & Parsley Producing L.P.
00167	PennzEnergy Company
02332	PennzEnergy Exploration and Production, L.L.C.
00167	Pennzoil Company
01853	Pennzoil Petroleum Pipeline Company
00167	Pennzoil United, Inc.
00481	Petrofina Delaware, Incorporated

Companies Categorized as Majors (Continued)

MMS Number	Business Name
00907	Petrofina Exploration, Inc.
01247	Petrolex USA, Inc.
02279	PG&E Texas Pipeline, L.P.
01570	Pioneer Natural Resources (GPC) Inc.
02223	Pioneer Natural Resources Company
01935	Pioneer Natural Resources USA, Inc.
02681	Pioneer Natural Resources Alaska, Inc.
02847	Pioneer Shelf Properties Incorporated
02702	Plains Exploration & Production Company
03199	Plains Offshore Inc.
03200	Plains Offshore Operations Inc.
01671	Plains Petroleum Operating Company
02885	Plains Pipeline, L.P.
01220	Plains Resources Inc.
03017	Pogo Producing Company LLC
02882	PXP Deepwater L.L.C.
02701	PXP Gulf Coast Inc.
03263	PXP Offshore LLC
03178	PXP Offshore LLC
03017	PXP Producing Company LLC
03192	PXP Resources Inc.
03192	PXP Resources LLC
02805	Repsol E&P USA Inc.
02805	Repsol E&P USA Inc.
02936	Repsol Offshore E&P USA Inc.
02936	Repsol Offshore E&P USA Inc.
00148	RME Petroleum Company
00325	Santa Fe Energy Company
01035	Santa Fe Energy Operating Partners, L.P.
01551	Santa Fe Energy Resources, Inc.
00977	Santa Fe International Company
00977	Santa Fe International Corporation
00784	Santa Fe International Corporation
00407	Santa Fe Minerals Co.- U.S.
00877	Santa Fe Minerals, Inc.
00407	Santa Fe Minerals, Inc.
01815	Santa Fe Minerals, Inc.
01551	Santa Fe Snyder Corporation
00899	Santa Fe-Andover Oil Company

MMS Number	Business Name
00672	Seagull Energy Corporation
02242	Seagull Energy Corporation
00672	Seagull Energy E&P Inc.
01956	Seagull Energy E&P Inc.
00932	Seagull Interstate Corporation
01712	Seagull Natural Gas Company
01790	Snyder Oil Corporation
01117	Snyder Oil Partners L.P.
01118	Snyder Operating Partnership L.P.
00748	SOCO Offshore, Inc.
02169	Spinnaker Exploration Company, L.L.C.
01247	Statoil Exploration (US) Inc.
02114	Statoil Exploration (US) Inc.
02748	Statoil Gulf of Mexico LLC
02748	Statoil Gulf of Mexico LLC
03019	Statoil Gulf Properties Inc.
15033	Statoil North America, Inc.
02528	Statoil USA E&P Inc.
03019	StatoilHydro Gulf Properties Inc.
02528	StatoilHydro USA E&P, Inc.
03285	Talisman GOM L.P.
01606	TEPCO Offshore, Inc.
00235	Texas Eastern Exploration Co.
00876	Texas Eastern Hydrocarbon Company
00875	Texas Eastern Petroleum Company
00176	Texas Eastern Transmission Corporation
00176	Texas Eastern Transmission, LP
00115	The Ohio Oil Company
01241	Thermal Exploration, Inc.
00313	Total American, Inc.
01500	TOTAL E&P USA, INC.
02280	TOTAL Exploration Production USA, Inc.
02754	TOTAL Holdings USA, Inc.
00312	Total Leonard, Inc.
00898	TOTAL MINATOME CORPORATION
00898	TOTAL Minatome Corporation
00312	Total Petroleum, Inc.
01500	TotalFinaElf E&P USA, Inc.
00818	Trafalgar House Oil and Gas Inc.
01221	Union Pacific Oil and Gas Company
00148	Union Pacific Resources Company

Companies Categorized as Majors (Continued)

MMS Number	Business Name
02239	Union Pacific Resources Group Inc.
00724	USS Holdings Company
00159	Valero Energy Corporation
00710	Valero Producing Company
02279	Valero Transmission, L.P.
01241	Washington Energy Exploration, Inc.
02148	Westport Oil and Gas Company, Inc.
02148	Westport Oil and Gas Company, L.P.
02401	Westport Resources Corporation
02525	Westport Resources Corporation
03165	ZMZ Acquisitions LLC

Appendix D– Terms and Conditions

The general terms and conditions pertaining to the value conclusion(s) stated in this Report are summarized below. If applicable, "special assumptions" are cited elsewhere in this Report.

1) To the best of our knowledge and belief, the statements of facts contained in this Report, upon which the analysis and conclusion(s) expressed are based, are true and correct. Information, estimates and opinions furnished to us and contained in the Report or utilized in the formation of the value conclusion(s) were obtained from sources considered reliable and believed to be true and correct. However, no representation, liability or warranty for the accuracy of such items is assumed by or imposed on us, and is subject to corrections, errors, omissions and withdrawal without notice.
2) The legal description of the appraised business, if exhibited in the Report, is assumed correct.
3) This valuation may not be used in conjunction with any other appraisal or study. The value conclusion(s) stated in this appraisal is based on the program of utilization described in the Report, and may not be separated into parts. The appraisal was prepared solely for the purpose, function and party so identified in the Report. Unless specifically stated, the appraisal Report may not be reproduced, in

whole or in part, and the findings of the Report may not be utilized by a third party for any purpose, without the express written consent of Opportune LLP.

4) No change of any item in any of the appraisal Report shall be made by anyone other than Opportune and we shall have no responsibility for any such unauthorized change.
5) We are not required to give testimony or be in attendance at any court or administrative proceeding with reference to the property appraised unless additional compensation is agreed to and priorarrangements have been made.
6) The working papers for this engagement are being retained in our files and are available for your reference. We would be available to support our valuation conclusion(s) should this be required. Those services would be performed for an additional fee.
7) Neither all nor any part of the contents of this Report shall be disseminated or referred to the public through advertising, public relations, news or sales media, or any other public means of communication or referenced in any publication, including any private or public offerings including but not limited to those filed with Securities and Exchange Commission or other governmental agency, without the prior written consent and approval of and review by Opportune.
8) Good and marketable title to the Company is assumed. We are not qualified to render an "opinion of title," and no responsibility is assumed or accepted for matters of a legal nature affecting the Company. No formal investigation of legal title was made, and we render no opinion as to ownership of Subsidiaries or condition of their title.
9) Management is assumed to be competent, and the ownership to be in responsible hands. The quality of Management can have a direct effect on a business's economic viability and value. The financial projections contained in the appraisal assume both responsible ownership and competent Management. Any variance from this assumption could have a significant impact on the final value estimate.
10) We take no responsibility for any events, conditions or circumstances affecting the subject asset(s) or its value, that take place subsequent to either the effective date of value cited in the appraisal or the date of our field inspection, whichever occurs first.

11) This valuation is based on historical and prospective financial statements. Some assumptions or projections inevitably will not materialize and unanticipated events and circumstances may occur during the forecast period. These could include major changes in the economic environs;

Appendix E – Valuation Certification

We certify that, to the best of our knowledge and belief:

- The statements of fact contained in this Report are true and correct.
- We have no present or prospective interest in the business or property that is the subject of this Report, and we have no personal interest or bias with respect to the parties involved.
- Our compensation is not contingent upon the reporting of a predetermined value or direction in value that favors the cause of the client, the amount of the value estimate, the attainment of a stipulated result, or the occurrence of a subsequent event.
- The engagement was not based on a requested minimum valuation, a specific valuation, or the approval of a loan.
- The analyses and conclusions are limited only by the reported assumptions and limiting conditions and represents our unbiased professional analyses and conclusions.
- This analysis and Report were prepared under the direction of Josh Sherman, with significant professional assistance provided by Steve Hendrickson, John Beaird, Darren Busch, Paul Legoudes, and Virginia Chan.

Appendix F– Sources of Information

During our valuation analysis, we relied upon financial and other information from Management and various public, financial, and industry sources. Our conclusion is dependent on such information being complete and accurate in all material respects. The principal sources of information utilized in performing our analysis include:

- S&P Capital IQ financial database
- Kroll Cost of Capital Navigator
- FASB ASC 805: Business Combinations
- FASB ASC 820: Fair Value Measurements andDisclosures
- IBISWorld Industry Report: Oil Drilling & Gas Extraction in the U.S., February 2022
- June 2022 Livingston Survey
- ProPublica: History of U.S. Gov't Bailouts, Updated April 15, 2009
- Bureau of Safety and Environmental Enforcement (BSEE) Formal solicitation for Temporary Abandonment Decommissioning Services - https://www.bsee.gov/formal-solicitation-issued-for- temporary-abandonment-decommissioning-services-virtual-pre
- Analysis of other facts and data resulting in our conclusions of value
- Historical production data retrieved from Enverus
- Rims II multipliers from the Department of Commerce

U.S. SMALL BUSINESS ADMINISTRATION
OFFICE OF ADVOCACY
REGULATION • RESEARCH • OUTREACH

August 28, 2023

VIA ELECTRONIC SUBMISSION

Liz Klein, Director
Bureau of Ocean Energy Management
Department of the Interior

Re: Risk Management and Financial Assurance for OCS Lease and Grant Obligations, RIN 1010–AE14.

Dear Director Klein:

On June 29, 2023, the Department of Interior's Bureau of Ocean Energy Management (BOEM) published a proposed rule entitled Risk Management and Financial Assurance for Outer Continental Shelf (OCS) Lease and Grant Obligations.[1] This letter constitutes the Office of Advocacy's (Advocacy) public comments on the proposed rule.

This proposed rule harms small businesses holding current leases for oil and gas extraction in the OCS by ignoring the joint and several liability of large predecessor leaseholders in Department of the Interior (DOI) regulations and written into the leases. It requires small businesses that have already paid for assurance bonds agreed as part of the sale of the lease to purchase duplicative assurance bonds for the federal government. BOEM should narrowly tailor this rule to cover only those liabilities for which there are no predecessor leaseholders that BOEM considers credit worthy.

I. Background

A. The Office of Advocacy

Congress established the Office of Advocacy under Pub. L. 94-305 to represent the views of small entities before Federal agencies and Congress. Advocacy is an independent office within the U.S. Small Business Administration (SBA). As such, the views expressed by Advocacy do not necessarily reflect the views of the SBA or the Administration. The Regulatory Flexibility Act (RFA),[2] as amended by the Small Business Regulatory Enforcement Fairness Act

---

[1] 88 Fed. Reg. 42136 (June 29, 2023).
[2] 5 U.S.C. §601 et seq.

409 3rd Street SW / MC 3110 / Washington, DC 20416
Ph 202-205-6533 / advocacy.sba.gov

SBA U.S. Small Business Administration

(SBREFA),[3] gives small entities a voice in the rulemaking process. For all rules that are expected to have a significant economic impact on a substantial number of small entities, the RFA requires federal agencies to assess the impact of the proposed rule on small entities and to consider less burdensome alternatives.

The Small Business Jobs Act of 2010 requires agencies to give every appropriate consideration to comments provided by Advocacy.[4] The agency must include a response to these written comments in any explanation or discussion accompanying the final rule's publication in the *Federal Register*, unless the agency certifies that the public interest is not served by doing so.[5]

Advocacy's comments are consistent with Congressional intent underlying the RFA, that "[w]hen adopting regulations to protect the health, safety, and economic welfare of the nation, federal agencies should seek to achieve statutory goals as effectively and efficiently as possible without imposing unnecessary burdens on the public."[6]

B. The Proposed Rule

On June 29, 2023, BOEM published a proposed rule to modify its criteria for determining when oil and gas lessees in the Outer Continental Shelf (OCS) would be required to provide bonds or other financial assurances to ensure their ability to comply with lease conditions, including decommissioning. BOEM issued this proposal in response to concerns about bankruptcies by leaseholders exposing taxpayers to significant unfunded liabilities for the decommissioning of OCS operations. Citing several recent bankruptcies, BOEM estimates the total offshore decommissioning liability is approximately $7.5 billion. However, BOEM also states that the actual financial risk to the United States is significantly less than that figure because multiple other private parties share responsibly for decommissioning costs.[7]

BOEM proposes that leaseholders provide additional financial assurance bonds to the federal government. BOEM estimates that there is a total offshore decommissioning liability of over $42 billion and believes that this rule would help protect taxpayers from this liability. As part of the effort to reduce the impact of this rule, BOEM proposes that companies with sufficiently high credit ratings or sufficiently high reserve ratios would be exempt from this requirement. BOEM recognizes that this exemption would apply to few small businesses. Excluding these large businesses, BOEM estimates that there will be an additional $9.6 billion in new financial assurance provided, with an annual premium cost of $379 million. These costs would be borne mostly by small businesses. Importantly, BOEM proposes that for these purposes it would not recognize the joint and several liability borne predecessor leaseholders and would require financial assurances from current leaseholders only.

---

[3] Pub. L. 104-121, Title II, 110 Stat. 857 (1996) (codified in various sections of 5 U.S.C. §601 et seq.).
[4] Small Business Jobs Act of 2010 (PL. 111-240) §1601.
[5] *Id.*
[6] *Id.*
[7] 88 Fed. Reg. at 42,139. Although there is no official figure representing the actual liability, Advocacy understands that BOEM provided a preliminary estimate to Congress that the unfunded taxpayer liability was less than one percent of this value.

BOEM prepared an Initial Regulatory Flexibility Analysis for this proposed rule and discussed a less stringent regulatory alternative that would reduce the burden on small businesses. This alternative would recognize the credit worthiness of predecessor leaseholders in determining when additional financial assurance would be required. The agency rejected that option, stating "consideration of predecessor lessees and grantees encourages moral hazard by incentivizing current lessees to pass risk to predecessors rather than proactively prepare for decommissioning and related obligations."[8]

II. Advocacy's Small Business Concerns

BOEM's proposed rule distorts the marketplace by ignoring a key element of the leases, joint and several liability. All participants in the market know and understand this liability in their business dealings. The proposed rule would require small businesses to pay twice to protect against decommissioning liabilities, once through the sales contract and again to the federal government. This proposed rule explicitly favors large businesses over small businesses without providing the taxpayer the intended protection from unfunded liabilities.

A. Joint and several liability protects taxpayers by ensuring that predecessors cannot shed liability and will instead engage in due diligence.

Joint and several liability is a fundamental characteristic of OCS leases. It has been a part of leases and in DOI regulations for decades. As a result, all market participants, past and current, have incorporated joint and several liability into their business decision, including the contractual sale to subsequent leaseholders. Decommission liabilities and the risk of default by subsequent leaseholders are well understood.

In response to this risk, leaseholders are rational market participants and seek their own assurances. These include not just financial bonding, of which there are billions of dollars outstanding. Sellers perform due diligence and establish long-term relationships to protect their financial interests. Leases are not sold like commodities, between faceless actors by price-takers. Each lease sale is negotiated, and joint and several liability is an ever-present factor. Taxpayers are protected by leaseholders understanding their liabilities and making sound business decisions to protect themselves.

However, BOEM treats these sellers as if they are naïve and lack experience in OCS leasing. In rejecting the regulatory alternative that would recognize predecessor leaseholders, BOEM cites moral hazard. Moral hazard is a situation in which a party carrying some kind of insurance, like a financial assurance bond, lacks incentives to avoid risks because they are insulated from the consequences.

Moral hazard is a symptom of asymmetric information because the insured party has more information about themselves and their future intentions than the insurer. This proposal does not solve a moral hazard problem, nor would the alternative create a new one. First, these markets are dominated by sophisticated sellers who do their due diligence and carefully negotiate the terms of sale and assurance. Second, this industry has repeated interactions and reputational

[8] *Id.* at 42,159.

consequences. Businesses and their agents that take unreasonable risks face consequences in higher premiums and loss of business. While bankruptcies can occur, this is not evidence of moral hazard nor of a failing of the existing allocation of liabilities through contract. Moral hazard is not a reasonable concern here. It does not justify a wholesale discarding of the market arrangements to allocate risk and liability that the free market itself has developed in response to the joint and several liability of the lease.

To the contrary, by creating a system that requires bonding for only for current leaseholders, BOEM is insulating predecessor leaseholders from joint and several liability and relieving the sellers of the need to perform due diligence on the subsequent leaseholder. This jeopardizes taxpayers and the environment by making future abandonments and bankruptcy more likely.

B. The proposed rule protects large companies at the expense of small current leaseholders.

The negotiated free market sale of leases, in the light of joint and several liability, includes the distribution liability through private assurance bonds. Small businesses have already paid for billions in private bonds to protect prior leaseholders. All parties are fully aware of the lease conditions and the risks to be shared.

In this proposal, BOEM ignore the liability of predecessor leaseholders in setting the bonding requirement, and BOEM recognizes that it is doing so by rejecting less stringent regulatory alternatives. As a result, this rule interferes with the market's allocation of risk. It shifts liability away from those best able to handle it, i.e., large businesses that were the initial leaseholders, and imposes it directly on small businesses with weaker credit ratings who are less able to afford it. Even worse, these small businesses had already paid assurance bonds for the benefit of the predecessor leaseholders, so these small businesses are paying to cover decommission liability twice.

BOEM asserts that this imposition of additional obligations on these small leaseholders is necessary to protect the taxpayer from their bankruptcies. However, it is unclear why BOEM believes that the taxpayer needs protection for the full value of these liabilities given its recognition that the actual financial risk is significantly less.

C. BOEM should narrowly tailor its rule to the unfunded decommission liability of actual concern.

BOEM has proposed that its requirements for additional bonding would not apply to companies with sufficiently strong credit rating. These requirements imply that BOEM is only worried about unfunded decommissioning liabilities held by small companies that do not have a sufficiently strong credit rating. But BOEM's analysis intentionally ignores joint and several liability as the way that the taxpayers are protected from these unfunded liabilities. It presumes that these small businesses impose a significant risk to the taxpayer despite the full backing of companies that BOEM has exempted. As a result, small businesses are not only disproportionately harmed by the proposal, but only small businesses are harmed by the proposal.

In its effort to protect the taxpayer, the agency should not ignore the role of the market in allocated risk. BOEM should instead focus its efforts on those risks for which there is no predecessor leaseholder that it trusts to have the resources necessary to make the taxpayer whole.

For this reason, the "first best" economic solution would be to require bonding only by the first leaseholder and allow rational economic actors to allocate risk through contracts. Recognizing that BOEM cannot often reach back to the first lessee, BOEM should include a waiver for leases with predecessor leaseholders who meet the proposed creditworthiness standard.

III. Conclusion

This proposed rule imposes an unreasonable burden for financial assurances on small businesses. Most of the burden is unnecessary because it ignores a core element underlying the commercial arrangements among leaseholders since the inception of OCS leasing, joint and several liability. BOEM should include a waiver from this requirement for leases with credit worthy predecessor leaseholders. Small businesses should not be required to stand alone and assume the full responsibility for making the taxpayer whole, especially where large businesses are equally liable under DOI regulations and the terms of the lease.

If you have any questions or require additional information, please contact me or Assistant Chief Counsel Dave Rostker at (202) 205-6966 or by email at david.rostker@sba.gov.

Sincerely,

/s/

Major L. Clark, III
Deputy Chief Counsel
Office of Advocacy
U.S. Small Business Administration

/s/

Dave Rostker
Assistant Chief Counsel
Office of Advocacy
U.S. Small Business Administration

Copy to: The Honorable Richard L. Revesz, Administrator
Office of Information and Regulatory Affairs
Office of Management and Budget

Chapter 2

The Biden Administration's Executive Overreach and Its Impact on American Energy Independence*

Committee on Natural Resources

Thursday, May 11, 2023
U.S. House of Representatives
Subcommittee on Oversight and Investigations
Committee on Natural Resources
Washington, DC

The Subcommittee met, pursuant to notice, at 2:16 p.m. in Room 1334, Longworth House Office Building, Hon. Paul Gosar [Chairwoman of the Subcommittee] presiding.

Present: Representatives Gosar, Rosendale, Hunt, Luna, Westerman; Stansbury, and Lee.

Also present: Representative Huffman.

Dr. GOSAR. Without objection, the Chair is authorized to declare a recess of the Subcommittee at any time.

We are now convening the Subcommittee on Oversight and Investigations, and it will come to order.

The Subcommittee is meeting today to hear testimony on the Biden administration's executive overreach and its impact on American energy independence.

---

* This is an edited, reformatted and augmented version of Oversight Hearing Before the Subcommittee on Oversight and Investigations of the Committee on Natural Resources, U.S. House of Representatives, Serial No. 118–26, dated May 11, 2023.

In: The Struggle for Energy Independence in America
Editor: Victoria Whitehead
ISBN: 979-8-89530-558-4

I ask unanimous consent that all Members testifying today be allowed to sit with the Subcommittee, giving their testimony, and participate in the hearing from the dais.

I ask that the gentleman, Mr. Huffman, be allowed to sit with the Subcommittee and participate in the hearing.

No objection, so ordered.

Without objection, the Chair is authorized to declare a recess of the Subcommittee at any time.

Under Committee Rule 4(f), any oral opening statements at the hearings are limited to the Chairman and the Ranking Minority Member. I therefore ask unanimous consent that all other Members' opening statements be made part of the hearing record if they are submitted in accordance with Committee Rule 3(o).

Without objection, so ordered.

Statement of the Hon. Paul Gosar, a Representative in Congress from the State of Arizona

Dr. GOSAR. I want to thank everyone for joining us today for this critically important hearing, especially the witnesses who have taken time out of their day and their jobs just to be here.

Speaking of showing up for your job today, I would like to point out that we are missing a government witness today, CEQ Chairman Brenda Mallory, who declined to attend today's hearing, citing official business travel, appears to be working from Hawaii on a taxpayer-funded junket. While I agree that soaking up the sun and making grant announcements in Hawaii sounds nicer than answering to Congress, sometimes you just need to face the music. Today, we are here to discuss the themes of the executive branch, specifically with the current Administration, and what can happen when that branch of government expands its power beyond the scope of its original intent.

America's founders were gravely aware of the dangers of concentrated power, and intentionally designed a system of separating government powers into the legislative, executive, and judicial. In the 20th century, as our nation grew, the size, scope, and the intrusion of the government activity increased exponentially. Today, the size and the breadth of the Federal Government would be unrecognizable to our nation's founders. The growth of the executive branch in both size and power has far outpaced the growth of other branches, and has resulted in more concentrated power and less accountability to the public.

The Biden administration has continued on a relentless march towards expansion, particularly by trying to reimagine the energy sector as a puppet of environmental activism. Instead of attempting to work with Congress and pass legislation, President Biden has ruled by fiat, abusing executive orders to radically restrict America's energy independence. These actions include canceling the construction of an energy pipeline and halting leasing on large swaths of Federal lands for oil and gas production on his first day in office; issuing an executive order that utilized the executive branch as forces for social and cultural movements for the left, rather than good governance; and issuing executive orders that inject principles of environmental justice into nearly all Federal actions.

Time and time again, we have seen President Biden abuse executive authority to promote his radical eco-agenda created by academics and social activists, regardless of the reality on the ground for the American public. The results of President Biden's radical eco-agenda have been catastrophic. We have seen the highest inflation in 40 years, crushing energy prices for American consumers and increased reliance on our foreign adversaries, notably China.

I am certain that my friends on the other side of the aisle will extoll the virtues of these executive orders today, in particular as it relates to climate change and environmental justice. But I ask you: at what cost? The left's race to achieve a clean energy standard is simply unsustainable, at least on the timeline that the Biden administration has set forth. Experts say a quantum leap in science and technology would be necessary to replace carbon-based energy sources on the timeline proposed by the Biden administration. I again ask you: at what cost?

In the United States, we utilize the best available practices, standards, and science in terms of extraction, manufacturing, and labor. The same cannot be said for our counterparts in China, Russia, Venezuela, and the African nations. We have lower greenhouse emissions, stringent labor laws, yet continually seek to move the goal posts, which results in nothing more than outsourcing our resources needs to countries that routinely violate human rights laws. The last time I checked, benevolent dictators don't really care about permits or anything closely resembling NEPA.

And what happens at home, the end result of all this nonsense? In my district, I can tell you. Some of the most vulnerable, those who identify as environmental justice communities, are still not being heard by this

Administration. They are paying more for the few services they do have, and losing out on opportunities for new income streams, like fully utilizing oil and gas allotments on their homelands.

Today, we take the first step at exposing the radical executive overreach of President Biden as we work to reverse course and inject a dose of restraint and reason into America's energy policies. And perhaps most important of all, restore the balance of powers between the executive and legislative branches.

With that, I turn and recognize Ranking Member Stansbury, my neighbor from the Southwest and neighboring New Mexico, for any statement she may have.

Statement of the Hon. Melanie A. Stansbury, a Representative in Congress from the State of New Mexico

Ms. STANSBURY. Good afternoon. Thank you, Mr. Chairman, and thank you to all of our witnesses for being here today.

I do want to just take a moment before I make a few statements about the content of today's hearing just to note that, while we are aware that the Administration is unable to join us today, we also want to be clear that the reason why they are unable to join us is there was not sufficient notice given in order for them to be here. So, we look forward to working with the Majority in order to make sure that they can attend a future hearing.

We are going to hear a lot today about domestic energy production. We are going to hear a lot today about permitting, about climate change, and about environmental justice. Why are these topics so important, and why do we continue to have hearings about these?

First and foremost, as we all know, the science is telling us and the latest report from the IPCC has indicated that if we do not take climate action now, our planet is facing a major nexus in terms of potential planetary disaster. And that is why we in the Majority, the Democrats during the last Congress, passed the most significant infrastructure and climate action bills ever in the history of the United States, including the Inflation Reduction Act, which is really aimed at reducing climate emissions by 40 percent by 2030.

But let's talk a little bit about domestic energy production. It is important to understand that over the last 20 years, the United States has actually become the second largest, not only consumer of energy in the world, but the second largest producer of energy in the world. In fact, we have the second highest crude oil production in America ever, and the highest natural gas production ever. And in my home state of New Mexico, in the Permian Basin, we are seeing the highest production ever in the history of our state. So, the oil and gas industry is not struggling in the United States, and certainly that is not what is driving prices at the pump, as our colleagues across the aisle would like to have us believe. We are, in fact, coming out of a major global pandemic in which our communities, like probably everyone who is sitting here in this hearing room, spent at least a year, if not more, inside our homes. We were not driving, we were not consuming petroleum products on the level we had historically. And as a result, oil and gas production actually reduced, due to the lack of demand. And also, that affected the price of oil and gas. We then saw, of course, last year Russia invaded Ukraine. That caused another spike in global oil and gas prices. And then there was an effort by OPEC+ and Russia to restrain oil production globally. All of these global forces have affected the global supply and demand of oil and gas, which drives the prices at the pump.

Our President, this Administration, and the Congress that is sitting before you have done everything we can within our toolbox to try to address these issues so that we can lower prices at the pump for Americans. But at the end of the day, much of this is outside of the power of the individuals who are sitting on this dais or are sitting at the other end of Pennsylvania Avenue.

I think it is crucial as part of this conversation to acknowledge we are going to hear a lot today, probably, about permitting and leasing, that because of the limitations on our ability to affect global oil and gas prices, the argument that, ultimately, at the end of the day, what is constraining prices at the pump is leasing on public lands is just factually untrue. Really, at the end of the day, what it is about and what is driving that argument is the desire by the oil and gas industry to make future oil and gas leases in order to keep their business plans afloat, especially as there has been consolidation in the industry and over-capitalization. It is not a permitting problem. It is not a leasing problem. So, it is crucial that we stay focused on the facts.

Now, the reason why the Administration has advanced multiple efforts in this Congress, this past Congress under Democratic leadership has advanced environmental justice initiatives is because our communities for far too long have borne the brunt of development of our oil and gas and other natural resources without seeing the benefits of those. So, it is crucial that we ensure, as we are investing in a more sustainable transition, a more climate-just future that our communities have the opportunity to get access to those resources, that they are able to plan for their own future, and to determine what they would like to see as their own economic future for their communities.

I want to thank our witnesses once again for being here. We are fortunate to be joined by one of our nation's foremost experts on NEPA, which is one of the environmental policy bills that we are going to discuss today.

And I just want to end by saying that we know that we have to take climate action now, and it is our job to ensure that we create a more sustainable, a more just, and a more equitable future for our communities.

Thank you, Mr. Chairman, and I yield back.

Dr. GOSAR. I thank the gentlelady from New Mexico. I have to correct the lady from New Mexico. We did give 2 weeks' notice as required. They just said that she had travel plans. So, they are not entitled to their facts; those are the facts. She was given ample time.

I now want to turn to the Full Committee Chair, Mr. Westerman, for his opening statement.

Statement of the Hon. Bruce Westerman, a Representative in Congress from the State of Arkansas

Mr. WESTERMAN. Good afternoon, everyone, and thank you to Subcommittee Chairman Gosar for holding this important hearing today, as well as the witnesses and Members for their time and for showing up.

And speaking of showing up, we have a key witness who declined to attend today. And again, that is Brenda Mallory, the Chair of the Council of Environmental Quality. And this is not the first time we have had issues with attendance from CEQ, although her office did note she looks forward to showing up soon. But I am not holding my breath. Apparently, she is tweeting from Hawaii today instead of being here with us. And all I can say is we are working closely with the Appropriations Committee and looking at their budget every chance we get. I am sure Hawaii is fun, but it is not a valid reason to skip coming here and taking care of the people's business.

Decisions made by CEQ have real impacts on American families, and it is time to face those facts. As of this morning, the average price of gasoline was $3.53 a gallon. That is nearly a full dollar higher than when President Trump left office. If last summer is any indication, this number will skyrocket right as families begin to head out on road trips. And speaking of road trips, the price of a new car is a whopping $48,000, and even used car prices average around $26,500. And I would like to note for the record that the average cost of a new vehicle right now is higher than the median household income in my hometown of Hot Springs, Arkansas.

So, why is this relevant to our discussion? The Biden administration continues repeating the same talking points about the need to invest in infrastructure, underserved communities, and social and environmental justice. Time and again, these lofty words result in more red tape, bigger bureaucracies, and higher prices for American families. It is all talk with not any action in the right direction.

Look no further than the dramatic growth of the Council on Environmental Quality, commonly known as CEQ, in both size and mission creep. From 2019 to 2023, CEQ's baseline budget increased by 63 percent, in addition to receiving a massive cash infusion from the Inflation Reduction Act to further inject radical environmental justice initiatives in the government. Those taxpayer dollars still haven't been spent, yet President Biden's latest budget request calls for even more money to CEQ.

Despite this Administration's best efforts, we can never spend our way out of the problems that America is facing. We need proactive, long-term solutions to get our economy back on track, to shore up our national security, and hold Federal agencies accountable. House Republicans have already advanced many of those solutions this Congress, and we remain committed to turning over every stone and ensuring taxpayer dollars aren't going to waste here in DC.

CEQ and our current permitting processes, primarily through NEPA, blanket the American energy sector in red tape. And I said the American energy sector, not just oil and gas, but the entire sector. It is keeping shovel-ready projects and good-paying jobs on the back burner, while bureaucrats twiddle their thumbs. Just last week, the editorial board of The New York Times, that is New York Times, not Wall Street Journal, described the red tape for energy transmission projects as "a jumble of approval processes" that "causes environmental damage." Given the length of time it takes to get new projects reviewed and approved to fix the mess, the Times editors called for decisive action to reform "unnecessarily cumbersome regulations," including NEPA.

I couldn't agree more. Our own regulations are killing us. I have often used the illustration of Aesop's fable of the eagle that was flying high overhead, and he was struck by an arrow, and as he fluttered to the ground with his lifeblood pouring out, he looked back at the arrow and said, "Alas, I have been mortally wounded by an arrow that is feathered with my own plume." That is what we are doing to ourselves in this country. We are killing our country with our own means, and I can't figure out for the life of me for what reason, because it is not doing anything to help the environment. It is certainly not helping the economy. And I don't understand the purpose or the reason behind these actions.

And that is why we passed H.R. 1, the Lower Energy Costs Act, and that is why we have continued advancing solutions to get our country back on track. Our businesses, families, our communities, and America deserve better, and that is what we are going to continue working to make sure that this country stays strong, that this country continues to lead the world in environmental stewardship, in human rights, and that we can lead the world in energy and manufacturing, and be an example instead of being a donor to countries like China and to OPEC that our current policies are making us dependent on.

With that, Mr. Chairman, I yield back.

Dr. GOSAR. I thank the Chairman for his opening statement, and now I am going to introduce our witnesses.

First, we have Ms. Diana Furchtgott-Roth. Did I say that right?

Ms. FURCHTGOTT-ROTH. Yes.

Dr. GOSAR. Bingo. Director, Center of Energy, Climate, and Environmental Policy, The Heritage Foundation.

A personal friend of mine, Mr. Jeremy Harrell, Chief Strategy Officer from ClearPath.

Good to see you, Jeremy.

Ms. Jamie Pleune, Associate Professor, S.J. Quinney College of Law, University of Utah.

Thank you.

And Mr. Kenny Stein, Director of Policy, Institute for Energy Research.

Let me remind the witnesses that under Committee Rules, they must limit their oral statements to 5 minutes, but their entire statement will appear in the hearing record.

To begin your testimony, please push the "on" button so we can hear you.

We use timing lights. When the light turns green, you have 5 minutes. When it turns yellow, you have 1 minute. And when it turns red, please summarize and try to complete your statement. I will also allow all witnesses to testify before Members' questioning.

I now recognize Ms. Furchtgott-Roth for 5 minutes.

Statement of Diana Furchtgott-Roth, Director, Center for Energy, Climate, and the Herbert and Joyce Morgan Fellow in Energy and Environmental Policy, the Heritage Foundation, Chevy Chase, Maryland

Ms. FURCHTGOTT-ROTH. Thank you very much. Chairman Westerman, Chairman Gosar, Ranking Member Stansbury, and members of the Subcommittee, I am honored to be invited to testify before you today.

Creeping executive overreach not envisaged by Congress in the original messages or purposes of cabinet agencies is driving up the price of energy and energy-related products. This is raising automotive and electricity costs for all Americans, and poor and middleclass Americans are disproportionately paying the price. This is not environmental justice or any kind of justice to have to pay higher prices for necessities such as electricity, gasoline, and cars. Led by the White House, mission creep can be found at many agencies, including the Environmental Protection Agency, the Council on Environmental Quality, the Office of the Comptroller of the Currency, the Securities and Exchange Commission, and the Federal Energy Regulatory Commission.

President Biden's higher prices for electricity and transportation bring no climate benefits. Even completely eliminating all fossil fuels from the United States would result in less than 2/10 of 1 degree Celsius in temperature mitigation by the year 2100, according to research using EPA's own climate models. So, these higher prices are all for nothing.

EPA's attempt to regulate regional emissions through a novel attempt of the Clean Air Act, known as the Clean Power Plan, was deemed executive overreach by the Supreme Court in West Virginia v. EPA. Creeping executive branch oversight of energy started on the first day in office, when President Biden revoked the permit for Keystone XL pipeline. In August 2021, the President issued an executive order that, by 2030, half of new vehicles sold in the United States must be electric, including fuel cell electric plug-in hybrids and full battery electric.

Last month, EPA proposed costly regulations on automobile emissions that would require new car sales to be 60 percent battery powered electric by 2030 and 67 percent by 2023, compared to fewer than 6 percent in 2022. This will make cars more expensive, less safe, and cost American auto workers jobs. United Auto Workers President Shawn Fain came out on April 26 saying, "Shame on Congress for doing this, shame on the President for doing this," because it is costing his workers jobs. Stellantis is laying off 3,500 people. They closed a plant in Illinois in December.

The Office of the Comptroller of the Currency, which regulates banks, has appointed a Chief Climate Risk Officer to assess and monitor climate-driven risks to banks. This is going to discourage investment in fossil fuels, and will allow the investigations of companies and banks that OCC believes are making the wrong investment.

The Council on Environmental Quality issued new guidance on January 9, 2023 requiring Federal agencies to use the NEPA process to reduce emissions from greenhouse gases. That means that projects that result in higher greenhouse gas emissions will find it harder to get NEPA approval. The guidance takes effect immediately, without even waiting for CEQ to address comments from the public and regulated industries. Some agencies are including in their compliance with NEPA factors, factors that Congress did not originally intend. This is an example of harmful arbitrary and capricious policy.

The Securities and Exchange Commission also wants to stick its fingers into energy production. SEC Chairman Gary Gensler has proposed rules to require companies to disclose information about government and management of climate risks, making it more difficult to get approval for fossil fuel investments. Meanwhile, China is increasing its construction of coal-fired power plants. America has 225 and China has 1,118. China has increased carbon emissions by over 5,000 million metric tons over the past 16 years; the United States has reduced it by about 1,000 million metric tons.

Americans, particularly the poor and middle class, are bearing the costs in higher electricity prices, higher food prices, and a forced switch to electric vehicles without benefits for the environment. This is not environmental justice, social justice, or any kind of justice.

Thank you very much.

[The prepared statement of Ms. Furchtgott-Roth follows:]

Prepared Statement of Diana Furchtgott-Roth, Director, Center on Energy, Climate, and Environment and the Herbert and Joyce Morgan Fellow in Energy and Environmental Policy, The Heritage Foundation

My name is Diana Furchtgott-Roth. I am the director of the Center for Energy, Climate, and Environment at The Heritage Foundation. The views I express in this testimony are my own and should not be construed as representing any official position of The Heritage Foundation.

Chairman Gosar, Ranking Member Stansbury, Members of the Subcommittee, I am honored to be invited to testify before you today on the subject of, "The Biden Administration's Executive Overreach and Its Effect on American Energy Independence."

In addition to my role at The Heritage Foundation, I am also an adjunct professor of economics at George Washington University. My professional training is in economics. From 2019 to 2021, I was deputy assistant secretary for research and technology at the U.S. Department of Transportation. Previous positions include acting assistant secretary for economic policy at the U.S. Department of the Treasury; chief economist at the U.S. Department of Labor; chief of staff of the Council of Economic Advisers under President George W. Bush; and deputy executive secretary of the Domestic Policy Council under President George H.W. Bush.

In the last State of the Union, President Biden said, "The climate crisis doesn't care if your state is red or blue. It [the climate crisis] is an existential threat."[144] The President has used the so-called existential climate crisis to expand dramatically the power of different executive branch agencies with the object of reducing greenhouse gas emissions.

This creeping overreach, not envisaged by Congress in the original missions or purposes of these agencies, has driven up the price of energy and energy-related products. A runaway government is trying to control our lives for no apparent benefit, but guaranteeing impoverishment of opportunity, safety and security. Government is sticking its fingers into every aspect of our lives, making it more difficult and more expensive to get ahead. By creeping over the limits on its power, government is switching off the power for growth.

---

[144] President Joe Biden, State of the Union Address, February 7, 2023, https://www.whitehouse.gov/briefing-room/speeches-remarks/2023/02/07/remarks-of-president-joe-biden- state-of-the-union-address-as-prepared-for-delivery/ (accessed May 8, 2023).

Some government agencies noted for mission creep include, but are not limited to: the Environmental Protection Agency, the Council on Environmental Quality, the Office of Comptroller of the Currency, the Securities and Exchange Commission, the Federal Energy Regulatory Commission, and the White House itself.

Overreach is sometimes used when Congress declines to pass a bill into law. Rather than accept the status quo, the president takes matters into his own hands with rules or guidance from executive branch agencies. The result is policies outside of congressional authorization. In this case, poorly-considered policies are forcing all Americans to pay more for electricity and transportation, for little or no benefit— because the new regulations will have a minimal effect on global temperatures. These costs are falling disproportionately on poor and middle-class Americans, many of whom voted for President Biden, who pay a higher share of their income in food and energy costs.

Sadly, there are many examples of regulatory overreach.

The Environmental Protection Agency's attempt to regulate regional emissions through a novel interpretation of the Clean Air Act, known as the Clean Power Plan,[145] proposed in 2015 under President Barack Obama, was deemed executive overreach by the Supreme Court in West Virginia v. Environmental Protection Agency. Ruling 6-3 on June 30, 2022, the Supreme Court decided that the Clean Air Act does not allow the EPA to move from regulating individual power plants to regulating regional emissions.

Although the Clean Air Act allows the EPA to set maximum levels of new and existing emissions sources, the Clean Power Plan[146] went further. If emissions exceeded the EPA's requirements, a state, or group of states, would be required to shut down power plants or to install renewable energy sources. The plan was similar to the American Clean Energy and Security Act, introduced by Democratic Congressmen Henry Waxman and Edward Markey in 2009, and the American Power Act, introduced by Senators John Kerry and Joe Lieberman in 2010. Neither bill became law, despite sizable Democratic majorities in both chambers.

The opinion in West Virginia v. Environmental Protection Agency, written by Chief Justice John Roberts, cited the major questions doctrine, according to which Congress must "speak clearly if it wishes to assign to an

---

[145] *Federal Register*, Vol. 80, No. 205 (October 23, 2015), pp. 64661–65120, https://www.gpo.gov/fdsys/pkg/FR-2015-10-23/pdf/2015-22842.pdf (accessed May 9, 2023).

[146] Ibid.

agency decisions of vast economic and political significance."[147] Writing for the majority, Chief Justice Roberts stated that "EPA claimed to discover an unheralded power representing a transformative expansion of its regulatory authority in the vague language of a long-extant, but rarely used, statute designed as a gap filler. That discovery allowed it to adopt a regulatory program that Congress had conspicuously declined to enact itself."[148] Justice Neil Gorsuch elaborated on the major questions doctrine in a concurring opinion, writing, "The framers believed that the power to make new laws regulating private conduct was a grave one that could, if not properly checked, pose a serious threat to individual liberty."[149]

The case has far-reaching implications for other agencies that could currently be exceeding their statutory limits. The Securities and Exchange Commission, for example, has proposed requirements for companies to disclose their exposure to climate risk and to provide details about the climate effects of their operations. The Office of the Comptroller of the Currency has appointed a chief climate risk officer who has decided to do the same.

Executive overreach could also be occurring in other areas. Courts are in the process of examining whether the Department of Education can unilaterally cancel or reduce student loans. Meantime, the National Labor Relations Board is considering making franchise businesses such as McDonald's accountable for the actions of local franchises. Such rules could find themselves on the wrong side of the Court's approach, which found the EPA's rulemaking to be an example of "agencies asserting highly consequential power beyond what Congress could reasonably be understood to have granted."[150] If the Court had found the other way, EPA could have changed the entire vehicle fleet to electric vehicles through a change in a regulation. When federal agencies exceed their authority, opportunities for ordinary people are stifled. Severe, government-imposed cuts in carbon emissions raise the cost of electricity and American-made goods. Under the Clean Power Plan, some states or groups of states would have had to meet EPA targets by ensuring plants cut emissions or by financing reductions in other ways, such as suppressing consumer demand or investing in more costly renewable energy. In any case, carbon emissions are declining naturally without the plan. Emissions of energy-related carbon dioxide declined by 18 percent from 2007 to 2021, according to the Energy Information

---

[147] *West Virginia et al. v. Environmental Protection Agency et al.*, 597 U.S. (2022).
[148] Ibid.
[149] Ibid.
[150] Ibid.

Administration.[151] Between 2014, when the Clean Power Plan was proposed, and 2021, four years after the Trump administration rescinded it, these emissions fell by 9 percent.[152]

Creeping executive branch oversight of energy started early in the Administration. On his first day in office, President Biden revoked the permit for the Keystone XL pipeline, which would have brought 850,000 barrels of oil per day from Canada to be refined in U.S. refineries.[153] This reduced energy independence, and higher gasoline prices and inflation soon followed. President Biden the following year asked Saudi Arabia and Venezuela to produce more oil. In addition to eliminating the Keystone XL pipeline, President Biden has reduced oil and gas production[154] by expanding the boundaries of the Grand Staircase-Escalante, Bears Ears, Northeast Canyons, and Seamounts Marine National Monuments, preventing oil and natural gas production in those areas.[155]

On August 5, 2021, President Biden issued an Executive Order[156] setting a goal that by 2030 half of new vehicles sold in the United States must be electric, including fuel cell electric, plug-in hybrids, and full battery electric. According to President Biden,[157] the Executive Order will "improve our economy and public health, boost energy security, secure consumer savings, advance environmental justice, and address the climate crisis." On the contrary, the Executive Order will raise costs for Americans, and poor and middle-class people disproportionately would pay the price.

---

[151] U.S. Energy Information Administration, "U.S. Energy-Related Carbon Dioxide Emissions, 2021," Figure 2, December 14, 2022, https://www.eia.gov/environment/emissions/carbon (accessed May 9, 2023).

[152] Ibid.

[153] Joseph R. Biden, Jr., "Protecting Public Health and the Environment and Restoring Science to Tackle the Climate Crisis," Executive Order 13990, January 20, 2021, https://www.whitehouse.gov/briefing-room/presidential-actions/2021/01/20/executive-order-protecting-public-health-and-environment-and-restoring-science-to-tackle-climate-crisis/ (accessed May 9, 2023).

[154] *Federal Register*, Vol. 86, No. 14 (January 25, 2021), pp. 7037–7043, http://www.federalregister.gov/documents/2021/01/25/2021-01765/protecting-public-health-and-the- environment-and-restoring-science-to-tackle-the-climate-crisis (accessed May 9, 2023).

[155] Ibid.

[156] Joseph R. Biden, Jr., "Strengthening American Leadership in Clean Cars and Trucks," Executive Order 14037, August 5, 2021, https://www.whitehouse.gov/briefing-room/presidential-actions/2021/08/05/executive-order-on-strengthening-american-leadership-in-clean-cars-and-trucks/ (accessed May 9, 2023).

[157] Ibid.

The president's 2021 announcement coincided with the August 11 meeting of the California Air Resources Board,[158] which discussed draft regulations to implement Governor Gavin Newsom's Executive Order[159] that all new vehicles sold in the Golden State be electric by 2035 and a ban on the sale of diesel trucks by 2036. Through regulations and executive orders, President Biden wants to follow the lead[160] of California and other states seeking to reduce emissions—laws that Congress would not pass.

The Environmental Protection Agency followed with new proposed regulations[161] on automobile emissions from the Environmental Protection Agency would require new car sales to be 60% battery powered electric by 2030 and 67% by 2032, compared to fewer than 6% in 2022. EPA is also planning new rules for power plants,[162] driving up the costs of the electricity needed to charge these vehicles. These rules again would raise driving costs for Americans, and poor and middle-class Americans disproportionately would pay the price.

New electric vehicles cost more than gasoline-powered vehicles. The electric version of the base version of the Ford 150 pickup truck, the best-selling vehicle in America, costs an additional $26,000.[163] Tesla's base prices

---

[158] California Air Resources Board, "Public Workshop on Advanced Clean Cars II," https://ww2.arb.ca.gov/events/public-workshop-advanced-clean-cars-ii-0 (accessed May 9, 2023).

[159] Executive Department State of California, "Executive Order N-79-20," https://www.gov.ca.gov/wp-content/uploads/2020/09/9.23.20-EO-N-79-20-Climate.pdf (accessed May 9, 2023).

[160] Press Release, "Fact Sheet: President Biden Announces Steps to Drive American Leadership Forward on Clean Cars and Trucks," The White House, August 5, 2021, https://www.whitehouse.gov/briefing-room/statements-releases/2021/08/05/fact-sheet-president-biden-announces-steps-to-drive-american-leadership-forward-on-clean-cars-and-trucks/ (accessed May 9, 2023).

[161] U.S. Environmental Protection Agency, ''Notice of Proposed Rulemaking: Multi-Pollutant Emissions Standards for Model Years 2027 and Later Light-Duty and Medium-Duty Vehicles,'' https://www.epa.gov/system/files/documents/2023-04/lmdv-multi-pollutant-emissions-my-2027-nprm-2023-04.pdf (accessed April 28, 2023).

[162] Valerie Volcovici, ''Biden EPA to Issue Power Plant Rules That Lean on Carbon Capture,'' Reuters, April 24, 2023, https://www.reuters.com/business/sustainable-business/biden-epa-issue-power-plant-rules-that-lean-carbon_capture-2023-04-23/ (accessed April 28, 2023).

[163] Ford Motor Company, Models & Specs, 2023 F–150 XL, https://www.ford.com/trucks/f150/models/?gnav=vhpnav-specs (accessed April 28, 2023); and Ford motor Company, Models & Specs, 2023 F–150 Lightning PRO, https://www.ford.com/trucks/f150/f150-lightning/models/?gnav=vhpnav-specs (accessed April 28, 2023).

start at about $40,000 for a Model 3 and go up to almost $100,000 for a Model X.[164] These are staggering costs to impose on American families.

Charging will also cost more. At the same time as EPA issuing new rules for automobile emissions, it is also planning new rules for emissions from power plants, The New York Times has reported.[165] According to the reports, EPA will regulate carbon dioxide and other so-called greenhouse gas emissions from both new and existing natural gas and coal-fired power plants, and require carbon capture systems or a switch to hydrogen fuels. These systems for capturing carbon are costly and will be passed on to consumers in the form of higher electricity rates. Drivers will find it more expensive to use electricity for all purposes, including charging their electric vehicles, harming poor and middle-class drivers the most.

Three-quarters of vehicles sold are previously owned cars.[166] In 2019, the last year for which complete data on used car sales are available, Americans bought 41 million used cars and 13 million new cars.[167] But people do not want to buy used electric vehicles, because it is difficult to evaluate how long the battery will last. Replacing an EV battery can cost anywhere from $5,000 to $20,000.[168] The poor and the middle class will suffer most from higher prices for used vehicles, because they cannot afford the new electric vehicles.

Mandating electric vehicles would reduce Americans' standard of living. Back in the early 1900s, when Henry Ford started producing cars, only rich Americans could afford them. Throughout the 20th century cars became less expensive, and many households could afford not one but two. Cars are already becoming more expensive, and the proposed rule accelerates that trend, taking America back a century, when new cars were only for the rich.

---

164 Tesla, Model 3, Purchase Price, https://www.tesla.com/model3/design#overview (accessed April 28, 2023); and Tesla, Model X, Purchase Price, https://www.tesla.com/modelx/design#overview (accessed April 28, 2023).

165 Coral Davenport and Lisa Friedman, "E.P.A. to Propose First Controls on Greenhouse Gases from Power Plants," *New York Times*, updated April 27, 2023, https://www.nytimes.com/2023/04/22/climate/epa-power-plants-pollution.html (accessed April 28, 2023).

166 Mathilde Carlier, Statista, New and Used Light Vehicle Sales in the United States, 2010 to 2021, https://www.statista.com/statistics/183713/value-of-us-passenger-cas-sales-and-leases-since-1990/ (accessed April 28, 2023).

167 U.S. Department of Transportation, Bureau of Transportation Statistics, New and Used Passenger Car and Light Truck Sales and Leases Data Set, 1990 to 2021, https://www.bts.gov/content/new-and-used-passenger-car-sales-and-leases-thousands-vehicles (accessed April 28, 2023).

168 Recurrent, "Updated: Electric Car Battery Replacement Costs," March 26, 2023, https://www.recurrentauto.com/research/costs-ev-battery-replacement (accessed April 28, 2023).

Recharging an electric vehicle from empty can take over an hour, compared to 5 minutes to fill up with gas.[169] If there is a line to use the charging station the wait can double. Manufacturers suggest not allowing EV batteries to go below 20%, and the charging rate goes down when it is charged over 80%.[170] Throughout America the poor rarely have access to indoor garages for overnight charging, and in most large cities, such as New York City, the middle-class also have no access to indoor charging. Using charging stations on the street, if available, risks theft of expensive charging cables.

Battery-powered vehicles lack sufficient range to satisfy most customers. Although 60 to 70 miles of range is enough for most trips, people buy cars for all circumstances, including long trips and cold weather. Batteries lose up to 40% of their range in cold climates and manufacturers suggest using heating systems.[171] A study by Autocar[172] shows that electric vehicles lose, on average, a third of their range in the winter, which reduces the typical 240-mile range to 160 miles. If a heat pump is added to the car, the loss is less, but still the 240-mile range would shrink to 180.

Car results varied. The Fiat 500 42kWh Icon lost 40% of its range in the winter.[173] The Ford Mustang Mach-E Extended Range RWD lost 35%, and the Porsche Taycan 4S Performance Battery Plus, with heat pump, lost 22% (the Taycan costs between $83,000 and $166,000).[174]

The loss of range in cold weather is one reason why, at the end of 2021, the latest full year available, North Dakota had 380 electric vehicle (EV) registrations, the fewest in the United States, according to the Energy Department.[175] North Dakota will receive $26 million for charging stations,

---

[169] Lazar, "How Long Does It Take to Refuel a Gasoline Car? GasAnswer, https://gasanswer.com/how-long-take-refuel-gasoline-car/ (accessed April 28, 2023).

[170] Sebastian Blanco, "How to Maximize EV Range," J.D. Power, July 20, 2022, https://www.jdpower.com/cars/shopping-guides/how-to-maximize-ev-range (accessed April 28, 2023).

[171] Ellen Edmonds, "Icy Temperatures Cut Electric Vehicle Range Nearly in Half," AAA News Room, February 7, 2019, https://newsroom.aaa.com/2019/02/cold-weather-reduces-electric-vehicle-range/(accessed April 28, 2023).

[172] Move Electric, "Electric Vehicle Range Test Reveals Up to 20% Drop in Winter," Autocar, March 17, 2022, https://www.autocar.co.uk/car-news/move-electric/electric-vehicle-range-test-reveals-20-drop-winter (accessed April 28, 2023).

[173] Ibid.

[174] Ibid.

[175] U.S. Department of Energy, Office of Energy Efficiency and Renewable Energy, Electric Vehicle Registrations by State Data Set, updated June 2022, https://afdc.energy.gov/data/10962 (accessed April 28, 2023).

according to the Department of Transportation,[176] or $68,000 per registered EV. Wyoming, with $27 million and 510 EVs, gets $53,000 per EV.[177] South Dakota, with $29 million, had 680 vehicles, and will collect $43,000 per vehicle.[178] Alaska had 1,290 registered electric vehicles (EVs). Alaska will get $52 million[179] of the $7.5 billion that the new Infrastructure Investment and Jobs Act allocates to states for electric charging stations.[180] That works out to more than $40,000 per electric vehicle. These funds could be saved or put to better use.

Minerals such as lithium and cobalt are essential for batteries. Mining for these minerals is energy-intensive, and the Chinese Communist Party (CCP) has substantial access to global mineral sources for battery production, resulting in a loss of American independence. Lithium is mined in western China's Qinghai Province, aided by government funding, and China purchases cobalt for electric batteries from Kisanfu, in the Democratic Republic of Congo.[181] Our federal government makes opening new mines in the United States virtually impossible, even though the jobs generated would help all Americans, particularly the poor and the middle class. Thus, the rule will result in a massive increase in mining in countries that have no respect for the environment or human welfare. The mining of minerals as a result of the rule will be bad for the environment and is frequently performed by child workers.

The Council on Environmental Quality, part of the Executive Office of the President, issued new guidance[182] on January 9, 2023, requiring federal agencies to use the National Environmental Policy Act (NEPA) process to reduce emissions from greenhouse gases. That means that projects that result

---

[176] U.S. Department of Transportation, Federal Highway Administration, 5-Year National Electric Vehicle Infrastructure Funding by State Data Set, FY 2022 to FY 2026, updated September 13, 2022,https://www.fhwa.dot.gov/bipartisan-infrastructure-law/evs_5year_nevi_funding_by_state.cfm (accessed April 28, 2023).

[177] Ibid.

[178] Ibid.

[179] Ibid.

[180] News Release, "President Biden, U.S. Department of Transportation Releases Toolkit to Help Rural Communities Build Out Electric Vehicle Charging Infrastructure," U.S. Department of Transportation, February 2, 2022, https://www.transportation.gov/briefing-room/president-biden-us-department-transportation-releases-toolkit-help-rural-communities (accessed April 28, 2023).

[181] Dionne Searcey, Michael Forsythe, and Eric Lipton, "A Power Struggle Over Cobalt Rattles the Clean Energy Revolution," *New York Times*, December 7, 2021, https://www.nytimes.com/2021/11/20/world/china-congo-cobalt.html (accessed April 28, 2023).

[182] *Federal Register*, Vol. 88, No. 5 (January 9, 2023), pp. 1196–1212, https://www.federalregister.gov/documents/2023/01/09/2023-00158/national-environmental-policy-act-guidance-on-consideration-of-greenhouse-gas-emissions-and-climate (accessed May 9, 2023).

in higher greenhouse gas emissions will find it harder to get NEPA approval. The guidance takes effect immediately, without waiting for CEQ to address the comments. Some agencies are including in their compliance with NEPA factors that Congress did not originally intend. This is an example of a harmful, arbitrary, and capricious policy.

Other Executive Branch agencies are deviating from traditional roles of approving investments and instead are slowing production and transportation of oil and natural gas. For instance, the Federal Energy Regulatory Commission proposed a new policy[183] on February 17, 2022, that would have made it even harder to put new pipelines in place to carry oil and gas from the interior of the country to the coasts, where it can be exported. FERC was intending to "consider a proposed project's impacts on existing pipelines" as well as the environmental effects of the new pipeline. The February 2022 policy statement was pulled back and deemed a "draft" in March 2022 due to public pressure.

Even though the policy statements have been downgraded to "draft" status, they have not been officially withdrawn, and regulatory uncertainty at FERC is slowing the development of pipelines. Last week FERC Commissioner James Danly said at a hearing before the Senate Committee on Energy and Natural Resources, "The consequences of premature retirements and resource scarcity are even more acute when you consider the constraints on natural gas supply resulting from the under-development of interstate natural gas infrastructure—again, driven by the FERC's maladministration of the Natural Gas Act. Although I am genuinely delighted that the Commission has recently increased the pace of natural gas pipeline reviews, the policies FERC recently sought to promulgate have had the very effects I predicted at last year's hearing: according to the Energy Information Administration, 2022 saw the lowest quantity of additional capacity added to the natural gas pipeline system since 1995, the obvious result of the FERC's slow walking natural gas pipeline applications over the last two years and the chilling effect of the regulatory uncertainty created by the Commission's issuances. Interstate natural gas infrastructure is absolutely critical: as coal, nuclear and

---

[183] Federal Energy Regulatory Commission, ''Fact Sheet: Updated Pipeline Certificate Policy Statement (PL18-1-000),'' updated February 17, 2022, http://www.ferc.gov/news-events/news/fact- sheet-updated-pipeline-certificate-policy-statement-pl18-1-000 (accessed May 9, 2023).

hydroelectric generators retire due to subsidies and public policy choices, the need for natural gas to ensure system reliability continues to grow.[184]

The Interior Department has issued a Report on the Federal Oil and Gas Leasing Program,[185]42 calling for fewer leases, higher royalties from oil and gas leases, and a more thorough bidding process to screen buyers. It proposed that oil and gas drilling not be a priority, and its recommendations would make it more difficult to drill. This interference with America's energy production makes it more difficult for companies to flourish and consumers to get affordable energy.

In the finance area, outside of its statutory authority, the Securities and Exchange Commission wants to stick its fingers into energy production. SEC Chairman Gary Gensler has proposed rules[186] to require companies to disclose information about the following: governance and management of climate-related risks; how climate related risks will affect companies' strategy and outlook; and the effects of climate events such as hurricanes and wildfires on financial statements. This rule would reduce America's energy independence and make it more difficult to get capital for fossil fuel investments.

The Office of the Comptroller of the Currency, which regulates banks, has appointed a Chief Climate Risk Officer to assess and to monitor climate-driven risks to banks. Dr. Yue Chen does not have a background in risk assessment. She is an engineer, with a bachelor's degree in chemical engineering from Tsinghua University in China and a Ph.D. in chemical engineering from the Massachusetts Institute of Technology.[187] Monitoring climate risks to bank lending and assets will have the effect of discouraging investments in fossil fuels and will allow the investigation of companies and banks that it believes are making the wrong investments.

The Defense Department, whose mission is to defend the United States, wants to use biofuels to make its military vehicles and jets more climate friendly. Such vehicles are less resilient and more costly, so the Pentagon would be able to afford fewer of them, with major effects on national security.

---

[184] James P. Danly, testimony before Committee on Energy and Natural Resources, U.S. Senate, May 4, 2023, https://www.energy.senate.gov/services/files/0A896B12-2895-4F68-A367-74009F2975C4 (accessed May 9, 2023).

[185] *Federal Register*, Vol. 86, No. 14.

[186] Press Release, "SEC Proposes Rules to Enhance and Standardize Climate-Related Disclosures for Investors," U.S. Securities and Exchange Commission, March 21, 2022, https:// www.sec.gov/news/press-release/2022-46 (accessed May 9, 2023).

[187] News Release, "OCC Announces Chief Climate Risk Officer," Office of the Comptroller of the Currency, September 12, 2022, https://www.occ.gov/news-issuances/news-releases/2022/nr-occ-2022-110.html (accessed May 9, 2023).

The Federal Trade Commission is considering investigating oil and gas companies for price gouging—even though people know that lower supply always leads to higher prices.

With prices of new and used cars higher, the biggest loser from runaway government is the American consumer, who will purchase fewer new cars, which have additional safety features and better fuel economy. More people will be injured or killed in car crashes. Mandating electrics won't solve problems of environmental justice. Requiring sales of electric vehicles discriminates against low-income Americans who cannot as easily afford expensive electric cars. Upper-income urban Americans buy Teslas; lower-income rural Americans buy pickup trucks and large vehicles. Mandating sales of electrics helps the rich and hurts the poor.

In April Stellantis announced that it would be offering buyouts to 33,500 hourly and white-collar workers in an attempt to cut 3,500 jobs due to its planned transition to electric vehicles.[188] It closed a plant in Illinois in December. General Motors and Ford are also laying off workers as part of their move to make more battery- powered vehicles.

United Auto Workers president Shawn Fain said in a statement on April 26, "Stellantis' push to cut thousands of jobs while raking in billions in profits is disgusting. This is a slap in the face to our members, their families, their communities, and the American people who saved this company 15 years ago. Even now, politicians and taxpayers are bankrolling the electric vehicle transition, and this is the thanks the working class gets. Shame on Stellantis."[189]

Americans' jobs in the oil and gas fields are being sacrificed to Chinese nationals (sometimes with forced labor) making wind turbines and solar panels. Americans' jobs in auto plants are being sacrificed to Chinese nationals (sometimes with forced labor) making batteries and electric vehicle components.

As well as taking away Americans' choice of cars, the Department of Energy wants to regulate a variety of other appliances, depriving Americans

---

[188] Ryan Felton and Nora Eckert, "Jeep Maker Stellantis to Offer Buyouts to Hourly, Salaried Workers," *The Wall Street Journal*, April 26, 2023, https://www.wsj.com/articles/jeep-maker- stellantis-to-offer-buyouts-to-hourly-salaried-workers-d3c71fdc (accessed May 1, 2023).

[189] United Auto Workers, UAW Statement on Job Cuts at Stellantis, https://uaw.org/uaw-statement-job-cuts-stellantis/#:?:text=%E2%80%9CStellantis'%20push%20to%20cut%20thousands,this%20company%2015%20years%20ago (accessed May 1, 2023).

of the ability to have a gas stove.[190] Poor and middle-class people will be disproportionately paying the price for these rules, both proposed and enacted.

Renewables raise the costs of electricity, which disproportionately affects the poor and middle class. President Biden's plan to transition away from fossil fuels is making the United States weaker and China stronger. This will have no noticeable effects on global temperatures, using the Model for the Assessment of Greenhouse Gas Induced Climate Change developed by researchers at the EPA, which is used for such calculations at the Heritage Foundation.

Government overreach is reducing America's energy independence and strengthening China, which makes nearly 80% of the world's electric batteries,[191] over 80% of global solar panels,[192] and almost 60% of wind turbines.[193]

This is especially troubling because the Chinese Communist Party (CCP) is a totalitarian regime which has a poor record both on the environment and on human rights. Beijing is engaged in genocide against the minority Uyghur people of Xinjiang and has imposed draconian restrictions on political freedoms in Hong Kong.[194] The CCP has reduced or eliminated religious liberties for Christians and Buddhist worshippers of the Dalai Lama throughout Tibet.[195] Empowering the Chinese government is fundamentally at odds with "good corporate governance."

Due to hypocritical, bureaucratic, executive branch mission creep, rather than using its own oil and natural gas resources, America will depend on energy from China. In order to produce supplies of renewables, China is increasing its construction of coal-fired power plants. America has 225 coal-fired power plants, and China has 1,118 (half of all the coal-fired plants in the world).[196] That is one reason why China has increased carbon emissions by

---

[190] *Federal Register*, Vol. 88, No. 21 (February 1, 2023) pp. 6818–6904.

[191] International Energy Agency, Global Supply Chains of EV Batteries, July 2022, p. 2, https://iea.blob.core.windows.net/assets/4eb8c252-76b1-4710-8f5e-867e751c8dda/GlobalSupplyChainsof EVBatteries.pdf (accessed May 9, 2023).

[192] International Energy Agency, Special Report on Solar PV Global Supply Chains, updated August 2022, p. 7, https://iea.blob.core.windows.net/assets/d2ee601d-6b1a-4cd2-a0e8-db02dc64332c/SpecialReportonSolarPVGlobalSupplyChains.pdf (accessed May 9, 2023).

[193] International Energy Agency, "Geographic Concentration by Supply Chain Segment, 2021," updated January 25, 2023, https://www.iea.org/data-and-statistics/charts/geographic-concentration-by-supply-chain-segment-2021 (accessed May 9, 2023).

[194] James J. Carafano et al., "Winning the New Cold War," p. 24.

[195] Ibid., p. 3.

[196] Jessica Aizarani, Statista, "Global operational coal-fired power stations by country 2022," January 30, 2023, https://www.statista.com/statistics/859266/number-of-coal-power-plants-by-country/ (accessed April 28, 2023).

over 5,000 million metric tons over the past 16 years.[197] In contrast, America's carbon emissions have declined by over 1,000 million metric tons over the same period due to the use of clean natural gas.[198]

A new report by the Heritage Foundation, Winning the Cold War: A Plan for Countering China,[199] shows how America's environmental policies benefit China and harm America.

Heritage issued the report on the same day that the House of Representatives introduced H.R. 1, The Lower Energy Costs Act,[200] sponsored by Majority Leader Steve Scalise (R-LA), Energy and Commerce Committee Chair Cathy McMorris Rodgers (R-WA), Natural Resources Committee Chair Bruce Westerman (R-AR), and Transportation and Infrastructure Committee Chair Sam Graves (R-MO). The bill emphasizes domestic energy production, lower energy costs, and reduction in emissions.

Rather than kowtowing to China, companies should reject environmental policies that raise the costs of doing business and favor the CCP. The rush to a green energy future, driven more by politics and virtue-signaling than economics and emissions reductions, will only enrich China at America's expense and place vital energy supply chains at mercy of Beijing.

China has not committed to reducing emissions until 2027. Research by Dr. Kevin Dayaratna, chief statistician and senior research fellow at The Heritage Foundation, has shown that even completely eliminating all fossil fuels from the United States would result in less than 0.2 degrees Celsius in temperature mitigation by 2100.[201] Americans, particularly poor and middle class, would be bearing major costs in higher electricity prices, higher food prices, and a forced switch to costly electric vehicles without benefits for the environment. They would pay the price for President Biden's energy agenda.

Cleaner air and efficient power generation are worthwhile goals. But so is the security that comes from the rule of law. The Supreme Court weighed in on that balance in 2022 and could weigh in further in the years ahead.

---

197 The Heritage Foundation, CO2 Emissions Trends in Key Countries Data Set, 2005–2021.

198 Ibid.

199 James J. Carafano et al., "Winning the New Cold War: A Plan for Countering China," Heritage Foundation *Special Report* No. 270, March 28, 2023, https://www.heritage.org/sites/default/files/2023-03/SR270_0.pdf (accessed April 28, 2023).

200 Lower Energy Costs Act, H.R. 1, 118th Congress, Session 1, (2023) https://www.congress.gov/118/bills/hr1/BILLS-118hr1ih.pdf (accessed April 28, 2023).

201 Kevin D. Dayaratna, PhD, Katie Tubb, and David Kreutzer, ''The Unsustainable Costs of President Biden's Climate Agenda,'' Heritage Foundation Backgrounder No. 3713, June 16, 2022, https://www.heritage.org/sites/default/files/2022-06/BG3713□0.pdf, (accessed May 1, 2023).

Dr. GOSAR. I thank the gentlelady.

The gentleman, Mr. Harrell, is recognized for 5 minutes.

Statement of Jeremy Harrell, Chief Strategy Officer, Clearpath, Washington, DC

Mr. HARRELL. Thank you, Mr. Chairman. Great to see you, and great to see many other familiar faces on both sides of the dais. My name is Jeremy Harrell, and I am the Chief Strategy Officer of ClearPath, and thank you for the opportunity to testify today.

America's energy demands are rapidly increasing. By some estimates, the United States will need to double the capacity of the grid by 2050 to meet expected energy demand. Unfortunately, building energy infrastructure at that pace is procedurally impossible in today's regulatory environment. Never has the phrase "time is money" been more appropriate. The combination of permitting delays and ping-pong decisions from administrations past and present make projects exceedingly expensive.

Fortunately, fixing this outdated system is at the top of the congressional agenda. This Committee has rightly put permitting reform front and center, passing with bipartisan support the Lower Energy Costs Act as H.R. 1. The bill addresses many bottlenecks that make the current system a quagmire: unnecessary duplication, a morass of reviews across multiple agencies, and superfluous legal action. Solving these challenges will create jobs, boost energy security, and reduce global emissions, all while providing safety and environmental protection for all our communities.

Project developers are ready to build today. The important thing is for policymakers to keep an eye on the prize. The next step in the process cannot be to water down H.R. 1 into something milque-toast that fails to change our broken system. As bipartisan efforts continue in both the House and Senate, I would like to highlight three pillars of reform that must be expanded upon to unlock our energy future: (1) restore predictability to the system; (2) provide more streamlined litigation; and (3) improve coordination between Federal, state, and local governments.

To restore predictability, we must flip the permitting paradigm from a system that favors stopping a project to one that expedites the approval of projects that bring net benefits. The status quo is overwhelmingly tilted toward those who seek to delay or block projects. That may have made sense four decades ago, but today it is actually resulting in increased emissions and environmental degradation. We need a system that promotes good outcomes, both economic and environmental. That means projects that do not have an environmental impact should be granted immediate approval. This approach, similar to permit-by-rule concepts some states have implemented, allow pre-qualified technologies proven to have minimal environmental impacts and immense positive outcomes to move forward without delay.

Additionally, designating a list of pre-qualified geographic areas such as brownfields is also reasonable. The environmental impacts of new development in these locations is minimal and, in many cases, they are near the communities that need the redevelopment the most. That would be a win-win for the environment and the economy. And if you take those no-brainer projects out of the way, you have more time and resources to focus on more complicated projects.

Critically, Federal action can no longer vacillate according to political whims, particularly when the Congress has acted. Developers must be able to rely on decisions from one administration to the next. We have seen this pointedly, particularly in mining, like the Resolution Copper mine in Arizona or the Twin Metals Project in Minnesota. This Administration's tortured approach increases U.S. reliance on minerals sourced from overseas, including from countries that lack basic environmental and human rights protections. Our system should create jobs here, promote American innovation, and foster better global environmental outcomes.

Second, once a project is approved, any further adjudication should move quickly. Litigation under NEPA has become the favored tools of those who seek to indefinitely delay projects. Such prominent examples include the Cape Wind Project off the coast of Massachusetts and the Atlantic Coast pipeline connecting the Utica and Marcellus gas fields to Appalachia, where excessive legal challenges to Federal and State permits forced developers to cut their losses.

Recent history has shown clean energy projects are not immune to these delay tactics. Any changes to judicial review must balance a plaintiff's right to be heard with the goal of reaching finality on a more predictable timeline. This could be accomplished by immediately elevating any legal challenge under NEPA to Federal appellate courts. Ultimately, legal disputes must be resolved in less than a year.

And finally, it is important to embrace federalism where appropriate, and return more permitting authority to the people who know the communities the best. One recent example is Class VI wells for carbon capture projects. While many states have long held enforcement authority for other well classes, only North Dakota and Wyoming have been granted primacy by the EPA for perma- nent CO2 storage. To gain primacy, a state must prove their standards are as rigorous as the Federal standards. Whereas the EPA has taken 6 years to permit a Class VI well, it took North Dakota only 5 months to do so after receiving primacy.

These Federal delays are particularly egregious, given that the Department of Energy is investing billions of dollars to deploy new carbon capture technologies that we need. These types of measures would improve efficiency with no different environmental outcome. So, why wouldn't we shift the authority to the people closest to the geology?

In conclusion, the challenges are numerous, but Republican and Democratic policy-makers have never been more closely aligned on the need for permitting reform. Whether the motivation is climate, economic growth, or energy security, it is well past time to fix what is broken. ClearPath looks forward to working with this Committee to push reform across the finish line. Thank you.

[The prepared statement of Mr. Harrell follows:]

Prepared Statement of Jeremy Harrell, ClearPath, Inc., Chief Strategy Officer

Good afternoon Chairman Gosar, Ranking Member Stansbury and members of the Committee. My name is Jeremy Harrell, and I am the Chief Strategy Officer of ClearPath, a 501(c)(3) organization that develops and advances policies that accelerate innovations to reduce and remove global energy emissions.

Thank you for the opportunity to testify today and for holding this important hearing. America's energy demands are rapidly increasing. Some estimates say the U.S. will need to double the capacity of the grid by 2050 to meet expected clean energy demand. To support that grid modernization and U.S. manufacturing competitiveness, America will simultaneously need to construct tens of thousands of miles of new pipelines carrying natural gas, hydrogen, and captured carbon dioxide from power plants and industrial facilities.

Financing and building enough energy infrastructure projects to meet our nation's need for reliable, affordable cleaner energy is an immense challenge.

Recent projections show that 1,300 gigawatts of new clean energy would need to be added by 2035. This would more than double the grid's current capacity within the next12 years. But under the current regulatory environment, this pace of deployment is procedurally impossible.

Never has the phrase "time is money" been more appropriate. Regulatory delays that can last nearly a decade are making projects more expensive, and impeding the U.S.' ability to deploy billions of dollars of capital that would create American jobs, enhance U.S. energy security, keep consumer costs affordable, and reduce emissions.

The Council on Environmental Quality's (CEQ) own data shows that on average it takes agencies 4.5 years to issue a Record of Decision for an Environmental Impact Statement (EIS).[202] But the average belies the real challenge. In reality, 10 percent of projects took 10 years or more to reach a Record of Decision. The projects most likely to be held up in permitting purgatory are those that have the potential to offer the greatest benefits to the United States, including reduced energy costs, enhanced energy independence, increased economic opportunity, and lower global emissions.

The current system is broken. The structures in place are overwhelmingly titled toward those who seek to delay or block projects as opposed to those who seek to build. While that dynamic may have made sense four decades ago when policy-makers enacted these laws as a response to environmental disasters, today, those laws are being used to block projects that will reduce emissions and improve environmental quality. We need a system that promotes good outcomes—both economic and environmental. The pace and scale necessary to build clean energy infrastructure projects to reliably meet America's energy demand and reduce emissions is not something the authors of the 1970s environmental laws could have imagined.

The energy infrastructure we need today is simply not getting built fast enough, and throwing federal money at the projects or the agencies reviewing them is not going to substantially change that problem. The combination of permitting delays and "ping-ponged" decisions from Administrations past and present have disrupted the U.S. ability to build to fulfill needs. As a result, it can now take six years to permit carbon dioxide storage locations needed to store billions of tons captured from industrial sites, 16 years to permit an offshore wind farm in Massachusetts, and up to 15 years for a new

[202]https://trumpwhitehouse.archives.gov/wp-content/uploads/2020/01/20200612CEQ_EIS_Length_Report_Update.pdf.

transmission line from Wyoming to Utah.[203,204,205] Another important example is the need for timely approval of a new LNG terminal as well as any necessary interstate natural gas pipelines to supply these new terminals. These are just a few of the hundreds of projects held up by the status quo of the current system.

Fortunately, fixing this outdated, broken system is at the top of the agenda this Congress. This Committee has rightly put permitting reform front and center this year, passing with bipartisan support its signature energy package, the Lower Energy Costs Act, as H.R. 1.

This bill addresses bottlenecks that make the current system a quagmire: unnecessary duplication, a morass of reviews across multiple agencies, and superfluous legal action. Solving these challenges will reduce emissions, increase production and boost U.S. energy security, all while providing safety and environmental protection for local communities.

Project developers are ready to build today. There is real opportunity for this Congress to work on a bipartisan basis to modernize the permitting process. The important thing is policymakers keep an eye on the prize. Senate action cannot simply water down H.R. 1 into something milquetoast that fails to fundamentally change the current regulatory regime.

This is underscored by recent proposals released this month, as leaders in the key Senate committees on both sides of the aisle have put forward their own proposals, including many concepts that match themes included in H.R. 1.

As the permitting reform effort continues in both the U.S. House and Senate, I will highlight three key solutions that have been identified by project developers, former federal officials, academics and environmental non-governmental organizations.

1) Restore predictability to the system;
2) Provide more streamlined litigation; and
3) Improve coordination between and among federal, state and local governments.

American entrepreneurs have the wind at their backs to deploy more energy projects now. Congress has come together in a bipartisan manner, with

203 https://www.reuters.com/business/energy/top-us-oil-states-vie-carbon-capture-oversight-speed- up-permits-2022-01-26/.
204 https://www.nytimes.com/2017/12/19/us/offshore-cape-wind-farm.html.
205 https://cowboystatedaily.com/2023/04/20/after-15-years-of-permitting-transwest-wind-trans mission-project-is-still-5-years-from-going-live/.

bills like the CHIPS and Science Act and the Infrastructure Investment and Jobs Act (IIJA), to bring new technologies to the market and invest in American supply chains. 2022 saw record industry investment in energy, with the largest boost in recent years coming from the power sector.[206]

But again, simply spending more money on new projects will not necessarily make them a reality. Without meaningful permitting reform, there is a real risk that these major investments in technologies that the globe needs, such as carbon capture, advanced nuclear, and geothermal will go unrealized. And the U.S. will miss out on an opportunity to lead a global energy transformation.

While these challenges are numerous, Republican and Democratic policymakers have never been more closely aligned on the need for significant permitting reform. Whether the motivation is climate, economic growth, more energy production, or energy supply chain security, it is well-past time to fix what is broken, as America's energy, environmental, and economic future depends on sweeping reform.

Restore Predictability to the System

Reform must flip the permitting paradigm from one that favors stopping a project to one that expedites the approval process for projects that bring net benefits and comply with the legal requirements meant to ensure clean water and clean air. This approach would rely on a three-pronged approach that automatically advances projects with significant net benefits, focuses environmental and permit review on uniquely local conditions of a project on an accelerated timeline review, and keeps the relevant agencies within the boundaries of the laws Congress has enacted. Many of these concepts were included in H.R. 1, and it is important a final bill doubles down on the concept and maximizes their impact.

First, projects that do not have an environmental impact should be granted immediate approval. For example, replacing a retiring power plant with a zero-emissions advanced nuclear generator at an existing site or building a solar project on a brownfield site should not require a yearslong permitting process. Advancing these types of projects without delay is a win-win. The economic and environmental benefits of these projects should not be delayed by unnecessary bureaucracy.

---

[206] https://www.iea.org/reports/world-energy-investment-2022/overview-and-key-findings.

There should be criteria to prequalify technologies that are proven to have minimal environmental impacts and immense positive outcomes—similar to "permit-by-rule" concepts some states have implemented. In other words, there should be a presumption of project approval so long as the specifics of a project satisfy certain predefined criteria. In many cases, this would alleviate the requirement to do unnecessary boilerplate re-analysis.

One starting point could be to automatically advance projects that have nationally significant outcomes, like enhancing resilience of the grid or a significant reduction of global emissions, where the environmental impacts of development are well known. For example, a carbon capture retrofit of an existing facility, the modernization of a grid substation, or powering of a non-powered dam. H.R. 1 took a similar tack for energy storage projects at existing facilities and maintenance or upgrades to existing transmission and distribution infrastructure.

Similarly, designating a list of prequalified geographic areas to encourage project sponsors to seek out specific locations, would go a long way toward accelerating projects with the lowest impact. Such areas could include previously disturbed locations or well categorized sites, such as brownfield sites that present opportunities to use existing electrical or mechanical infrastructure or former military bases. The environmental impacts to these locations related to energy deployment are minimal, and in many cases these locations are in or near communities that need the redevelopment most urgently. Congress could also consider regulatory incentives to direct investment toward areas where impacts are already well understood.

Another opportunity could be to pair existing financial incentives, such as the "Opportunity Zones" or "Energy Communities" classifications established by Congress, with a streamlined permitting process to further boost investment. Both Opportunity Zones and Energy Communities were established by Congress to drive investment in distressed areas and communities that would benefit the most from new energy investments. Matching financial incentives with regulatory certainty will create a clear signal to project developers during the site selection process. Coordinated incentives like these can help drive investment to previously under- served areas and ensure the benefits of clean energy reach these communities without unnecessary delays.

Some of the most egregious problems of our broken system would be solved by this type of reform. For example, nonsensical approaches to geothermal exploration inhibit our ability of scaling baseload clean energy at scale. The Department of Energy estimates that geothermal generation could double by 2035 if our immense potential was unleashed. But concurrently, the Department found that "because additional steps and NEPA analyses are required, confirming the resource is more costly and risky," translating to permitting timelines of 5–7 years, rather than a 1– 3 year period that would otherwise be available with a categorical exclusion.[207]

Geothermal energy uses similar technology as oil and gas exploration and drilling activities. When oil and gas uses this technology, these resource confirmation (e.g. exploration) activities benefit from statutory authority enacted by the Energy Policy Act of 2005 that expedites five types of development activities. However, when the same mechanisms are used to confirm a geothermal energy resource, the expedited pathway does not apply. As a result, two very similar methods to test for resource feasibility must undergo substantially different permitting reviews despite both having negligible environmental impact. The Bureau of Land Management has the authority to administratively grant this same expedited pathway for geothermal energy resources, yet has wrung its hands for years rather than simply updating its regulatory guidance. This system is clearly broken.

Further, departments should proactively consult with other agencies to identify existing NEPA categorical exclusions available to accelerate development of energy infrastructure projects. DOE's recent Request for Information (RFI) to adopt new Categorical Exclusions is a model that should be replicated across other federal agencies.

Second, reform must streamline the approval process for projects where there are unique environmental impacts. In these cases, the review process could focus specifically on issues of the highest impact, resulting in more efficient timelines that still ensure compliance with existing environmental laws.

There are several provisions that have earned broad bipartisan support, including applying the "One Federal Decision" framework to energy projects. Similar support exists to reuse existing environmental review documents when a project will have substantially similar impacts as one previously studied. These provisions are both included in H.R. 1 and other proposals that have been recently made public. Those principles should be expanded upon.

One immense opportunity that could be fostered by reforms like this is in new nuclear technologies. The U.S. Nuclear Regulatory Commission (NRC) has publicly stated it anticipates at least 13 applications for advanced reactors by 2027, technologies that could bring safe, flexible, and reliable clean energy

---

[207] https://www.energy.gov/sites/default/files/2019/06/f63/GeoVision-full-report-opt.pdf.

to our energy system.[208] Decades of operation have shown that nuclear energy has a minimal environmental impact. Future designs hold the same promise.

Since the dawn of the nuclear age in the 1950s, nuclear reactors have been supplying Americans with clean, reliable, and affordable energy. On a bipartisan basis in the 115th, 116th, and 117th Congresses, legislation has been passed that strengthens the U.S. nuclear industry. However, except for the Vogtle Unit 3 reactor that recently came online in Georgia, the vast majority of nuclear plants in the United States were constructed over 40 years ago.[209]

That is changing today. The advanced reactor market is at an inflection point. Investors and potential end-users are closely watching first-of-a-kind utility-scale projects eyeing the late 2020s and early 2030s for commercial operation. American electric utilities are projecting a need for 90 GW of new nuclear power by 2050, nearly doubling our nuclear energy capacity in the next 30 years.[210] Simplifying the permitting for projects like TerraPower's flagship project in Kemmerer, Wyoming, which is leveraging the infrastructure at a retiring coal plant, is a no brainer. A nuclear facility is different from a coal-fired power plant, but many of the environmental factors that must be considered are similar. Additionally, many advanced reactors are looking to develop alongside industrial facilities or existing nuclear sites, where previous environmental analysis and community engagement has been extensive. A rational permitting system would leverage that work to accelerate exciting projects, not force needless duplication.

Third, federal action can no longer vacillate according to political whims, particularly when the Congress has acted. Given long development timelines needed to bring a project from financing to construction, project developers need to be able to rely on regulatory certainty from one Administration to the next. This need is perhaps most acute for projects that seek to unlock critical minerals.

While the Administration has announced award selections worth a combined total of nearly $5 billion for critical minerals demonstration projects funded by the bipartisan infrastructure bill and other new programs, there remains one glaring omission in the critical minerals supply chain: none of these selected projects addresses our inability to extract new materials

208 https://adamswebsearch2.nrc.gov/webSearch2/main.jsp?AccessionNumber=ML22038A001.
209 https://www.eia.gov/tools/faqs/faq.php?id=228&t=3.
210 https://www.powermag.com/u-s-utility-survey-suggests-industry-mulling-additions-of-90-gw- of-new-nuclear/.

domestically.[211] The International Energy Agency (IEA) predicts that demand for energy-related minerals like lithium, cobalt, graphite, and nickel could grow by 20–40 times by 2040.[212] The U.S. will not be able to recycle its way out of this demand for critical minerals.

It is difficult to overstate the U.S. dependence on foreign supply chains, including reliance on China. According to the 2023 U.S. Geological Survey's Mineral Commodities Summary, the U.S. was 100 percent net import reliant for 12 of the 50 individually listed critical minerals and was more than 50 percent net import reliant for an additional 31 critical mineral commodities.[213] Meanwhile, China was the leading producing nation for 30 of the 50 critical minerals.[214] Regardless of where the minerals are mined, China exerts dominant control over the refining process for many of these critical minerals. Rising demand for minerals will place major stress on global supply chains and undermine the ability of the U.S. to deploy more clean energy.

One of the most prominent examples of America's inability to permit mines is Resolution Copper, which Congress explicitly authorized when the Southeast Arizona Land Exchange and Conservation Act was enacted into law with the Carl Levin and Howard P. "Buck" McKeon National Defense Authorization Act for Fiscal Year 2015 (P.L. 113-291). Once approved, the proposed mine is expected to become the largest copper mine in North America, capable of producing up to 25 percent of U.S. copper demand each year.[215] The proposal received a final EIS in January 2021, only to have it unpublished by the Biden administration two months later.[216] The Administration is explicitly subverting Congressional intent with this project. These unnecessary delays precede a decade of construction before operations can begin, bringing the project timeline to at least two full decades from its inception. This back and forth regulatory review is far too common. The Resolution Copper Mine is just one of many examples. And the regulatory overreach deters investors, increases capital costs, and delays the energy security benefits of developing a robust domestic supply chain for clean energy and related infrastructure.

[211] https://www.clearpath.org/clearpath-infrastructure-tracker/.

[212] https://www.iea.org/reports/the-role-of-critical-minerals-in-clean-energy-transitions/ executive-summary.

[213] https://pubs.er.usgs.gov/publication/mcs2023

[214] Ibid.

[215] https://resolutioncopper.com/project-overview/.

[216] https://www.fs.usda.gov/detail/r3/home/?cid=FSEPRD858166.

Absent a clear, predictable, and streamlined process, America will continue to rely on critical minerals sourced from overseas, including from countries that pose national security risks or those that lack basic environmental and human rights protections. The choice should be clear: producing American resources here at home creates jobs, promotes innovation, increases energy security, and leads to better global environmental outcomes.

Provide More Streamlined Litigation

Once a project is approved, any further adjudications should be addressed as expeditiously as possible to ensure that protracted litigation does not undermine project viability. Judicial review is the biggest wildcard in the current permitting system, and H.R. 1 appropriately recognized it as an area that could have the most meaningful impact toward efficient project deliverability. Establishing requirements where any legal disputes must be resolved in less than one year would meaningfully address this uncertainty.

In the spirit of the current system rewarding those who seek to delay rather than those who seek to build, litigation under NEPA has become the favored tool of those who seek to indefinitely delay projects through procedural lawsuits. Such prominent examples include the saga of the proposed Cape Wind project off the coast of Massachusetts, where protracted litigation, including more than 20 administrative and judicial challenges to both federal and state reviews, ultimately led utilities to cancel power purchase agreements, effectively killing the project.[217] While no single suit ever specifically terminated the project, the purposeful delay tactics requiring evermore environmental analysis ultimately led investors to pull the plug. This same playbook is now being used to protest the approval for the Vineyard Wind project, despite new state laws that mandate utilities to procure offshore wind energy.[218]

The Atlantic Coast Pipeline (ACP) is another prime example in which legal uncertainty contributed to an untenable business environment leading project developers to cancel the project and take a loss. ACP was intended to bring natural gas access to residential, commercial, defense, and industrial

---

[217] http://static.djlmgdigital.com.s3.amazonaws.com/cct/capecodonline/graphics/pdf/ GridFM Notice.pdf.

[218] https://www.eenews.net/articles/4-lawsuits-threaten-vineyard-wind/.

customers in Virginia and North Carolina, but legal challenges to federal and state permits contributed to more than three years of delays and increased project costs from $8 billion from an original estimate of $5 billion.

Additionally, the Mountain Valley Pipeline (MVP) from northwestern West Virginia to southern Virginia has also attracted extensive Congressional attention because of similar uncertainty. No doubt, the expected build out of gas, hydrogen, and CO2 pipelines needed to meet our future system demands requires a more predictable process for the private sector to deliver on these projects.

As more clean energy projects enter the permitting process, clean energy projects will increasingly find themselves subject to these delay tactics. Such actions too often delay significant economic and environmental benefits, like new clean energy generation from that wind farm or the net reduction in global emissions from the use of lower emissions U.S. gas relative to dirtier Russian supplies in Europe. These increases in emissions or environmental harm are the very outcomes that NEPA was enacted to avoid and prevent.

Any changes to judicial review must balance a plaintiff's right to have his or her day in court with the goal of reaching finality on a more predictable timeline. One such proposal would be to immediately elevate any legal challenge under NEPA to the federal appellate court where the project is to be constructed or alternatively the DC Circuit. This would be similar to the process already used to challenge agency decisions, including those made by FERC.

Improve Coordination with State and Local Governments

Finally, it is important to recognize and address, to the maximum extent practicable, challenging permitting projects at the state and local level—without trampling on federalism.

An example of an unpredictable regulatory environment is the prolonged delay to review and approve permits for Class VI underground injection control wells needed to permanently sequester carbon dioxide. Class VI wells are a necessary part of the carbon capture equation of preventing more emissions and are the only authorized method permitted by the Environmental Protection Agency (EPA) to sequester carbon dioxide in permanent geologic storage.

While many states have long held primary enforcement authority for other well classes, only North Dakota and Wyoming have received primacy for this newest well class established in 2010. Congress rightly included provisions in the 2021 Consolidated Appropriations Act and the IIJA directing the EPA to support states applying for Class VI primacy and to actively improve the Class VI permitting review process.[219]

The advantages of state primacy for Class VI wells are readily apparent in North Dakota. Whereas the EPA has taken an average of three years to permit Class VI wells, it took North Dakota only five months. The EPA currently has more than 70 pending applications across eight states awaiting regulatory approval.[220]

This backlog is a prime example of where this Administration is working against its own priorities. The Department of Energy is investing billions of dollars to deploy new carbon capture technologies now, while the EPA muddles through reviews of storage sites at a palatial pace and the Department of the Interior stands in the way of related infrastructure projects across regions prime for commercial scale up.

The most egregious example may be in Louisiana. After years of delay, the Administration finally issued the draft rule necessary to approve Louisiana's request for Class VI primacy earlier this month. This initiates a 60-day comment period and a subsequent EPA response period that historically can take upwards of a year for a state to be granted final authority.[221] Once final, the decision is likely to have an immediate impact as 10 of the current outstanding Class VI permits are located in the state, which could unlock up to 6 million tons of carbon dioxide per year in Louisiana alone. If Republican and Democratic policy makers did not lean into the federal agencies in recent oversight hearings, this rule would likely still be stuck in the bureaucracy.

It is clear that the time to transfer Class VI authority should be improved for the other states looking to obtain primacy such as Pennsylvania, Arizona, Texas, and West Virginia, which are preparing applications for Class VI primacy. To date, primacy is the number one tool to get these projects permitted quickly, while preserving the safety of local communities. Additionally, this would allow federal agencies to focus their energies on permits in states not-yet equipped to take on permitting primacy or accelerate

---

219 https://carboncapturecoalition.org/recently-enacted-infrastructure-investment-and-jobs-act-to-bolster-economywide-deployment-of-carbon-management-technologies-upon-full-implementation/.

220 https://www.epa.gov/uic/class-vi-wells-permitted-epa.

221 https://www.epa.gov/newsreleases/epa-opens-public-comment-proposal-granting-louisiana-primacy-carbon-sequestration-and.

review of storage opportunities on federal lands or the Outer Continental Shelf, which have immense potential to contribute to our long-term energy future.

Similar barriers exist for proposed transmission lines that can better connect both new and existing generation assets to load as timelines to get new transmission projects developed now routinely stretch to over a decade.

One example is the SunZia line, designed to move power from New Mexico to California. The 550 mile line required cooperation from 10 federal agencies, 5 state agencies, and 9 local authorities while incorporating input from a host of additional state, local, and federal stakeholders. Projected to come online in 2025, the 3.5 GW project, which would provide power for millions of customers, will have taken over 17 years from proposal to completion.[222] These timelines, complicated by the intersection of different requirements from federal, state, tribal, and local regulators, impede the ability of new projects to interconnect to the grid.

According to the Lawrence Berkeley National Lab, there are 2000 GW of electricity and storage waiting in the interconnection queue to be connected to the grid.[223] While not all of these projects will be built, this figure is nearly double the current system capacity as it exists today. This backlog is especially relevant as hundreds of gigawatts of clean energy projects spend years stuck in the interconnection process, awaiting evaluation by transmission providers to determine their impact on the broader system. An average completion rate of 21% and queue wait time of 4 years makes meeting any target for a reliable and affordable clean energy system infeasible.[224] It is also important to note that analysis recently conducted by the regional transmission organization (RTO) PJM estimates that 40 GW of baseload generation, more than 21 percent of current installed capacity, is at risk of retirement by 2030 without reliable generation lined up to replace it and keep up with demand growth.[225]

While there is no silver bullet to rapidly and reliably modernize the grid, a combination of process improvements, permitting reforms, and technological innovation will help avoid clean energy deployment from hitting a wall.

---

[222] https://www.power-grid.com/td/sunzia-transmission-line-is-a-win-but-also-a-lesson-in-what- not-to-do/#gref.

[223] https://emp.lbl.gov/sites/default/files/queued_up_2022_04-06-2023.pdf.

[224] https://emp.lbl.gov/queues.

[225] https://www.pjm.com/-/media/library/reports-notices/special-reports/2023/energy-transition-in-pjm-resource-retirements-replacements-and-risks.ashx.

Conclusion

The current permitting system stymies clean energy resources and broadly delays the highest impact projects from delivering benefits. It is imperative that Congress address both aspects of the permitting process to maximize public and private sector investments and put steel in the ground. These pillars of pre-qualification to expedite review, more streamlined litigation, and improved coordination with state and local governments are priorities that merit consideration as the process to reach a permitting deal moves forward.

These reforms are ambitious by design as half measures have failed to move the needle for more than two decades. Anything less will only prolong the inability of the U.S. to build big things.

We look forward to working with this Committee to both further legislative action on regulatory reform and to reign in executive branch overreach. I look forward to today's discussion.

Dr. GOSAR. I thank the gentleman. The Chair now recognizes Ms. Pleune for 5 minutes.

Did I say that right?

Ms. PLEUNE. You did.

Dr. GOSAR. OK.

Statement of Jamie Pleune, Associate Professor (Research), S.J. Quinney College of Law, University of Utah, Salt Lake City, Utah

Ms. PLEUNE. Thank you, Chairman Gosar, Ranking Member Stansbury, and members of the Committee for the opportunity to be here today. My name is Jamie Pleune. I am an Associate Professor at the University of Utah Law School. We have done extensive empirical research on NEPA and its implementation times.

Little is known about the National Environmental Policy Act and its process. The data that is available focuses exclusively on environmental impact statements. To address this shortage of information, my colleagues, John Ruple and Erik Heiny, and I undertook the most comprehensive analysis of NEPA decision making that has been conducted. We analyzed 16 years of Forest Service data, which was 41,000 NEPA decisions.

The first question that we sought to answer was how long does the NEPA process really take? We found that the median time to complete an EIS was 2.8 years. For an EA, it was 1.2 years. And for a CE, it was only 4 months. These evidence-based time frames are dramatically shorter than the anecdotal time frames that are often cited.

We also reaffirmed the GAO estimate. The EISs are an extremely small percentage of all NEPA decisions. Specifically, within the Forest Service, who conducts more EISs than any other agency, EISs constitute only 2 percent of all decisions. The other 98 percent face less rigorous review. To put this in perspective very roughly, when we looked at the evidence over 16 years, only 200 decisions took longer than 4 years, and 33,000 took less than 1 year. The median time for projects that have identified, known, well-understood, and insignificant impacts is 4 months.

We also looked for reasons for delay. We developed a regression analysis that was able to look at NEPA-specific factors, and we found that those could only predict 25 percent of the variation. This meant that the primary causes of delay were external to the NEPA regulatory process. We found that the primary causes of delay are a lack of agency capacity, specifically projects hit bottlenecks when there are insufficient staff members to review a permit, or when there are not enough staff members with the expertise that is necessary to review a permit.

Additionally, unstable budgets and outdated technology caused delays. There were also delays in waiting for information from the operator.

And finally, compliance with other laws and coordinating with other agencies or coordinating within a team that is approving a permit also caused delay. This finding was consistent with an observation made by the Congressional Research Service that NEPA often functions as an umbrella statute. That is, it serves asa framework for compliance with other laws and regulatory requirements. Delays caused by compliance with those other legal standards are reflected in the NEPA process, but NEPA itself is not the cause of delay.

This interplay is visible in a study conducted by Amanda Miner regarding Forest Service litigation. Her research recognized that NEPA litigation usually involves multiple legal claims. Focusing on the cases in which the Forest Service lost, her research showed that 69 percent of the time the Forest Service would have lost, even if NEPA did not exist. This is important because reforms that focus solely on speeding up NEPA completion times may compromise agency's ability to comply with the mandates of other laws, which would ultimately create more delay in implementing projects.

After completing the study of Forest Service decision making, we looked at the mine permitting process. We found the same three primary causes of delay have consistently been identified in research investigation since 1999. So, what does this tell us about permit reform?

First, the biggest source of delay is a lack of staff and unstable budgets. The most important thing to improve permit processing time is to bolster agency capacity. They must have sufficient staff and staff with relevant expertise. With expanded capacity, agencies can engage in pre-application meetings with project sponsors and encourage early engagement with stakeholders.

This will address the second cause of delay: waiting for information from operators.

Finally, encouraging coordination between permitting authorities is a way to streamline the permitting process and make it more predictable. The procedures incorporated through FAST-41 have been effective in achieving predictability, transparency, and improved timelines. Significantly, this is important because these are the most complex projects available. Thank you.

[The prepared statement of Ms. Pleune follows:]

Prepared Statement of Professor Jamie Pleune, University of Utah, S.J. Quinney College of Law

Thank you, Chairman Gosar and Ranking Member Stansbury for the opportunity to testify today. My name is Jamie Pleune. I am an Associate Professor of Law (Research) at the S.J. Quinney College of Law, University of Utah and a Wallace Stegner Center Fellow.

The Wallace Stegner Center provides objective and actionable research on contemporary environmental issues. The non-partisan Law and Policy Program at the Wallace Stegner Center has done extensive empirical research into different aspects of the National Environmental Policy Act and its implementation. My testimony today reflects the results of that research and my personal observations. I do not speak on behalf of the University of Utah, and the views I express do not necessarily represent the views of the state of Utah, or the University of Utah.

I. Research Does Not Substantiate the Claim That NEPA Causes Delays

The National Environmental Policy Act is often blamed for delays in the permitting process. However, research does not support that characterization. Multiple studies have found that the NEPA analysis is rarely the primary cause of delay, even though delays may be reflected in the NEPA process.[226] Analysis on a project may stop and restart for external reasons such as funding, engineering requirements, changes in agency priorities, delays in obtaining non-federal approvals, or political opposition to the project.[227] These delays create the appearance of a long NEPA process, because the NEPA process marks a public beginning and ending, even though the NEPA analysis did not cause the delay.[228]

Additionally, the NEPA process may provide the structure for multiple different and independent permitting decisions. As the Congressional Research Service observed, "Most agencies use NEPA as an umbrella statute—that is, a framework to coordinate or demonstrate compliance with any studies, reviews, or consultations required by any other environmental laws."[229] This can create confusion because the need to comply with another law may be identified during the NEPA process, but NEPA is not the source

---

[226] Ryan Sud, Sanjay Patnaik & Robert Glicksman, The Brookings Institute, How to Reform Federal Permitting to Accelerate Clean Energy Infrastructure: A Nonpartisan Way Forward 14 (Feb. 2023).

[227] Government Accountability Office, GAO-14-379, National Environmental Policy Act: Little Information Exists on NEPA Analyses 15 (Apr. 2014); Linda Luther, Cong. Res. Serv., R4279, The Role of the Environmental Review Process in Federally Funded Highway Projects: Background and Issues for Congress 9 (Apr. 11, 2012) ("The environmental review process may start, stop, and restart for reasons unrelated to environmental issues. Local and state issues have shown to have the most significant influence on whether a project moves forward relatively quickly or takes longer than anticipated.").

[228] Executive Office of the President, Council on Environmental Quality, Environmental Impact Statement Timelines (2010–2018) 2 (June 2020) ("For some EISs, the timeline does not represent continuous activity. Delays may be attributable to the agency, the applicant, Congress, the needs of cooperating agencies, States, Tribes, and local interests, or public controversy."); John C. Ruple, Jamie Pleune & Erik Heiny, *Evidence-Based Recommendations for Improving National Environmental Policy Act Implementation*, 46 Colum. J. Envt'l L. 273, 304 (2022) [*hereinafter* Ruple et al. *Evidence-Based Recommendations for Improving Implementation of NEPA*] (conducting a detailed analysis of NEPA decisionmaking times and observing that complex projects can be completed quickly and simple projects subject to a truncated NEPA analysis may encounter delays); Executive Office of the President, Council on Environmental Quality, Environmental Impact Statement Timelines (2010–2018) 8 (June 2020).

[229] Cong. Res. Serv, RL 33152, The National Environmental Policy Act (NEPA): Background and Implementation 1 (Jan. 10, 2011).

of the obligation.[230] In fact, there is some evidence that the structure provided by the NEPA analysis actually reduces decisionmaking times.[231] The NEPA process can also reduce costs by identifying design problems before implementation of a project begins. A study prepared for the Transportation Research Board emphasized this potential benefit. "Spending more monies during planning and design will reduce the time and cost required for construction by avoiding unforeseen conditions, reducing to a minimum design errors and omissions, and developing schemes that will support the most efficient approach to construction."[232] For these reasons, projects exempted from NEPA may not be faster or cheaper. Instead of achieving speed by exempting projects from NEPA, permit reform should distinguish between productive and unproductive delays in the permitting process,[233] and focus on eliminating the latter.

II. NEPA Decisionmaking Times Vary Significantly and Projects That Encounter Extensive Delays Are the Exception, Not the Norm

Working with my colleague, John Ruple, and another colleague, Erik Heiny, at Utah Valley University, we analyzed 41,000 NEPA decisions made by the Forest Service at all levels of review.[234] This is one of the most comprehensive investigations into NEPA decisionmaking times that has been undertaken.[235]

---

[230] *Id.*; *see also* Ruple et al. *Evidence-Based Recommendations for Improving Implementation of NEPA supra* note 3 at 317–322 (exploring this dynamic with the National Forest Management Act).

[231] John C. Ruple et al., *Does NEPA Help or Harm ESA Critical Habitat Designations? An Assessment of Over 600 Critical Habitat Rules*, 46 Ecology L.Q. 829, 842 (2019) (finding that critical habitat designations subject to NEPA review were completed an average of 93 days faster than those that were not subject to NEPA review).

[232] Linda Luther, Cong. Res. Serv. R.42479, The Role of the Environmental Review Process in Federally Funded Highway Projects: Background and Issues for Congress 36 (Apr. 2012) (Citing H.R. Thomas and R.D. Ellis, *Avoiding Delays During the Construction Phase of Highway Projects*, Transportation Research Board, National Research Council, NCHRP 20–24 (Oct. 2001).

[233] Jamie Pleune, *Playing the Long Game: Expediting Permitting Without Compromising Protections*, 52 Env. L. Rep. 10893, 10896–07 (2022) [*hereinafter* Pleune, *Playing the Long Game*].

[234] Ruple et al., *Evidence-Based Recommendations for Improving NEPA Implementation supra* note 3 at 294.

[235] See also Forrest Fleischman et al., U.S. Forest Service Implementation of the National Environmental Policy Act: Fast, Variable, Rarely Litigated, and Declining, 118 J. of Forestry 403, 408 (2020) (conducting a descriptive analysis of a slightly different set of data from the Forest Service MYTR database); Exec. Office of the President, Council on Env't

The first question we sought to answer was, how long does the NEPA process actually take?" We found that the median time to complete an EIS was 2.8 years. For an EA, it was 1.2 years. And for a CE, the median was only 4 months.[236]

More importantly, only 2 percent of all decisions were made through an EIS.[237] The other 98% of NEPA decisions faced less rigorous review. The median time for projects with well-understood and insignificant impacts was only 4 months.

This approach makes sense. Projects with insignificant and well-understood impacts should receive quick approval, and the research shows that they usually do. On the other hand, projects that will impose significant impacts on communities, create hazards to health, threaten clean water, compromise clean air, or destroy natural resources deserve more rigorous review. The permitting process ensures that each project adheres to basic environmental and safety standards. The NEPA process creates a mechanism to explore whether a potentially harmful project can be adjusted to avoid, reduce, or mitigate harmful consequences. This approach has protected communities and resources from poorly contemplated projects for over 50 years.

While there are ways, which I will discuss, to improve the efficiency of permitting and the NEPA process, eliminating environmental standards or reducing analytical rigor is a poor choice that does not address the true sources of unproductive delay.

III. True Causes of Delay Can Be Addressed Without Compromising Environmental or Safety Standards

Permit reform should be driven by accurate data that defines the scope of the problem. The available data indicates that only a small percentage of NEPA decisions encounter excessive delays of the type that are commonly relied upon as anecdotal evidence.

Quality, Environmental Impact Statement Timelines (2010–2018) 1 (June 12, 2020) (providing description of government-wide EIS decisionmaking times).

236 *Id.* at 293.

237 *Id.* at 289. *See also* Government Accountability Office, GAO-14-379, National Environmental Policy Act: Little Information Exists on NEPA Analyses 8 (Apr. 2014) (estimating that government-wide, less than 1% of NEPA decisions are EISs, 5% are EAs, and 95% are CEs).

For example, in 2016, the Government Accountability Office studied processing times for mine permit applications.[238] Between 2010 and 2014, the BLM and the Forest Service approved 68 mine plans of operations. The majority (55%) were processed in less than 18 months, and 63% were processed in under two years. The remaining 37% were spread across a wide timeframe, with only six applications (less than 10%) taking longer than four years.

A similar trend is visible in the Forest Service data. Only projects in the 75th percentile of EISs took longer than four years.[239] Because there were only 870 EISs total, that means roughly 217 documents took longer than four years over 16 years. In contrast, at least 28,552 decisions were made in less than 1.2 years.[240] In other words, efficiency is possible, and it happens. It is also important to keep the big picture in focus. The commonly cited statistic that permits take six years represents less than 1% of all NEPA decisions.

Moreover, analytical rigor does not appear to be the primary cause of delay.[241] When we used a regression model to identify causes of delay, we learned that NEPA specific factors could not predict whether a project would encounter a delay.[242] Instead, the most common causes of delay were functional and external to the NEPA analysis. Those were: (1) agency capacity, which includes both staff availability and appropriate expertise; (2) delays attributable to the operator including waiting for information, changed plans of operation, and shifting priorities; and (3) compliance with other laws, which includes coordination with other permitting authorities.[243]

Notably, both the GAO and the National Research Council made the same observations regarding delays in the mine permit application process.[244] Three prominent causes of delay were: (1) insufficient resources, including staff, expertise, funding, or technology; (2) waiting for operator responses following vague applications or changes to a mine plan; and (3) compliance with other

---

[238] GAO, Hardrock Mining: BLM and Forest Service Have Taken Some Actions to Expedite the Mine Plan Review Process but Could Do More 6–7 (2016).

[239] Ruple et al., Evidence Based Recommendations for Improving NEPA supra note 3 at 297.

[240] *Id.* at 289 and 297 (showing that there were 6,881 EAs, 50% of which were completed in less than 1.2 years and there were 33,143 CEs, at least 75% of which were completed in less than one year).

[241] *Id.* at 302–303.

[242] *Id.* at 300–306.

[243] *Id.* at 306–322.

[244] Gov't Accountability Off., GAO-16-165, Hardrock Mining: BLM and Forest Service Have Taken Some Actions to Expedite the Mine Plan Review Process but Could Do More (2016); National Research Council, Hardrock Mining on Federal Lands (1999).

legal standards and/ or ineffective agency coordination during the mine plan review process.[245]

Even permits that can be processed quickly are affected by these factors. A 2014 investigation by the Office of Inspector General into BLM permit processing times for oil and gas wells emphasizes this point.[246] The BLM receives approximately 5,000 new APDs each year, which are processed at 33 different field offices. According to the BLM, the average processing time in 2012 was 228 days, but this number only tells part of the story. Even though each field office is governed by the same legal standard, the permit processing times varied widely. Buffalo, WY and Miles City, MT took more than 300 days to process permits. In contrast, five field offices took less than 100 days. Anchorage, AK averaged 37 days.

This dramatic variation in permit processing times cannot be blamed on NEPA or environmental standards because each field office was applying the same legal standard to the same activity. Sources of delay were a lack of staff, poor data management, and weaknesses in oversight and accountability.[247] Even though BLM had repeatedly identified staffing shortages as a problem, limited budgets combined with a high cost of living made it difficult to attract and retain employees. Field offices with staff shortages experienced prolonged review times. Permit coordination and management also mattered. Most field offices did not assign a manager to oversee the APD process, resulting in an open-ended process where applications languished and no one could predict when the application would be finished. Field offices that did assign field managers processed permits in less than half the average time.[248] In other words, improving permit coordination resulted in faster permit processing times.

The consistency of these findings across time, agencies, and practice constitute reliable evidence as to the real causes of delay in permit processing. These delays can be summarized as: (1) agency capacity; (2) delays attributable to the operator; and (3) permitting coordination. These findings demonstrate that the choice between speed and environmental standards is a false dilemma. Each of the true sources of delay can be addressed without

---

245 *See* Pleune, *Playing the Long Game supra* note 8 at 10900–10906 (discussing these studies).

246 Office of Inspector General, Department of the Interior, Onshore Oil and Gas Permitting, U.S. Dept. of Int., Report No. CR-EV-MOA-0003-2013 (June 2014).

247 *Id.* at 6.

248 The Field Office in Silt, CO which uses supervisors, averaged 108 days in FY2012 versus the bureau wide average of 228 days. The field office in Carlsbad NM uses a field manager, as well as an in-house automated tracking system. Its processing times average 110 days. *Id.* at 7.

compromising environmental standards that protect safe, healthy, and clean communities.

IV. Permit Reform That Addresses the True Causes of Delay

We can improve permit processing times by bolstering agency capacity, fostering early communication with permit applicants, and improving permit coordination. Notably, improved communication and better coordination depend on sufficient agency capacity. So the first step for permit reform should be to focus on agency capacity.

1. Building Agency Capacity Requires Long-Term Funding and Strategic Workforce Planning

Wisely, Congress has already taken a step in the right direction. The Inflation Reduction Act contained roughly $1 billion directed toward improving agencies' environmental review processes and NEPA implementation. These funds are a critical—and encouraging—first step; however, the journey is not over. Agencies must rebuild from chronic shortages. Additionally, the funding must be reliable and sustained to allow agencies to implement strategic workforce plans, retain experienced staff, and engage in proactive planning that can facilitate faster decisionmaking. Finally, agencies must have flexibility to spend the funds in the area of greatest need.[249]

Many agencies that were already understaffed suffered extreme losses of staff under the last Administration. The Bureau of Land Management offers an instructive case study.[250] Since 2011, it has been on the GAO's list of programs at high risk and vulnerable to waste, fraud and abuse due in part to a lack of staff. This problem was further exacerbated in July 2020, when the

---

[249] *See* Office of Inspector General, Department of the Interior, Onshore Oil and Gas Permitting, U.S. Dept. of Int., Report No. CR-EV-MOA-0003-2013, 10 (June 2014) (noting that Congress had allocated funding to seven field offices as a pilot project for improving oil and gas processing times through expanded capacity; however those field offices no longer carried the heaviest workload due to shifting development patterns and the act did not allow BLM to transfer the special funding to offices with the greatest need).

[250] This discussion draws heavily from the following article. Jamie Pleune & Ted Boling, *This Permit Reform Works. Why Aren't Mine Projects Using It?* 53 Envt'l L. Rep. (forthcoming June 2023) [*hereinafter* Pleune & Boling, *This Permit Reform Works*]. *See also* Gov't Accountability Off., GAO-20-379R, Bureau of Land Management: Agency Reorganization Efforts Did Not Substantially Address Key Practices for Effective Reforms (Mar. 6, 2020) and Gov't Accountability Off., GAO-22-104247, BLM, Better Workforce Planning and Data Would Help Mitigate the Effects of Recent Staff Vacancies (Nov. 2021).

last Administration abruptly decided relocate BLM's headquarters from Washington D.C. to Grand Junction, Colorado. The Headquarters Office, which develops guidance and regulations, should be staffed by 311 career positions. However, it was already severely understaffed with 132 vacant positions before the relocation announcement. In response to the relocation announcement, 81 more staff left, leaving the leadership at 31 percent capacity. The remaining leadership team were dispersed among multiple offices.

Numbers do not tell the whole story. The BLM also suffered a loss of experienced staff. Every BLM staff member interviewed reported that the loss of experienced staff negatively affected their offices' ability to conduct its duties. For example, the loss of institutional knowledge about laws and regulations meant that the BLM could not provide knowledgeable input on proposed rules and legislation. Other staff admitted that the rapid loss of experienced staff hindered knowledge transfer. In a follow-up report, a year later, all BLM staff interviewed by the GAO reported challenges in completing their duties due to headquarters vacancies. As a result of delays in creating or clarifying guidance or policies, some staff relied on outdated policy guidance to make decisions. Other staff reported delays implementing upgrades to information technology systems, which GAO had previously recommended be updated. Obviously these institutional challenges would affect permitting times. Applicants could not receive good guidance from experienced staff, and staff members processing permits had little instruction on how to proceed effectively. The BLM is not alone. Multiple agencies with permitting or infrastructure responsibilities, are short-staffed and underfunded.

Filling vacancies requires strategic workforce planning, but few agencies have engaged in that process. When the GAO investigated the BLM's workforce planning in 2020, it found that the BLM had no way of tracking vacancies and no recruitment plan for filling vacancies.[251] When asking for data on the total number of positions and vacancies agency wide, the GAO was told that BLM does not maintain a list of vacancies for state offices. As a result, it was not possible to determine the proportion of positions that are vacant at any given time or the specific positions that are vacant. This lack of information obviously creates a problem for improving capacity in a way that results in improved efficiency. Strategic workforce planning is critical to

---

[251] Government Accountability Office, Bureau of Land Management: Agency Reorganization Efforts Did Not Substantially Address Key Practices for Effective Reforms GAO-20-379R, 10 (Mar. 6, 2020).

ensuring that agencies spend wisely and build a workforce capable of fulfilling agency missions. Additionally, agencies must be confident that the funding will not disappear. Unstable budgets do not build durable workforces.

2. Pre-Application Meetings, Early Stakeholder Engagement, Permit Sequencing, And Transparent Schedules Are Proven Methods for Improving Efficiency Without Compromising Environmental Standards Or Public Participation

With expanded capacity, agencies can address other sources of delay such as communication with permit applicants, and interagency coordination. The best practices and procedural requirements of FAST-41 target both problems and serve as a valuable test case for their efficacy.[252] Wisely, Congress has also appropriated funding to support the Permitting Council, which should help propagate and further implement these practices.

Pre-application meetings with project proponents speed permitting by avoiding delays later in the process.[253] Meeting with regulators and stakeholders early allows project sponsors to incorporate environmental and social sensitivities at the design phase, when impact avoidance is still feasible and cost-effective. Additionally, project sponsors can benefit from agencies' experience with addressing controversial or complex impacts in other similar projects. This results in fewer modifications later in the process. It also reduces permit review times by minimizing the amount of additional information agencies must request during the review process. Especially in complex areas with conflicting or overlapping jurisdictions, the analytical and public comment procedures imposed by the NEPA process can actually speed project approval by providing a framework for analysis and bringing multiple permitting authorities to the table.

For example, the NEXUS Gas Transmission Project, was a 250-mile natural gas pipeline traversing Pennsylvania, West Virginia, Ohio, and Michigan.[254] During the pre-application process, which included extensive public participation, the project sponsors incorporated 239 route alternatives and variations in the pipeline design to address landowner requests, avoid sensitive resources, or respond to engineering restraints. This feedback

---

[252] For a thorough discussion of these issues, *see* Pleune & Boling, *This Permit Reform Works supra* note 25.

[253] *Id.* at Section IV.

[254] Office of the Executive Director, Federal Permitting Improvement Steering Council, Recommended Best Practices for Project Review and Permitting for Infrastructure Projects for Fiscal Year 2018, at 17 (2017).

resulted in a 91 percent change from the original proposed route design—a number of modifications that would have been prohibitively expensive at the end of the review process. Using this information at the beginning of the process improved efficiency and arguably resulted in a better end-result and a final application that was processed more expeditiously.[255]

Early stakeholder engagement is also consistent with the results of recent research conducted by MIT investigating sources of delay for renewable energy projects.[256] Two critical sources of delay were opposition from affected landowners due to real or perceived harms that the project would bring, and inconsistency between overlapping authorities, such as local, state, tribal, and federal jurisdictions. Based on their empirical research, they concluded that "incorporating all stakeholder perspectives from the outset of a siting process will probably save time and money.[257]

This is not to say that early stakeholder engagement will eliminate all opposition. The NEXUS Gas Transmission Project demonstrates a conundrum with finding permitting success stories. The ideal result of implementing a best practice, like preapplication meetings, is the avoidance of a bad result, like project delays from unexpected impacts or local opposition. If the best practice works, the bad result will not occur. Which means that that a success story must prove a negative. It makes sense intuitively that such substantial route alterations would address many concerns and reduce opposition, but there is no way to prove what would have happened if the pipeline design had proceeded as originally designed. Moreover, the absence of opposition is an unrealistic standard to demonstrate success. With large projects, like the NEXUS Transmission Gas Line, it would be virtually impossible to avoid all opposition.[258] Thus, recognizing success requires enough familiarity with the process to understand what could have happened in a given scenario.

---

[255] This is not to say that early engagement eliminated local opposition in every community. See Heidi Gorovitz Robertson, Home Rule Symposium: Cities and Citizens Seethe: A Case Study of Local Efforts to Influence Natural Gas Pipeline Routing Decisions, 122 W. Va. L. Rev. 881, 907–934 (Spring 2020) [hereinafter Gorovitz, Cities and Towns Seethe] (describing FERC's extensive public engagement and local opposition in three Ohio towns).

[256] Lawrence Susskind et al., *Sources of Opposition to Renewable Energy Projects in the United States*, 165 Energy Policy 112922 (2022).

[257] *Id.* at 13.

[258] Gorovitz, *Cities and Towns Seethe supra* note 30 at 907–934 (detailing some local opposition to the pipeline).

FAST-41 procedures also focus heavily on avoiding delays caused by poor interagency coordination.[259] In early stakeholder meetings, agencies develop a Coordinated Project Plan that encourages the development of concurrent, rather than sequential, analyses. Early and coordinated stakeholder engagement enhances the efficiency of this process. The Permitting Dashboard creates accountability, reducing the likelihood that a decision will linger on the back of someone's desk. Finally, enhanced oversight from the Executive Director of the Permitting Council creates an opportunity for conflicts between jurisdictions to be addressed early, and in a coordinated manner. The Permitting Council's 2020 Annual Report to Congress offered both quantitative and qualitative evidence of the program's success. Between 2010 and 2018, the average time across all agencies for a project to complete an EIS was 4.5 years. In contrast, the average time to complete an EIS for projects that went through the FAST-41 process by 2020 was 2.5 years. Perhaps more persuasively, the report included testimonials from project proponents praising the transparency and efficiency of the process.

A GAO investigation, also found evidence of the program's success that extended beyond federal agencies.[260] Although FAST-41 only directly affects federal agencies, the opportunities for coordination extend to state, local, and tribal permitting authorities. Early engagement creates opportunities for permitting authorities to enter into Memorandums of Understanding establishing roles and responsibilities As an example, the GAO report discussed the Mid-Barataria Sediment Diversion project, which was a complex FAST-41 project focused on restoring ecosystems damaged by coastal erosion and the Deepwater Horizon Oil spill, which oiled over 684 miles of wetlands across the Gulf of Mexico, particularly in the Barataria Bay.[261] With the Army Corps of Engineers as lead agency, six federal agencies entered into a memorandum of understanding with several Louisiana state agencies, accelerating the environmental review and permitting process by nearly two years.[262] Due to the complexity of this project, many project participants believe that it would not have been possible without the coordination procedures afforded through the FAST-41 process. The GAO

---

[259] For a thorough discussion of these issues, *see* Pleune & Boling, *This Permit Reform Works supra* note 25.

[260] Gov't Accountability Office, GAO-20-19, Infrastructure Projects: Actions Needed to Fully Develop Performance Schedules for Environmental Reviews 20–21 (Oct. 2019) [*hereinafter* GAO, Actions Needed to Fully Develop Performance Schedules].

[261] *Id.* at 22.

[262] *Id.*

attributed this efficiency to enhanced interagency coordination.[263] In summary, efficient permitting is possible without compromising environmental standards.

V. Renewable Energy Projects Face Permitting Delays Unrelated to Environmental Standards Due to Serious Problems with the Interconnect Queue and Regional Transmission Organizations

A recent study by the Lawrence Berkeley National Laboratory found that there are over 2,000 GW of total generation and storage capacity waiting for approval to connect to the grid, 95 percent of which are solar, wind, or battery storage.[264] The combined wind and solar capacity actively seeking connection to the grid is approximately 1,250 GW, which is roughly equal to the current installed capacity of the entire U.S. power plant fleet and consistent with what is necessary to achieve the 2030 target.[265]

The hurdles facing these projects involve issues with the interconnect queue, regional control of transmission, and a lack of high power transmission lines. The wait time for projects in the interconnect queue is steadily rising.[266] Between 2000– 2007, the time between an initial connection request and a fully built, operational plant was typically less than two years. Between 2018–2022, that timeframe doubled to an average of almost 4 years, with an increasing trend. By 2022, the median time between an interconnection request to commercial operations date reached almost 5 years.[267] The increased volume of proposed renewable projects sitting in the queue promises

---

[263] *Id.* at 22; *see also* Mark Schleifstein, *Louisiana Granted Final Funds for Unprecedented Coastal Restoration Project* NOLA.com (Mar. 9, 2023) (reporting that Louisiana was granted the final necessary funds to build the unprecedented Mid-Barataria Sediment Diversion project aimed at helping slow land loss devastating the coast).

[264] Berkeley Lab, Queued Up: Characteristics of Power Plants Seeking Transmission Interconnection, https://emp.lbl.gov/queues (last visited May 9, 2023); Energy Technologies Area, *Grid Connection Requests grow by 40% in 2022 As Clean Energy Surges, Despite Backlogs and Uncertainty* (Apr. 6, 2023) https://energy.lbl.gov/news/grid-connection-requests-grow-40-2022b(last visited May 9, 2023).

[265] *Id.*; *see also* Emma Penrod, *Why the Energy Transition Broke the U.S. Interconnection System* Utility Dive (Aug. 22, 2022) (quoting Bhaskar Ray, vice president of interconnection and development engineering for energy developer Qcells USA as estimating that if all the capacity currently waiting in the interconnection queue today were built by 2030, the U.S. would meet the 80% clean energy share milestone).

[266] Berkeley Lab, Queued Up: Characteristics of Power Plants Seeking Transmission Interconnection, https://emp.lbl.gov/queues (last visited May 9, 2023).

[267] *Id.*

to continue amplifying this problem. For example, at least two regional transmission organizations, the entities responsible for approving requests to connect to the grid, have announced pauses on accepting new projects until they can process their backlogs. The nation's largest electric grid operator, PJM Interconnection, coordinates electricity movement in 13 states and the District of Columbia.[268] It has announced that it will not process any new applications until the end of 2025.[269] Similarly, CAISO,[270] a California grid operator, declined to accept any new projects in 2022 while they processed their backlog. Both entities are looking for systemic solutions to improve the grid connection process. The interconnect queue, regional control of transmission, and a lack of high power transmission lines are serious problems delaying implementation of green technology. These problems are worthy of permit reform and require Congressional attention. Strengthening federal authority to facilitate interstate transmission power transmission may be one solution. Regardless of the solution to these complex problems, delays caused by the interconnect queue should not be conflated with NEPA and environmental standards.

VI. Good Decisions Are More Important Than Rushed Decisions

The discussion around permit reform often focuses exclusively on decisionmaking times. A legitimate sense of urgency to implement clean energy projects drives many to seek shortcuts. However, rushed decisions have consequences. For example, the GAO recently investigated the Department of Energy's management of $1.1 billion in funding for carbon capture sequestration projects.[271] It found that between 2009 to 2022, the DOE provided $684 million to eight coal projects. Every project went over budget and exceeded timelines. Only one project went into operation, and even that

[268] Delaware, Illinois, Indiana, Kentucky, Maryland, Michigan, New Jersey, North Carolina, Ohio, Pennsylvania, Tennessee, Virginia, West Virginia. https://pjm.com/about-pjm/who-we-are/ territory-served (last visited Apr. 24, 2023).

[269] Institute for Energy Research, PJM Plans for a Two-Year Pause on Reviewing Project Applications (Feb. 22, 2022). https://www.instituteforenergyresearch.org/the-grid/pjm-plans-for-a- two-year-pause-on-reviewing-project-applications/#:?:text=To% 20implement %20it%2C%20PJM%20is%20proposing%20an%20interim,on%20those%20coming% 20at%20the%20end%20of%202027.

[270] California Independent System Operator https://www.caiso.com/Pages/default.aspx.

[271] Gov't Accountability Off., GAO-22-105111, Carbon Capture and Storage: Actions Needed to Improve DOE Management of Demonstration Projects (Dec. 2021).

project eventually shut down because it was not economically viable. In summary, none of the projects succeeded.

Rushed decisionmaking contributed to the DOE's poor judgment and money management, according to the GAO. Eager to spend funding from the American Recovery and Reinvestment Act of 2009, the DOE bypassed its usual selection procedures and committed to coal projects at their initial selection. Additionally, DOE used expedited timeframes for project negotiations—shortening them from one year to three months. Bypassing procedures reduced DOE's ability to identify and mitigate risks, resulting in a $684 million loss for the American people.

Like mismanaged money, environmental resources, communities, and human health are not replaceable. The BP Deepwater Horizon Oil Spill offers a cautionary tale against expediting permitting by eliminating analytical rigor and environ- mental analysis.[272] The Macondo well never underwent NEPA's hard look requirement due to a broad categorical exclusion.[273] Additionally, staff within the managing agencies were subject to 30-day deadlines and instructed to approve permits as quickly as possible, without creating unnecessary delays.[274] Furthermore the industry had grown more quickly than the agency, resulting in a offices that were short-staffed and lacked training or expertise to understand the implications of the technology being used.[275]

Due to NEPA streamlining procedures, short review deadlines, and insufficient agency capacity, risks that might have been exposed through the NEPA process went undetected. For example, BP's emergency response plan for a subsea blow-out was to drill a relief well.[276] When this became the only option for containing the spill, BP revealed that drilling the relief well would take at least three months. Meanwhile, the spill kept gushing over 50,000 barrels of oil per day into the Gulf.[277] This risk should have been caught by agency staff at the permitting stage, but it wasn't. Because the permit was not subject to public comment or circulated to other agencies, no one else saw it either. We cannot know if a more rigorous permitting process would have avoided the spill. But we can learn from the past.

---

[272] National Commission on the BP Deepwater Horizon Oil Spill and Offshore Drilling, Report to the President, Deep Water: The Gulf Oil Disaster and the Future of Offshore Drilling (January 2011) [*hereinafter* Deepwater Horizon Commission Report].

[273] *Id.* at 82.

[274] *Id.*

[275] *Id.* at 73–74.

[276] *Id.* at 132.

[277] *Id.* at 167.

Good decisions are more important than rushed decisions. Many reform proponents emphasize statutory exemptions for favored projects, expanding the use of categorical exclusions, imposing short deadlines and page limits, restricting public comment opportunities, and reducing the rigor of environmental analysis.[278] As the Deepwater Horizon example demonstrates, these reform proposals are likely to underestimate risks and overlook problems.

Conclusion

Despite its ubiquity, the phrase "permit reform" is misleading. There is no single "permit law" that can be amended to eradicate delays. Complex projects, like transmission lines, mine permits, and renewable energy projects implicate a variety of legal standards and permitting authorities, each focused on protecting different resources such as clean air, clean water, endangered species, and cultural resources. Used properly, the NEPA process can facilitate coordinated information gathering and decisionmaking and streamline the permitting process. Improved inter-agency coordination, shared data management, and strategic permit sequencing facilitate concurrent review between permitting authorities. Enhanced communication with the project sponsor, implementation of permitting best practices, and greater accountability for permitting authorities avoid inefficient delays. Early stakeholder engagement creates an opportunity to identify, avoid, and mitigate harms at the most cost-effective phase of a project's life cycle. All of these improvements depend on robust agency capacity. These reforms are not easily encapsulated in a pithy soundbite, but they do address the true causes of delay in permitting without compromising environmental or safety standards.

---

278 *See e.g.*, The Building U.S. Infrastructure through Limited Delays and Efficient Reviews (BUILDER) Act (H.R. 2515) *available at* https://transportation.house.gov/builder-act/default.aspx (last visited Mar. 17, 2023).

Questions Submitted for the Record to Professor Jamie Pleune, University of Utah, S.J. Quinney College of Law

Questions Submitted by Representative Westerman

Question 1. Can you explain how Congress can modernize and clarify the CEQ regulations to facilitate more efficient, effective, and timely NEPA reviews by Federal agencies by simplifying regulatory requirements, codifying certain guidance and caselaw relevant to those proposed regulations, revising the regulations to reflect current technologies and agency practices, eliminating obsolete provisions, and improving the format and readability of the regulations?

Answer. There has been intense focus on revising CEQ's NEPA regulations and codifying certain changes to NEPA in order to improve decisionmaking timelines. In addition to CEQ's ongoing efforts to revise its NEPA implementing regulations,[279] each agency also has its own implementing regulations tailored to the practices and challenges of implementing NEPA within the context of work accomplished by each agency.[280] While there may be some regulatory provisions that merit revision, research suggests that the regulatory requirements of NEPA are only a small factor in the variation between decisionmaking times. External factors such as agency capacity, budgets, technology, project management, compliance with

---

[279] Council on Environmental Quality, Update to the Regulations Implementing the Procedural Provisions of the National Environmental Policy Act, 85 Fed. Reg. 43,304 (July 16, 2020) (''2020 Regulations); Council on Environmental Quality, National Environmental Policy Act Implementing Regulations Revisions, 87 Fed. Reg. 23,453 (Apr. 20, 2022) (''Phase 1 Final Rule''); Fall 2022 Unified Agenda of Regulatory and Deregulatory Actions, RIN 0331-AA07 (describing scope of Phase 2 revisions to NEPA implementing regulations). *See also*, John Ruple et al., *Evidence-Based Recommendations for Improving National Environmental Policy Act Implementation*, 46 Columbia J. Envt'l L. 274, 283-284 (2022) [*hereinafter* Ruple et al., *Evidence- Based Recommendations for Improving NEPA Implementation*] (explaining regulatory structure and procedural history of amendments to NEPA's implementing regulations).

[280] 40 C.F.R. 1507.3 (2020) (instructing agencies to develop or revise procedures to implement NEPA); Council on Environmental Quality, Deadline for Agencies to Propose Updates to National Environmental Policy Act Procedures, 86 Fed. Reg. 34,154 (June 29, 2021) (extending deadline for agencies to revise NEPA regulations); U.S. Forest Service, National Environmental Policy Act (NEPA) Compliance 85 Fed. Reg. 73,620 (Nov. 19, 2020) (finalizing updates to Forest Service Regulations implementing NEPA); Ruple et al., *Evidence-Based Recommendations for Improving NEPA Implementation supra* note 1 at 286-287 (describing potential conflict where the Forest Service had already initiated a procedure for updating its implementing regulations for NEPA prior to the issuance of the 2020 Rule).

other laws, changes to the scope of the project, and litigation aversion appear to have a heavier influence on efficient and timely NEPA reviews.[281]

Rather than focusing on regulatory changes that reduce analytical rigor and environmental protections without offering a substantive benefit, a more productive approach to improving NEPA efficiency would focus on improving agency capacity, promoting strategically-sized analyses for long-term efficiency, using NEPA as a framework for structured inter-agency (and intergovernmental) collaboration, and utilizing the NEPA process to develop consensus.[282] The procedures adopted through FAST-41 appear to improve predictability, transparency and timeliness for complex projects.[283]

Additional promising practices have been publicized in the annual Best Practices Reports issued by the Federal Permitting Improvement Steering Council.[284] The early iterations of these reports (2017 & 2018) were particularly creative and identified specific practices that had been implemented by different agencies with positive results for efficiency. Many of the best practices identified in these reports could be replicated to improve efficiency across agencies. For example, the Army Corps of Engineers improved its web-based application for a general permit, including creating an online permit application, with a video tutorial on how to fill it out, and specific contact information for assistance. This reduced the frequency of incomplete or inaccurate applications, which reduced processing times. It also freed up staff members to focus on more complex permits.[285] These reports identify manageable, affordable, and replicable practices that improve efficiency. Unfortunately, later iterations of the best practices report focus less on success stories. In particular, the format adopted in 2022, which requires a Quarterly Agency Performance Report, imposes an additional workload on the Permitting Council and agencies without providing the same overview of

---

[281] Gov't Accountability Off., GAO-14-370, National Environmental Policy Act: Little Information Exists on NEPA Analyses 1, 15 (2014) [*hereinafter* GAO, NEPA: Little Information Exists] (noting that for non-federal projects requiring a federal permit, delays in obtaining project funding, changes to a proposal that occur during the NEPA process, and non-federal approvals all may delay a NEPA analysis); Ruple et al., *Evidence-Based Recommendations for Improving NEPA Implementation supra* note 1 at 299, 327-333).

[282] *Id.* at 335-340.

[283] Jamie Pleune & Edward Boling, *This Permit Reform Already Works. Why Aren't More Mining Projects Using It?* 53 Env. L. Rep. 10463, 10468 (June 2023) [*hereinafter* Pleune & Boling, *This Permit Reform Already Works*].

[284] https: // www.permits.performance.gov/fpisc-content / reports-and-publications#annual reportto congress.

[285] Federal Permitting Improvement Steering Council, Recommended Best Practices for Environmental Reviews and Authorizations for Infrastructure Projects for Fiscal Year 2018 (Dec. 2017).

creative measures adopted by different agencies that test improved practices and assess their value.

Question 2. Yes or No—the study on NEPA implementation you co-authored, Evidence-Based Recommendations for Improving National Environmental Policy Act Implementation, only analyzed NEPA decisions completed by the U.S. Forest Service. If yes, please explain the limits of extrapolating data and conclusions from a study on one government agency and applying those conclusions to other government agencies.

Answer. Yes. The article, Evidence-Based Recommendations for Improving National Environmental Policy Act Implementation, used a database of NEPA decisions issued by the U.S. Forest Service. To our knowledge, the Multi-Year Trend Report database compiled by the Forest Service is the most comprehensive, detailed, and reliable set of data regarding NEPA decisions gathered by any agency.[286] It would be valuable for other agencies to develop similar databases in order to compare practices.[287] The MYTR database also has limitations. It was designed as a tracking system to facilitate compliance with public disclosure duties.[288] The information that it contains is specific to NEPA decision documents, which are distinct from the time required to implement a project following its approval. Additionally, the database does not track when work on a NEPA decision document is paused due to changes in the scope of the project, political priorities, or budgetary limitations. Finally, the MYTR database offers information regarding decisionmaking times, but it does not provide a way to test whether the NEPA process produces better projects through the twin aims of meaningful public engagement and careful consideration of environmental impacts. Valuable information about avoided impacts, improved time to implementation, and reduction of community opposition to a project are not visible in this database.

Despite these limitations, there are two reasons to believe that the information we obtained regarding Forest Service NEPA practice is informative for the practices of other agencies. First, the Forest Service conducts more EISs than any other agency.[289] Second, when we turned our attention to the mine permitting process, multiple reports identified the same underlying causes of delay in the mine permitting process that we observed in

---

[286] Ruple et al., Evidence-Based Recommendations for Improving NEPA Implementation supra note 1 at 288, 333.

[287] Id. at 333.

[288] Id. at 289.

[289] Executive Office of the President, Council on Environmental Quality, Length of Environmental Impact Statements (2013-2018).

the Forest Service's NEPA practices.[290] For these reasons, we believe that the information regarding Forest Service practice is informative, even if it is not perfect.

Question 3. The study you co-authored on NEPA implementation in the U.S. Forest Service, Evidence-Based Recommendations for Improving National Environmental Policy Act Implementation, found that the administrative region had a "significant influence" on decision-making times. Can you explain the regional differences that affected the variation in completion times and why further research is necessary to explain the regional differences in decision-making times?

Answer. The regression model revealed that the Forest Service administrative region responsible for overseeing a NEPA analysis has a significant influence on decisionmaking times. This finding surprised use because each Forest Service region is implementing the same laws, subject to the same regulations, pursuant to the same administrative guidance, involving the same activities. If delays in decision- making times were caused solely by the NEPA process, we would expect similar mean completion times across regions. The regional variation suggests that factors external to the NEPA process was affecting decisionmaking times. We posited some potential influences including ecological differences, cultural differences, and different budgetary structures. However, these were simply ideas. We have no way to test these hypotheses, which is why further research is necessary. Understanding why some regions complete the NEPA process more quickly may reveal administrative and management efficiencies that could be replicated. Conversely, understanding why some regions tend toward slower decisions could identify barriers to efficiency that can be eliminated.

Question 4. Regarding the study you co-authored on NEPA implementation in the U.S. Forest Service, Evidence-Based Recommendations for Improving National Environmental Policy Act Implementation, can you explain the correlation between regional decision-making time and wildfire suppression costs?

Answer. Throughout our study period, fire borrowing affected the staff and resources available to complete NEPA projects and thereby increased NEPA compliance times.[291] Additionally, the uncertainty caused by wildfire suppression activities was identified by Forest Service staff and stakeholders

---

[290] Jamie Pleune, Playing the Long Game: Expediting Permitting Without Compromising Protections, 52 Env. L. Rep. 10893 (Nov. 2022).

[291] Ruple et al., Evidence-Based Recommendations for Improving NEPA Implementation supra note 1 at 328-29.

as a cause of delay that complicated NEPA compliance.[292] Sources of delay included unstable budgets as well as staff reductions and shifting staff from project management to wildfire duties.[293] Additionally, according to a 2006 report by the Office of Inspector General, some regions bore an "inequitable wildfire protection burden" because wildland fire protection agreements between the Forest Service and other agencies had not been renegotiated to reflect appropriate WUI protection responsibilities.[294] Finally, due to ecological differences, some regions have higher wildfire hazards than others. There appeared to be some correlation between regions with longer decisionmaking times and those with greater wildfire burdens. However, other than observing the overlap, we had no way to test the relationship. It is worth noting Congress stabilized funding for wildfire suppression costs in 2018. However, the effect of this legislation was not visible during the period of study for our research.[295] This could be a productive area of study.

Question 5. Would additional resources for wildfire management help reduce NEPA decision-making times [for] areas with higher wildfire suppression costs?

Answer. According to a 2019 report from the Congressional Research Service, "Fire expenditures continue to climb, affecting the implementation of other programs . . . through personnel and funds transferred to fire control."[296] A series of roundtables conducted with stakeholders engaged with the Forest Service NEPA process also described how the high priority of wildfire suppression activities affect decisionmaking times. "Budget shortfalls and statutory mandates on funding for fire response, combined with a shortage of trained employees in areas other than fire and/or a frequent diversion of staff to emergency response or shifting priorities, hamper the ability of the Agency to make progress on other important forest and grassland resource management efforts." They also noted that "staffing levels are not adequate to meet the current demand'' and that "timelines are often lengthened due to the need for hiring or onboarding additional staff, including 'holes' in

---

292 Id. at 329-330.

293 Id. at 330.

294 Ruple et al., Evidence-Based Recommendations for Improving NEPA Implementation supra note 1 at 326.

295 Id. at 327-328.

296 Kate Hoover & Anne A. Riddle, Cong. Res. Serv., R43872, National Forest System Management: Overview, Appropriations, and Issues for Congress (2019).

interdisciplinary team specialist representation"[297] Based on these reports, it appears likely that stabilizing budgets and bolstering agency capacity in non-fire suppression roles would improve decisionmaking times and efficiency in the NEPA process.

Question 6. Can you provide more information on how America needs permitting reform for transmission lines?

Answer. A recent study by the Lawrence Berkeley National Laboratory found that there are over 2,000 GW of total generation and storage capacity waiting for approval to connect to the grid, 95 percent of which are solar, wind, or battery storage.[298] However, these projects face long wait times and uncertainty when attempting to connect to the grid. Between 2000–2007, the time between an initial connection request and a fully built, operational plant was typically less than 2 years. Between 2018–2022, that timeframe doubled to an average of almost 4 years, with an increasing trend. By 2022, the median between an interconnection request to commercial operations date reached almost 5 years. The increased volume of proposed renewable projects sitting in the queue promises to amplify this problem. For example, at least two regional transmission organizations, the entities responsible for approving requests to connect to the grid, have announced pauses on accepting new projects until they can process their backlogs. The nation's largest electric grid operator, PJM Interconnection, coordinates electricity movement in 13 states and the District of Columbia.[299] It has announced that it will not process any new applications until the end of 2025.[300] Similarly, CAISO,[301] a California grid operator, declined to accept any new projects in 2022 while they processed their backlog. Both entities are looking for systemic solutions to improve the grid connection process. In addition to these challenges, transmission lines also face siting challenges, especially within communities

---

[297] Nat'l Forest Found., EADM, Environmental Analysis and Decisionmaking, Regional Partner Roundtables: National Findings and Leverage Points 18 (2018) https://www.nationalforests.org/assets/pdfs/National-EADM-Report.pdf.

[298] Berkeley Lab, Energy Technologies Area, *Grid Connection Requests grow by 40% in 2022 As Clean Energy Surges, Despite Backlogs and Uncertainty* (Apr. 6, 2023) https://energy.lbl.gov/news/grid-connection-requests-grow-40-2022.

[299] Delaware, Illinois, Indiana, Kentucky, Maryland, Michigan, New Jersey, North Carolina, Ohio, Pennsylvania, Tennessee, Virginia, West Virginia. https://pjm.com/about-pjm/who-we-are/ territory-served (last visited Apr. 24, 2023).

[300] Institute for Energy Research, PJM Plans for a Two-Year Pause on Reviewing Project Applications (Feb. 22, 2022). https://www.instituteforenergyresearch.org/the-grid/pjm-plans-for-a-two-year-pause-on-reviewing-project-applications/#:_:text=To%20implement%20it%2C%20PJM%20is%20proposing%20an%20interim,on%20those%20coming%20at%20the%20end%20of%202027.

[301] California Independent System Operator https://www.caiso.com/Pages/default.aspx.

that do not benefit from the power line. The interconnect queue, regional control of transmission, and siting high power transmission lines are serious problems that deserve a national strategy and are worthy of permit reform.

Question 7. How would having a lead federal agency for permitting benefit getting more energy projects up and running.

Answer. Experience within the FAST-41 program demonstrates that identifying a lead federal agency and tasking it with the responsibility to convene stakeholders early in the permitting process can improve the transparency, predictability, and timeliness of permitting.[302] Tasking a lead agency with the responsibility for coordinating multiple different permitting authorities, sequencing permitting decisions, identifying information that must be gathered, developing data management protocols, and coordinating stakeholder engagement can promote efficiency. For example, the Federal Railroad Administration (FRA) utilized the NEPA process to overcome the challenge of inter-agency variance in decisionmaking for multiple federal, state, and local entities affected by a proposal to improve intercity passenger rail service in the Northeast Corridor.[303] By engaging multiple agencies early, and identifying points of contact within each agency, the FRA ensured that partner agencies could provide timely information that the technical team utilized, avoiding conflict down the road. Communication protocols enabled the creation of an interactive dataset encompassing multiple local and state jurisdictions, transportation authorities and watersheds that could be used for subsequent environmental analyses. This created a framework for collaboration that would foster continued efficiencies beyond project implementation because future projects can utilize the established interjurisdictional database and communication protocols.

Question 8. Can you further explain permitting delays caused by litigation aversion of agency staff?

Answer. In 2018, the Forest Service launched an agency-wide effort to improve processes related to Environmental Analysis and Decision Making (EADM).[304] The acronym EADM includes the NEPA processes, as well as underlying environmental decisions such as forest planning, issuance of special use permits, implementation of forest management activities, and fulfilment of other statutory obligations, including compliance with the Clean

---

[302] Pleune & Boling, This Permit Reform Already Works supra note 5 at 1046-1047.

[303] Ruple et al., Evidence-Based Recommendations for Improving NEPA Implementation supra note 1 at 337-338.

[304] https://www.nationalforests.org/collaboration-resources/environmental-analysis-and-decision-making-roundtables.

Water Act, the Endangered Species Act, and the National Historic Preservation Act. As part of that effort, the Forest Service asked the National Forest Foundation to assist in hosting ten regional partner roundtables across the country with the objective of collecting diverse feedback to inform ways to improve decisionmaking processes. During the roundtables, concern over litigation aversion featured prominently in every region.[305] According to participants in the roundtables, Forest Service staff avoid making controversial decisions for fear of affecting opportunities for promotion.[306] A controversial decision may sit on the back of someone's desk until that person is promoted or sent on detail, leaving someone else to bear the political or professional backlash of signing a decision that gets litigated. Additionally, litigation aversion leads to unwieldy, bulky, time-consuming documents. Participants in the roundtables explained, "Risk aversion and a history of legal challenges to USFS decisions have led to the 'bullet-proofing of environmental analysis documents and specialist reports" resulting in documents where "the complexity and size of analysis is often inconsistent with the complexity and size of the project."[307] These observations are consistent with external research on Forest Service NEPA practice, which found that the threat of litigation had more influence on the decision to prepare an EIS or an EA than the degree of environmental impacts.[308] Practitioners also recognize the problem. As one observed, "[i]t has been the author's frequent experience that BLM and Forest Service delay decision-making in order to prepare more and lengthier documents in an effort to bulletproof their decisions from appeal. As a result, the diversion of agency resources and attention to the preparation of up-front disclosures under NEPA means less attention and resources are devoted to on the ground efforts such as monitoring the effects of agency decisions."[309] Helen Leanne Serassio, a lawyer with more than 14 years in the Department of Transportation, suggested that "the most effective action agencies can take

---

[305] Ruple et al., Evidence-Based Recommendations for Improving NEPA Implementation supra note 1 at 330.

[306] *Id.* at 331.

[307] *Id.* (quoting Michael J. Mortimer et al., *Environmental and Social Risks: efensive National Environmental Policy Act in the US Forest Service*, 109 J. Forestry 27, 29-30 (2011).

[308] *Id.*

[309] *Id.* at 332 (quoting Laura Lindley, *NEPA Streamlining: Some Observations on Its Use in the Context of BLM and Forest Service Oil and Gas Program*, in Rocky Mt. Min. L. Found, Natural Resources and Environmental Administrative Land and Procedure II (2004).

to increase efficiencies in the NEPA review process is to get back to the basics with NEPA and halt efforts to make NEPA documents litigation-proof."[310]

Litigation aversion, which is a cultural problem that affects an untold number of decisions, is different from delays caused by actual litigation. Government-wide, only about two-tenths of one percent of more than 50,000 NEPA decisions that are documented each year result in litigation.[311] An investigation by the GAO regarding Forest Service fuel reduction projects from fiscal years 2006–2008 revealed that only 29 out of 1,415 decisions were litigated and the litigation impacted only 1% of lands slated for fuel reduction projects.[312] In summary, the fear of litigation appears to create far more delay than litigation itself.

Question 9. How can litigation reform reduce delays caused by litigation aversion by agency staff?

Answer. According to participants in the EADM roundtable discussions, Forest Service staff fear that their opportunities for promotion will be reduced if they sign a NEPA decision that is litigated.[313] Assuming the accuracy of these reports, the best way to reduce delays caused by litigation aversion is to reward agency officials who make prompt, well-supported decisions, regardless of whether the decision is litigated. This shift in focus would enable agencies like the Forest Service to encourage field officers to act decisively and exercise discretion to focus the NEPA analysis on significant issues identified during the scoping process.[314] As Helen Leanne Serassio observed, "Agencies must recognize and use their discretion to determine the necessary length of their NEPA documents, the methodologies to use, and the depth of the analysis necessary to make an informed decision. If the agency's decision is not 'arbitrary, capricious, an abuse of discretion, or otherwise not in accordance with law,' it will withstand judicial review."[315] It is also worth noting that the NEPA process itself can help an agency avoid litigation by addressing stakeholder concerns through impact avoidance, reduction or

---

310 Helen Leanne Serassio, *Legislative and Executive Efforts to Modernize NEPA and Create Efficiencies in Environmental Review*, 45 Tex. Envtl. L.J. 317, 333 (2015).

311 *Id.* at 333-334.

312 Gov. Accountability Off., GAO-10-337, Forest Service: Information on Appeals, Objections, and Litigation Involving Fuel Reduction Activities, Fiscal Year 2006 through 2008 1 (2010).

313 Ruple et al., Evidence-Based Recommendations for Improving NEPA Implementation supranote 1 at 330.

314 Id. at 342.

315 Helen Leanne Serassio, Legislative and Executive Efforts to Modernize NEPA and Create Efficiencies in Environmental Review, 45 Tex. Envtl. L.J. 317, 334 & 335-341 (2015).

mitigation.[316] It can also help ensure that a contentious agency decision is defensible. Through NEPA's public participation procedures, the agency has an early opportunity to identify issues that may be litigated and justify its decisions regarding those issues. "Courts do not typically overturn NEPA decisions when the administrative record demonstrates that the agency has followed NEPA's procedural steps, when there are minor deficiencies in the NEPA document, or when an agency documents why it has chosen to exclude information."[317] Without the public participation process afforded through NEPA, an agency may not understand the weaknesses in its decision until it is too late. Finally, eliminating a cause of action under NEPA will not protect a weak or unjustified agency decision from litigation. A study analyzing 20 years of Forest Service land management litigation recognized that most lawsuits involve multiple claims arising under different statutes.[318] In cases involving multiple statutes, the majority of the time, the Forest Service would have lost even if NEPA did not exist.[319] The importance of this observation comes into sharper focus when one considers the proliferation of local, regional, and state regulations that can also provide a cause of action to challenge a controversial action.

Question 10. Yes or no—do you agree that the Biden administration push to electric vehicles, and other so-called "clean energy" technologies, will require an increase in mineral production, i.e. mining?

Answer. Yes. According to the Biden Administration's Report, Building Resilient Supply Chains, Revitalizing American Manufacturing, and Fostering Broad-Based Growth, demand for critical minerals and materials "is projected to surge over the next two decades, particularly as the world moves to eliminate net carbon emissions by 2050."[320] The report provided the following examples, "global demand for lithium and graphite, two of the most important materials for electric vehicle batteries, is estimated to grow by more than 4000 percent by 2040 in a scenario where the world achieves its climate goals."

---

[316] Id. at 340-341.

[317] Id. at 335; Ruple et al., Evidence-Based Recommendations for Improving NEPA Imple mentation supra note 1 at 342-343.

[318] Amanda M.A. Miner et al., *Twenty Years of Forest Service Land Management Litigation*, 112 J. Forestry 32, 36 (2014).

[319] *Id.*; Ruple et al., *Evidence-Based Recommendations for Improving NEPA Implementation supra* note 1 at 319.

[320] The White House, Building Resilient Supply Chains, Revitalizing American Manufacturing, and Fostering Broad-Based Growth: 100-Day Reviews Under Executive Order 14017, 9 (June 2021).

Question 11. Yes or no—Do you agree that the United States is currently reliant on foreign countries for critical minerals needed to transition to electric vehicles?

Answer. Yes. According to the Biden Administration's Report, Building Resilient Supply Chains, Revitalizing American Manufacturing, and Fostering Broad-Based Growth, "Global production for lithium, cobalt, and graphite are primarily dependent on a single nation . . . for each of these materials, a single country controls over 60 percent of the global production."[321]

Question 12. Yes or No—Are there steps we can take to mine and process more critical minerals in America?

Answer. Yes. On May 10, 2023, the Biden Administration identified several steps "to expand and accelerate responsible domestic production of critical minerals in a manner that upholds strong environmental, labor, safety, Tribal consultation, and community engagement standards."[322] Promising reforms include updating the General Mining Law and providing legal clarification for laws affecting remining and remediation projects. These are two examples of arenas where legal ambiguities caused by complexities in the law create delay.[323]

Question 13. Yes or no—the main producers of critical minerals—such as China, Congo, and Indonesia—don't have nearly the same environmental standards on mining as the United States, correct?

Answer. I am not an expert on international environmental law. Based on my limited knowledge, it appears that China, Congo, and Indonesia have lower environmental standards. Other main producers, including Canada and Australia, appear to have comparable environmental regulatory regimes.

Question 14. Yes or no—the main producers of critical minerals—such as China, Congo, and Indonesia—do not have as many protections for human rights or labor as the United States, correct?

Answer. I am not an expert in international human rights or labor practices. Based on my limited knowledge, it appears that China, Congo, and Indonesia have fewer human rights protections. Other main producers,

---

[321] *Id.* at 120.

[322] White House, Fact Sheet: Biden-Harris Administration Outlines Priorities for Building America's Energy Infrastructure Faster, Safer, and Cleaner (May 10, 2023) https://www.whitehouse.gov/briefing-room/statements-releases/2023/05/10/fact-sheet-biden-harris-administration-outlines-priorities-for-building-americas-energy-infrastructure-faster-safer-and-cleaner/.

[323] Pleune, *Playing the Long Game supra* note 12 at 10901-10905.

including Canada and Australia, appear to have comparable human rights and labor regimes.

Questions Submitted by Representative Grijalva

Question 1. Republicans are demanding 22% across-the-board cuts to annual appropriations. How would cuts like those affect the speed of permit processing, which they claim they are trying to improve?

Answer. Common causes of delay in the permitting process can be summarised into three categories: (1) a lack of agency capacity, which includes insufficient allocation of resources (e.g., number of staff, staff expertise, funding, infrastructure, training, and or computer technology); (2) waiting for information from an applicant, particularly following a permit application that was vague or incomplete, or following a change to a proposed plan; (3) compliance with other legal requirements and/or ineffective agency coordination or collaboration during the permitting process.[324] Notably, these three categories are not independent. The first category— agency capacity— affects the other two. Without sufficient staff or expertise, an agency cannot provide support or training to assist operators in submitting complete applications with the required information. It is also unlikely that they will effectively engage in proactive coordination. Thus, a lack of agency capacity exacerbates the other two causes of delay.

Question 2. What impact does Executive Order 14096, Revitalizing Our Nation's Commitment to Environmental Justice for All, have on energy independence?

Answer. Executive Order 14096 reduces the risk that we will repeat mistakes of the past by failing to account for the disproportionate impacts that often fall to minority, underrepresented, and socially or economically deprived communities.

Question 3. Would extending or even expanding the use of fossil fuels make us more competitive or less competitive when compared to countries that are racing toward a renewable energy economy?

Answer. This is outside my area of expertise.

Question 4. Would extending or even expanding the use of fossil fuels increase our national security or weaken it?

---

[324] Pleune, Playing the Long Game supra note 12 at 10475.

Answer. According to the Secretary of Defense, "No country can find lasting security without tackling the climate crisis."[325] Expanding the use of fossil fuels will exacerbate the climate crisis and consequently weaken our national security.

Question 5. What is the connection between NEPA and energy independence?

Answer. One way to achieve energy independence is transitioning to a renewable energy economy, which means building massive infrastructure. It also will require cooperation between agencies with different jurisdictional duties, states, local communities, and tribes. Without a framework for coordinating analysis, considering stakeholder input, identifying potential hazards, and avoiding, reducing, or mitigating those impacts, this build out of infrastructure would be practically impossible. One barrier to the deployment of renewable energy projects is opposition from affected landowners due to real or perceived harms that the project would bring, and inconsistency between local, state, tribal, and federal laws.[326] A research team from MIT concluded "incorporating all stakeholder perspectives from the outset of a siting process will probably save time and money" by addressing concerns early and avoiding sustained political opposition. The NEPA process is a familiar tool that can be used to engage stakeholders early and streamline renewable energy deployment. The NEPA process also serves as a tool for decisionmakers to defend justified decisions. It offers a public process for deliberation. It helps decisionmakers identify issues of concern. It provides a forum to justify use of agency discretion. It provides a preview potential sources of conflict and a mechanism for avoiding, reducing, or mitigating impacts. And it helps agencies understand legal vulnerabilities of a decision before it's made.

Question 6. What would weakening NEPA mean for frontline communities, communities that have historically borne a disproportionate burden of environmental harms?

Answer. Weakening NEPA would harm these communities. NEPA's "look before you leap" mandate requires agencies to disclose environmental impacts, weigh alternatives, and consider public comment before committing

---

[325] Department of Defense, Office of the Undersecretary of Defense (Acquisition and Sustainment), Department of Defense Climate Adaptation Plan 2022 Progress Report. Report Submitted to National Climate Task Force and Federal Chief Sustainability Officer (Oct. 2022).

[326] Lawrence Susskind et al., Sources of Opposition to Renewable Energy Projects in the United States, 165 Energy Policy 112922 (2022).

public resources to a course of action.[327] Looking to the past is highly relevant to the future. Prior to NEPA's enactment, agencies were free to implement decisions without regard to the collateral damage on communities or natural resources. For example, the Federal Highway Act of 1956 initiated construction of the interstate highway system.[328] Focused on speedy implementation, the Department of Transportation routed highways through low cost, low opposition lands, which tended to be parks, historic sites, recreation areas, and working class or low-income neighborhoods. As a result, low-income communities were disproportionately harmed during the build out of the interstate highway system. One example is a segment of I-95 that cut through an innercity community outside of Miami, Florida. Overtown, known at the time as the "Harlem of the South," was a thriving black community. The selected route for I- 95 cut straight through the community, even though a less destructive route along a nearby abandoned railroad corridor was available.[329] In a 2009 report, the Federal Highway Administration acknowledged that the selected route "had a disastrous impact on the economic and social structure of the community" with lingering effects that to this day fuel anger, resentment, and distrust toward the Department of Transportation.[330] Many other communities suffered similar fates due to one-sided planning focused on finding the cheapest route, without regard to the collateral impacts. As one group of scholars described, "Take any major American city, and you're likely to find a historically Black neighborhood demolished, gashed in two, or cut

---

[327] *Robertson v. Methow Valley Citizens Council*, 490 U.S. 332, 349 (1989) (''The statutory requirement that a federal agency contemplating a major action prepare such an environmental impact statement serves NEPA's action-forcing purpose in two important respects. It ensures that the agency, in reaching its decision, will have available, and will carefully consider, detailed information concerning significant environmental impacts; it also guarantees that the relevant information will be made available to the larger audience that may also play a role in both the decisionmaking process and the implementation of that decision.'' (internal citations and quotations omitted)); *Baltimore Gas & Elec. v. Nat. Res. Def. Council*, 462 U.S. 87, 97 (1983) (''NEPA has twin aims. First it places upon an agency the obligation to consider every significant aspect of the environmental impact of a proposed action. Second it ensures that the agency will inform the public that it has indeed considered environmental concerns in its decision- making process.'' (internal citations and quotations omitted)).

[328] Linda Luther, Cong. Res. Serv., The Role of the Environmental Review Process in Federally Funded Highway Projects: Background and Issues 6-7 (Apr. 2012).

[329] Hillary Simmons, *The Heart of ''Harlem of the South,'' The Beacon* (Dec. 28, 2020) *available at* https://mastthebeacon.wordpress.com/2020/12/28/the-heart-of-harlem-of-the-south/ (last visited May 17, 2023).

[330] Linda Luther, Cong. Res. Serv., The Role of the Environmental Review Process in Federally Funded Highway Projects: Background and Issues 6-7 (Apr. 2012).

off from the rest of the city by a highway."[331] The disclosure requirements imposed through NEPA were intended to avoid similar ill-advised and harmful uses of federal power and funding. As the nation considers how to build a new interstate energy infrastructure system, the lessons of the past should serve as a cautionary tale.

Question 7. Under the Trump administration, the Bureau of Land Management lost significant numbers of experienced staff when they tried to move the agency's headquarters out of D.C. While the Biden administration has taken important steps to address the bureau's workforce challenges. Based on your research, would you please explain the expected impact of the Trump BLM relocation on permitting times, and why it is important to provide sufficient and stable funding.

Answer. The Bureau of Land Management is one of several agencies that suffered extreme losses of staff under the prior administration.[332] Since 2011, it has been on the GAO's list of programs at high risk and vulnerable to waste, fraud and abuse due in part to a lack of staff. This problem was further exacerbated in July 2020, when the Trump Administration abruptly decided relocate BLM's headquarters from Washington D.C. to Grand Junction, Colorado. The Headquarters Office, which develops guidance and regulations, should be staffed by 311 career positions. However, it was already severely understaffed with 132 vacant positions before the relocation announcement. In response to the relocation announcement, 81 more staff left, leaving the leadership at 31 percent capacity. The remaining leadership team were dispersed among multiple offices.

Numbers do not tell the whole story. The BLM also suffered a loss of experienced staff. Every BLM staff member interviewed reported that the loss of experienced staff negatively affected their offices' ability to conduct its duties. For example, the loss of institutional knowledge about laws and regulations meant that the BLM could not provide knowledgeable input on proposed rules and legislation. Other staff admitted that the rapid loss of experienced staff hindered knowledge transfer. In a follow-up report, a year later, all BLM staff interviewed by the GAO reported challenges in completing

---

[331] Rachael Dottle, Laura Bliss, and Pablo Robles, *What It Looks Like to Reconnect Black Communities Torn Apart by Highways*, Bloomberg (July 28, 2021).

[332] Pleune & Boling, *This Permit Reform Works supra* note 5 at 10476; *see also* Gov't Accountability Off., GAO-20-379R, Bureau of Land Management: Agency Reorganization Efforts Did Not Substantially Address Key Practices for Effective Reforms (Mar. 6, 2020) and Gov't Accountability Off., GAO-22-104247, BLM, Better Workforce Planning and Data Would Help Mitigate the Effects of Recent Staff Vacancies (Nov. 2021).

their duties due to headquarters vacancies. As a result of delays in creating or clarifying guidance or policies, some staff relied on outdated policy guidance to make decisions. Other staff reported delays implementing upgrades to information technology systems, which GAO had previously recommended be updated. Obviously, these institutional challenges would affect permitting times. Applicants could not receive good guidance from experienced staff, and staff members processing permits had little instruction on how to proceed effectively. The BLM is not alone. Multiple agencies with permitting or infrastructure responsibilities, are short-staffed and underfunded.

When the GAO investigated the BLM's workforce planning in 2020, it found that the BLM had no way of tracking vacancies and no recruitment plan for filling vacancies.[333] When asking for data on the total number of positions and vacancies agency wide, the GAO was told that BLM does not maintain a list of vacancies for state offices. As a result, it was not possible to determine the proportion of positions that are vacant at any given time or the specific positions that are vacant. This lack of information obviously creates a problem for improving capacity in a way that results in improved efficiency.

In order to resolve workforce challenges, agencies require stable funding that will enable strategic workforce development. Additionally, agencies must be confident that the funding will not disappear. Unstable budgets do not build durable workforces. Shortchanging agencies of staff and funding will cause delays in the permitting process.

Question 8. You've talked about the importance of staff but have emphasized that certain staff attributes are important as well. What do we need to see in those staff, and how can we make that happen?

Answer. Strategic workforce planning is critical to ensuring that agencies spend wisely and build a workforce capable of fulfilling agency missions. The Army Corps of Engineers developed a comprehensive strategic workforce plan in 2017 that demonstrates how this type of planning can achieve long-term efficiencies at all levels of agency operations.[334] The Corps developed planning strategies to align human capital with changing workloads and missions using real-time workforce data for decision-making.[335] For example,

---

[333] Government Accountability Office, Bureau of Land Management: Agency Reorganization Efforts Did Not Substantially Address Key Practices for Effective Reforms GAO-20-379R, 10 (Mar. 6, 2020).

[334] Government Accountability Office, Army Corps of Engineers: Workforce Planning Follows Most Leading Practices but Could Be Enhanced with Additional Actions GAO-22-104053 (Dec. 2021).

[335] *Id.* at 9.

they initiated a Workload to Workforce Assessment, which is an annual planning activity conducted by workforce managers agency-wide to assess the capacity, competency, and balance of the workforce to meet the projected workload in the next 1 to 3 years.[336] Based on the results, managers prepare action plans to address any potential workforce capacity, competency, or balance gaps. They also created Civilian Workforce Dashboards, which are an interactive online display that provide workforce managers at all levels of the organization with direct access to the most current human capital information for planning purposes.

The Corps also created a strategy to address all four stages of the human capital life cycle. The recruiting stage involves sourcing and acquiring top talent to accomplish current and future missions, shaping the workforce to meet mission needs, and marketing missions to attract the workforce of the future.[337] The Corps human resources managers created a talent acquisition team to support the recruitment program. They also utilized tools to attract talent including direct-hire authorities. Use of direct hire authorities reduced vacancy fill time and allowed managers to quickly hire top talent.[338]60 Additional tools included policy guidance for managers to offer monetary and non-monetary incentives as tools to attract and retain personnel. Incentives include student loan repayment; recruitment, relocation, retention, and enhanced retention incentives; credit for prior non-federal work experience and certain military experience for determining annual leave accrual rate; and superior qualifications appointment and special needs pay-setting authority.[339]

The second stage of the workforce life cycle is the developing stage, which focuses on ensuring a culture of continuous skill improvement throughout the organization and fostering technical and leader development.[340] To address career development challenges, the Corps provides career specific training, development, and mentoring activities to its workforce. This includes job-related training, an Emerging Leaders Program, and a Senior Leader Academy.[341] In addition to training new staff, the mentoring and leadership training programs have the added benefit of preserving and passing along institutional knowledge within the agency.

---

[336] *Id.* at 10.
[337] *Id.* at 11.
[338] *Id.* at 12.
[339] *Id.* at 13.
[340] *Id.*
[341] *Id.* at 14-15.

The final stage is the sustaining, or retention stage, which focuses on preventing critical talent loss and improving organizational performance by valuing and engaging employees at all levels. Some strategies utilized at this stage include an online exit survey tool for departing employees to capture the underlying causes of talent loss. Some offices also implemented a "stay survey" to measure employee engagement during their tenure as a proactive alternative to surveying staff that are leaving. The Corps also utilizes the annual Federal Employee Viewpoint Survey as a tool to make improvements. The Federal Employee Viewpoint Survey is an existing tool with valuable information about workforce problems. The Corp directs managers at all levels to analyze the annual survey results for their workforce and develop action plans to address areas of concerns. Since adopting this strategy, survey response rates and scores have steadily risen. For example, employees satisfied with the organization rose from 55 percent in 2013 to 71 percent in 2019.

Using these strategies, the Corps has been able to determine critical skills and competencies needed, and align its workforce to those needs. One strategy is through employee performance reviews. The missions of the Army and the Corps are incorporated into each employee's performance plan and evaluation. Employee progress reviews link back to the mission and goals of the organization. Managers provide feedback and recommend training to assist the employee in reaching individual goals.[342]

Another strategy is agency-wide Workload to Workforce Assessments, which assess the status of the workforce—such as vacancy fill rates—to forecast the ability to meet the future workload and the type and experience level of employees that will be needed.[343] Use of the Federal Employee Viewpoint Surveys provide further insight into areas for improvement in employee engagement and working groups focus on ways to use the information to make the Corps a better place to work. Finally, the Corps utilizes ongoing real-time data to evaluate the effectiveness of these strategies.[344]

In summary, each agency will have different staffing needs. Strategic workforce planning is an existing tool that agencies can use to ensure sufficient staff, expertise, and capacity to achieve the agency's mission.

---

[342] *Id.* at 31.
[343] *Id.* at 31.
[344] *Id.* at 33.

Question 9. What is the biggest barrier to renewable energy transmission projects, and what is the evidence for that conclusion?

Answer. A recent study by the Lawrence Berkeley National Laboratory found that there are over 2,000 GW of total generation and storage capacity waiting for approval to connect to the grid, 95 percent of which are solar, wind, or battery storage.[345] However, these projects face long wait times and uncertainty when attempting to connect to the grid. Between 2000–2007, the time between an initial connection request and a fully built, operational plant was typically less than 2 years. Between 2018–2022, that time frame doubled to an average of almost 4 years, with an increasing trend. By 2022, the median between an interconnection request to commercial operations date reached almost 5 years. This is a major barrier to energy transmission. Additional hurdles include siting challenges, and cost-benefit allocation for transmission lines that pass through communities without offering a benefit.

Question 10. My friends on the other side of the aisle have claimed that litigation slows energy projects. What effect does NEPA have on litigation of major energy projects?

Answer. The NEPA process creates an opportunity to discover and mitigate concerns with a proposed project before finalizing an action, which can often avoid litigation that could delay implementation of an action.[346] The NEPA process can also reduce costs by identifying design problems before implementation of a project begins. A study prepared for the Transportation Research Board emphasized this potential benefit. "Spending more monies during planning and design will reduce the time and cost required for construction by avoiding unforeseen conditions, reducing to a minimum design errors and omissions, and developing schemes that will support the most efficient approach to construction."[347]

---

[345] Berkeley Lab, Energy Technologies Area, *Grid Connection Requests grow by 40% in 2022 As Clean Energy Surges, Despite Backlogs and Uncertainty* (Apr. 6, 2023) https://energy.lbl.gov/news/grid-connection-requests-grow-40-2022.

[346] *Id.* at 340-341; Pleune, *Playing the Long Game*, 52 Env. L. Rep. at 10905-10904 (''without providing an opportunity to raise concerns during the [NEPA] scoping process, stakeholders may raise concerns late in the process or through litigation . . . [and] some of those concerns may require collecting additional baseline data that may have been easily collected at the beginning of the permitting process'').

[347] Linda Luther, Cong. Res. Serv. R.42479, The Role of the Environmental Review Process in Federally Funded Highway Projects: Background and Issues for Congress 36 (Apr. 2012) (citing H.R. Thomas and R.D. Ellis, *Avoiding Delays During the Construction Phase of Highway Projects*, Transportation Research Board, National Research Council, NCHRP 20-24 (Oct. 2001).

For example, the NEXUS Gas Transmission Project, was a 250-mile natural gas pipeline traversing Pennsylvania, West Virginia, Ohio, and Michigan.[348] During the pre-application process, which included extensive public participation, the project sponsors incorporated 239 route alternatives and variations in the pipeline design to address landowner requests, avoid sensitive resources, or respond to engineering restraints. This feedback resulted in a 91 percent change from the original proposed route design—a number of modifications that would have been prohibitively expensive at the end of the review process. Using this information at the beginning of the process improved efficiency and arguably resulted in a better end-result and a final application that was processed more expeditiously.[349] Similar efficiencies could be achieved with other major energy projects.

The NEPA process can also provide a mechanism to build consensus, which can reduce the risk of litigation. For example, in 2012, the Forest Service completed the 4FRI EIS. The project goal was to restore the ponderosa pine forest stretching across northern Arizona, while reducing communities' exposure to wildfire threats, rehabilitating ecosystems, and sustaining the forest industry operating in local communities.[350] The EIS analyzed the largest number of acres in Forest Service history, stretching across four different national forests, for restoration-based mechanical forest treatments. Despite its ambitious scale, the EIS was completed more quickly than the average time frame for EISs completed that year, and when it came time for implementation, the Forest Service was not delayed by litigation. Using the NEPA process as an opportunity for collaborative decisionmaking developed consensus among diverse stakeholders that had long-term benefits and ultimately sped up implementation of the project.

This brings up another important distinction. Although NEPA's detractors often blame litigation for delay, the evidence shows that litigation is rare.

---

[348] Office of the Executive Director, Federal Permitting Improvement Steering Council, Recommended Best Practices for Project Review and Permitting for Infrastructure Projects for Fiscal Year 2018, at 17 (2017).

[349] This is not to say that early engagement eliminated local opposition in every community. See Heidi Gorovitz Robertson, Home Rule Symposium: Cities and Citizens Seethe: A Case Study of Local Efforts to Influence Natural Gas Pipeline Routing Decisions, 122 W. Va. L. Rev. 881, 907-934 (Spring 2020) [hereinafter Gorovitz, Cities and Towns Seethe] (describing FERC's extensive public engagement and local opposition in three Ohio towns).

[350] Ruple et al., Evidence-Based Recommendations for Improving Implementation of NEPA supra note 20 at 338.

Governmentwide, only an estimated 0.22% of NEPA decisions are litigated.[351] An investigation by the Government Accountability Office regarding Forest Service fuel reduction projects from fiscal years 2006–2008 revealed that only 29 out of 1,415 decisions were litigated, and litigation impacted only 1% of lands slated for fuel reduction projects.[352] In other words, used properly, the NEPA process is more likely to avoid potential litigation than cause it. Proposed reforms like short deadlines and page limits threaten to undermine NEPA's capacity to serve as a flexible tool for structured and transparent deliberation.

Dr. GOSAR. I thank the gentlelady.

I now recognize Mr. Kenny Stein for 5 minutes.

Statement of Kenny Stein, Director of Policy, Institute for Energy Research, Washington, DC

Mr. STEIN. Mr. Chairman, thank you for the opportunity to testify at this hearing.

For nearly 50 years a constant over-riding concern of United States energy policy was the shortage and scarcity of oil and natural gas. Dependence on foreign oil was seen as a national security crisis. Numerous half-baked policy ideas, regulations, and subsidies were spawned to address this perceived crisis, many of which survive to this day, continuing to distort markets and increase energy costs: fuel efficiency mandates, ethanol subsidies and mandates, solar and wind subsidies, export controls, subsidies for domestic production. All of this and more was done in the name of energy independence and security, and all to little effect.

---

351 John C. Ruple & Kayla M. Race, Measuring the NEPA Litigation Burden: A Review of 1,499 Federal Court Cases, 50 Envtl. L. 479, 497-99 (2020); David Adelman & Robert L. Glicksman, Presidential and Judicial Politics in Environmental Litigation, 50 Ariz. St. L.J. 3, 7 (2018) (conducting an empirical study of NEPA litigation during the presidencies of George W. Bush and Barack Obama and observing, "[w]e find little evidence that litigation under NEPA is out of control or that NEPA's processes are overly burdensome"); John C. Ruple & Heather Tanana, Debunking the Myths Behind the NEPA Review Process, 35 Nat. Res. & Env't 14, 15 (2020); Forrest Fleischman et al., U.S. Forest Service Implementation of the National Environmental Policy Act: Fast, Variable, Rarely Litigated, and Declining, 118 J. Forestry 403, 404 (2020).

352 Gov't Accountability Off., GAO 10-227, Forest Service, Information on Appeals, Objections, and Litigation Involving Fuel Reduction Activities, Fiscal Years 2006 through 2008 1 (2010).

But today, we have just about achieved that long-sought-after energy security. Since lows in the mid-2000s, the United States has more than doubled domestic oil production, and almost doubled natural gas production. In 2019, the United States became a net exporter of petroleum. The United States became a net exporter of natural gas in 2017, and the United States is also a net exporter of coal and refined products. The United States is the largest oil and natural gas producer in the world.

This position has completely changed this country's energy posture, making the United States the most energy secure that it has been since at least the first half of the 20th century. This success did not come from government. It took years of experimenting with and perfecting hydraulic fracturing and directional drilling. It took companies taking the risk of investing in new areas and new formations and new depths with no guarantees of success. It took government actually getting out of the way of exporting LNG and crude oil, providing new markets for U.S. production. And it needed a lack of Federal Government hostility.

For all its environmental radicalism, the Obama administration recognized the need for domestic energy production. They somewhat reduced production on Federal lands through policy interference, but mostly left production on private and state lands to its own devices. In short, the many decades of government interference intervention did not solve the energy security problem; the private sector did when it was allowed to escape from the help of the Federal Government.

But just as we have just about achieved this long-sought energy security, the Biden administration has launched a full spectrum assault on domestic energy production, trying to crush production on Federal lands while looking to regulate non-Federal production into oblivion, all while subsidizing promoting energy sources whose inputs and supply chains are controlled by China and Chinese state-owned companies. It is an agenda to replace independence with dependence, and dependence on one of the countries in the world most hostile to the United States, no less.

The Administration has grasped for any lever it can to pursue its promotion of Chinese energy and suppression of U.S. energy. This includes many of the laws and regulations passed over the decades attempting to address the old crisis of scarcity, which the Administration is now seeking to repurpose in often illegal ways to support the crusade to eliminate the new abundance.

As just a few examples of the Administration's war on domestic energy, the Administration is actively seeking to halt oil and gas production on Federal lands by any means in direct contravention of congressional intent; the Council on Environmental Quality is writing new guidelines for the National Environmental Policy Act designed to turn an already burdensome NEPA process into a cudgel to crush domestic energy production; the Administration has participated in numerous sue-and-settle deals with environmental groups so that they can use the settlements to prevent oil and gas production without going through Congress or the normal regulatory process; in a brazen and illegal move, the Administration, for the first time in our history, has refused to finalize an offshore leasing plan entirely; the Administration is attempting to use old fuel efficiency mandates to mandate electric vehicles; and finally, the Administration has, without statutory mandate from Congress, sought to inject vague notions of environmental justice into every decision-making process. This concept is not scientific or measurable. It is entirely in the eye of the beholder. It is simply a tool for arbitrarily halting development that the Administration does not like.

These actions, and the hundreds of other additional and related actions are cumulative. Every action making it harder to produce energy domestically reduces the desire of companies to invest domestically. Oil and gas production requires ongoing investment just to keep production stable, much less increase it. Reduced investment means reduced production down the line, and that reduced production erodes our hard-won energy security. It ends with the spectacle of the President of the United States, leader of the world's largest oil-producing country, going hat-in-hand to beg Saudi Arabia for more production, or cowardly seeking to relax sanctions against one of the most brutal regimes in the Americas to get more oil supply from Venezuela.

U.S. oil production still has not reached the level seen at the end of 2019, just before the start of the pandemic. That is not because the oil isn't there or there isn't demand. It is because the companies are concerned about making the investments needed when there is a hostile Administration willing to use any means to destroy their industry. These investments then get made in other countries, with the United States left having to import to meet our future needs.

The Biden administration's regulatory assault on domestic energy is an assault on this nation's security and prosperity. It is a dependence agenda that makes energy more expensive and less available, while at the same time making our energy system dependent on foreign supplies from China. That this is being done by regulatory fiat using questionable authorities, or even in direct conflict with laws passed by Congress is, frankly, a crisis of our democratic system.

Thank you, and I look forward to taking your questions.

[The prepared statement of Mr. Stein follows:]

Prepared Statement of Kenneth Stein, Policy Director, Institute for Energy Research

Mr. Chairman, thank you for the opportunity to testify at this hearing.

My name is Kenny Stein, I am the Policy Director for the Institute for Energy Research, a free-market organization that conducts research and analysis on the function, operation, and regulation of energy markets.

For nearly 50 years, a constant overriding concern of United States energy policy was shortage and scarcity of oil and natural gas. Dependence on foreign oil was seen as a national security crisis. Numerous half-baked policy ideas, regulations and subsidies were spawned to address this perceived crisis, many of which survive to this day continuing to distort markets and increase energy costs. Fuel efficiency mandates, ethanol subsidies and mandates, solar and wind subsidies, export controls, subsidies for domestic production; all of this and more was done in the name of energy independence and security. And all to little effect.

But today, we have just about achieved that long sought after energy security. Since lows in the mid 2000s, the US has more than doubled domestic oil production[353] and almost doubled natural gas production.[354] In 2019, the US became a net exporter of petroleum.[355] The US became a net exporter of natural gas in 2017. The US is also a net exporter of coal and refined products. The US is the largest oil and natural gas producer in the world. This position has completely changed this country's energy posture, making the US the most energy secure it has been since at least the first half of the 20th century.

This success did not come from government. It took years experimenting with and perfecting hydraulic fracturing and directional drilling. It took companies taking the risk of investing in new areas, new formations, and new depths with no guarantee of success. It took government getting out of the way of exporting LNG and crude oil, providing new markets for US production. And it needed a lack of federal government hostility. For all its environmental radicalism, the Obama administration recognized the need for domestic energy production, somewhat reducing production on federal lands through policy

---

[353] https://www.eia.gov/dnav/pet/hist/LeafHandler.ashx?n=PET&s=MCRFPUS2&f=M.
[354] https://www.eia.gov/energyexplained/natural-gas/where-our-natural-gas-comes-from.php#.
[355] https://www.eia.gov/energyexplained/oil-and-petroleum-products/imports-and-exports.php.

interference, but mostly leaving production on private and state land to its own devices. In short, the many decades of government interference and intervention did not solve the energy security problem, the private sector did when it was able to escape from the "help" of the federal government. But just as we have just about achieved this long-sought energy security, the Biden administration has launched a full spectrum assault on domestic energy production, trying to crush production on federal lands while looking to regulate non-federal production into oblivion, all while subsidizing and promoting energy sources whose inputs and supply chains are controlled by China and Chinese state-owned companies.[356] It is an agenda to replace independence with dependence, and dependence on one of the countries in the world most hostile to the United States no less. The administration has grasped for any lever it can in pursuit of its promotion of Chinese energy and suppression of US energy. This includes many of the laws and regulations passed over the decades attempting to address the old crisis of scarcity, which the administration is now seeking to repurpose, often in illegal ways, to support their crusade to eliminate abundance.

To list just a few examples of the Biden administration's war on domestic energy:

- The administration is actively seeking to halt oil and gas production on federal lands by any means, in direction contravention of Congressional intent.
- The Council of Environmental Quality (CEQ) is writing new guidelines for the National Environmental Policy Act (NEPA) designed to turn the already burdensome NEPA process into a cudgel to crush domestic energy production.
- The administration has participated in numerous sue-and-settle deals with environmental groups so they can use the settlements to prevent oil and gas production without going through Congress or the normal regulatory process.
- In a brazen and illegal move, the administration for the first time in our country's history has refused to finalize an offshore leasing plan.
- The administration is attempting to use old fuel efficiency mandates to mandate electric vehicles.

[356] Institute for Energy Research, The Economic and Strategic Importance of Domestic Mineral Production, April 2023 https://www.instituteforenergyresearch.org/wp-content/uploads/2023/04/The-Economic-and-Strategic-Importance-of-Domestic-Mineral-Production.pdf.

- And the administration has, without statutory mandate from Congress, sought to inject vague notions of "environmental justice" into every decision- making process. This concept is not scientific or measurable, it is entirely in eye of the beholder. It is simply a tool for arbitrarily halting development that the administration does not like.

These actions and the hundreds of other additional and related actions are cumulative. Every action making it harder to produce energy domestically reduces the desire of companies to invest domestically. Oil and gas production requires ongoing investment just to keep production stable, much less increase it. Reduced investment means reduced production down the line. And that reduced production erodes our hard-won energy security. It ends with the spectacle of President of the US, leader of the world's largest oil producing country, going hat in hand to beg Saudi Arabia for more oil production, or cowardly seeking to relax sanctions against one of the most brutal regimes in the Americas to get more oil supply from Venezuela.

US oil production still has not reached the level seen at the end of 2019 just before the start of the pandemic.[357] That is not because the oil isn't there or that there isn't demand, it's because companies are concerned about making the investments needed when there is a hostile administration willing to use any means to destroy their industry. These investments then get made in other countries, with the US left having to import to meet our future needs. The Biden administration's regulatory assault on domestic energy is an assault on this nation's security and prosperity. It is a dependence agenda that makes energy more expensive and less available, while at the same time making our energy system dependent on foreign supplies from China. That this is being done by regulatory fiat, using questionable authorities or even in direct conflict with laws passed by Congress, is frankly a crisis of our democratic system.

Dr. GOSAR. Thank you, Mr. Stein. I thank all the witnesses for your testimony. I will now recognize Members for 5 minutes for their questions. I am going to start with the Chairman for the Full Committee, Mr. Westerman.

Mr. WESTERMAN. Thank you again, Chairman Gosar, and thank you to the witnesses for being here today.

Ms. Pleune, you made a pretty solid argument about the efficiency of NEPA. And if I understand from your testimony, your position would be that NEPA is OK like it is, and it doesn't need any reforms. Is that correct?

---

[357] https://www.eia.gov/dnav/pet/hist/LeafHandler.ashx?n=PET&s=WCRFPUS2&f=W.

Ms. PLEUNE. NEPA is capable of efficiency, and it is done efficiently under the existing regulations. However, inefficiency also happens. What we found, however, is that the sources of inefficiency are not the regulatory problems. If they were, then we wouldn't see quick decisions.

In fact, what we are finding is that the primary sources of inefficiency are a lack of staff capacity, a lack of regulatory people with relevant expertise in order to make a decision. I will just give you a quick example.

Mr. WESTERMAN. I have some more questions. Ms. PLEUNE. OK.

Mr. WESTERMAN. Did you see any changes in the data when administrations would change, where maybe permits were issued quicker under one administration than another administration?

Ms. PLEUNE. We looked for that trend, and we did not see that trend. But one, I think, really poignant example was a study that the GAO did of application for permit-to-drill processing times. And what they found is that, even though the different field offices had significantly different times in processing them, ranging from 300 days in Miles City to 37 days in Anchorage—exact same standard, exact same law—and the difference was a lack of capacity and how well the permitting process was managed.

Mr. WESTERMAN. Has that process increased since the—I think it was in the IRA, or maybe the Infrastructure and Jobs Act—there was a huge amount of money put out for agencies for permitting. Have you seen any change in the numbers since there has been a flood of money into the agencies?

Ms. PLEUNE. We have. And in fact, one of the things that is showing quite significant improvements in permit processing times are the reforms, and also the attention and support that has been given to projects that are going through the FAST-41 process. And those improved processing times are quite remarkable, first because the projects that go through FAST-41 are chosen because they are complex, they involve multiple agencies, lots of different laws, and the permit processing times within there are much shorter.

For example, in 2020, the average time for projects that went through FAST-41 was about 2.5 years for EISs. Other projects are going even more quickly, notably——

Mr. WESTERMAN. Having done engineering work before I came to Congress, when I hear 2.8 years or 2.5 years, that causes people real problems that are actually trying to do projects, to have it delayed that long where you are waiting for permitting. Plus, it is not just NEPA. If you are building something, you end up with clean air permits, water discharge permits. So, these permits get stacked on top of each other.

But my real question here is, do you think that permitting reform is necessary for America to meet the Administration's clean energy goals?

Ms. PLEUNE. There are things that need to be reformed, and some examples are the interconnect queue, the process for approving transmission lines. The interconnect queue—right now projects are waiting an average of 5 years because it is so long to get through the regional transmission organizations.

Mr. WESTERMAN. Is there anything in H.R. 1, the Lower Energy Costs Act that we proposed on permitting reform, such as putting one Federal agency in charge, limiting the length of the permit applications, would any of that be beneficial, or do you think that is just some kind of political gamesmanship?

Ms. PLEUNE. Identifying a lead Federal agency is a productive practice that was implemented through the FAST-41 Act, and it does work.

You have multiple different permits that you have to obtain, and the NEPA structure can provide a way for agencies to coordinate that decision-making process, and that is what they use the FAST- 41 for.

Imposing time limits on NEPA process creates a possibility that you won't be able to use the process in order to address all of those different permits, especially where there are also state, local, or tribal interests or permit requirements imposed.

Mr. WESTERMAN. You think there is any problem with litigation on NEPA, litigation that may not be justified, that it is used for more political purposes than to actually achieve the objectives of NEPA?

Ms. PLEUNE. I think there are two elements of litigation that are important to recognize. The first is that the amount of litigation is extremely small. It is an estimated 0.22 percent of all NEPA decisions, and secondly, I don't think NEPA litigation itself slows down very many projects.

However, there is a distinction between delays that are caused by litigation and delays that are caused by litigation aversion. In our research, we found that agency staff members are concerned that their opportunities for promotion will be limited if they sign a document that gets litigated. As a result, the document sits on the back of the desk. It doesn't get signed until that person gets promoted or they go on detail. And that is a real significant delay that doesn't benefit anyone.

So, creating a system where we reward agency members for quick decisions, even if they are litigated, would be one way to avoid delays that are caused by the fear of litigation.

Mr. WESTERMAN. Thank you, Mr. Chair. I am out of time, even though this is a fun discussion. Thank you.

Dr. GOSAR. It was a fun discussion. Thank you, Chairman. I now recognize Mrs. Lee from Nevada.

Mrs. LEE. Thank you, Mr. Chair. I want to thank all of the witnesses today.

This hearing today can feel like it is tailor-made to highlight the disagreements that those of us on this dais have had and continue to have regarding energy policy. But I always try to highlight areas where we can be in agreement. And I will begin by asking you all a simple yes-or-no question.

Would you agree that, as a general rule, Americans get the biggest bang for our buck when we prioritize energy development and exploration in non-sensitive areas where there is also a high likelihood of finding and harnessing significant energy resources?

And I will start with you. Just a yes-or-no answer. Do you agree with that statement?

Ms. FURCHTGOTT-ROTH. Energy companies have their own ways of deciding which are the best investments, and sensitive or less sensitive is not an objective manner of describing a particular area.

Mr. HARRELL. Yes, we need to produce more American energy.

Ms. PLEUNE. Yes, I think producing energy in a way that reduces harms, avoids harms is a good idea.

Mr. STEIN. Again, repeating the previous statement, the sensitive versus not sensitive is a very subjective term, so it is a little hard to agree or disagree with that——

Mrs. LEE. What about in areas with a higher likelihood of producing energy?

Mr. STEIN. Well, yes, but I think that is where investments go.

Mrs. LEE. All right. Well, I represent a state that has an incredibly wide range of energy sources. I come from Nevada. We have critical minerals like lithium and, obviously, our sunshine, a source of renewable. But what we don't have are great quantities of oil and natural gas. Nonetheless, the GAO reports that a shocking two-thirds of total acreage nominated for onshore oil and gas leasing nationwide from 2009 to 2019 was in my home state of Nevada. That is 60.7 million out of 86.8 million acres.

BLM officials in Nevada have said that it took their staff over 5 years just to review and process 28 million acres nominated in my state in 2014, most of which did not result in any energy production. In fact, of the 61 million acres nominated in Nevada across this 10-year period—get this—out of 61 million, only 3.5 million acres were ever actually leased, including more than 2 million acres leased non-competitively due to a lack of industry interest.

Professor Pleune, I would like to ask you, is it fair to say that American taxpayers and our public lands both stand to benefit from bipartisan policy-making that would shift focus away from speculative and ultimately unproductive oil and gas leasing in places like Nevada toward projects with a more promising return on investment?

Ms. PLEUNE. That certainly sounds reasonable to me. The last time I looked at the statistics, a significant number of parcels do not get bid on.

And moreover, a large number of parcels do not go into production. A GAO investigation looked at the number of APD permits that were processed, whether or not they actually went into production. And roughly half of all APD permits that are processed never actually go into production, which is actually quite a bit of waste of time, given that we have limited staff capacity.

Furthermore, there is no system for regularly tracking lease suspensions. So, the amount of leases that are currently being held, even though they haven't gone into production, they are in a suspended state and there is no systemic way of identifying where they are, how long they have been suspended, the reasons for suspension, and whether those reasons are still justified.

Mrs. LEE. Yes. So, not only is this an incredible time suck on behalf of staff, it also costs us taxpayers a lot of money.

I have just a few seconds remaining. I just wanted to ask you, with NEPA, can you explain how prioritizing early stakeholder consultation during the permitting process would help the process along, and help industry build and avoid costly changes and delays?

Ms. PLEUNE. Absolutely. One of the most productive methods that we have seen toward improving the timelines of permit processing is early engagement with stakeholders. That includes all of the agencies that need to issue a permit on a project, state and local governments, as well as people who are likely to be affected by the project.

Agencies like FERC have even engaged in pre-application meetings. This allows the permit applicant to identify at the design phase sensitivities and regulatory standards that they need to when it is least expensive in a project's life cycle. A good example of this is the Nexus pipeline. They engaged in early stakeholder engagement and, through a large process of stakeholder engagement with the public, they ended up with almost a 91 percent change in the course of the pipeline, which would not have been something you could have done if it was later in the cycle of the project. But because it was so early, they were able to do it and avoid both sensitivities, as well as opposition.

Mrs. LEE. Thank you. I am over my time.

Dr. GOSAR. I thank the gentlewoman from Nevada. Great conversation. I recognize the gentleman from Montana, Mr. Rosendale, for his 5 minutes.

Mr. ROSENDALE. Thank you very much, Mr. Chair.

Wow, this is fascinating. Ninety percent of the land in Nevada didn't have oil and gas on it. I can tell you, spending a little bit of time in the oil and gas industry myself, that there is nothing more disappointing for someone who goes out there and drills than to find a dry hole.

There is so much investment and research and development that takes place to try to develop an oil field. And I assure you that the people that are funding that are not expending funds to enrich the state of Nevada; they are doing so because they think that possibly that there is a resource underneath the surface that they can develop. Unfortunately, most of the oil and gas resources that they are tapping into are about 2 miles below the surface, so it is sort of hard to use a divining rod and try to identify where they are. Moving on to other permitting problems, which is not anecdotal, I can tell you, coming from Montana, I testified at the first public hearings to simply permit a pipeline, the safest, most monitored pipeline that had been proposed in our nation's history, and it was called the Keystone XL. And it took 10 years from the time that those first hearings began until it was finally permitted through an executive order, in order to make sure that we could get those additional resources, roughly 850,000 barrels a day of crude oil into our country—150,000 barrels a day, which would have come from Montana and North Dakota, so that we could increase our domestic production.

And yet with the stroke of a pen, on his first day in office, President Biden killed that pipeline, a pipeline. We are not talking about drilling. We are talking about a pipeline to transport these products. Again, the safest, most monitored pipeline that had been proposed in the nation's history. That cost the state of Montana $60 million a year in tax revenue losses. And it was going through some of the poorest counties in the state, places where that would have taken care of the expense of the roads, and the schools, and the medical facilities for those counties with populations of 1,500 to 2,000 people.

We haven't had a new mine permitted for 20 years. A lot of this is our fault, right here, because Congress has continued to write legislation that is so subjective, and then turn the rulemaking authority over to the agencies to do. So, that is our fault.

But I will also tell you that there is an awful lot of litigation that takes place. We have activist judges that tie these projects up, and there are attorneys and law firms that get enriched because of it. And then agencies that overstep their bounds, as evidenced by the most recent case that was decided by the

U.S. Supreme Court, West Virginia v. EPA, where they did get outside of their lane, and they were told that, no, you can't make those decisions, that the U.S. Congress does.

All that being said, I would like to ask a couple of questions to Ms. Furchtgott-Roth.

Ms. Furchtgott-Roth, who would you say, as far as a specific demographic, is affected most by the Biden administration's aggressive, climate justice-focused agenda, and how, when it comes to energy?

Ms. FURCHTGOTT-ROTH. The lowest income are the most affected, because they pay the highest proportion of their income in electricity, motor fuel, and automotive costs. So, the bottom 20 percent pay, on average, around 10 percent of their income. The top 20 percent pay about 1 percent. So, when you increase electricity and automotive costs, it is the poor who are hurt most. This is not justice, not social justice, not environmental justice.

Mr. ROSENDALE. Thank you. And in your testimony you speak a lot on the Administration's push for electric vehicles. Can you please delve deeper into the potential consequences this push has had for the middle- and lower-income Americans if they are forced to transition to electric vehicles, especially in the more rural states and colder states like Montana?

Ms. FURCHTGOTT-ROTH. Right. Well, only 6 percent of new vehicle sales last year were electric. It is because Americans are smart, they know that electric vehicles are more expensive, they have smaller range, they lose 25 to 40 percent of their battery power in cold climates. It is hard to heat them in cold climates.

That is why, for example, Wyoming has about 500 electric vehicle registrations, even fewer in Montana, North Dakota, South Dakota. They are just not practical. When people have them, it is generally as a second, third, or fourth car. They have an SUV, or a minivan, or a pickup truck if they want to go on long trips. So, it is more expensive. It is a toy for the upper-income groups. And having 60 percent of vehicles be electric would impose a huge burden on low- income Americans.

Plus, we don't have the charging infrastructure for it yet, especially since the Environmental Protection Agency has come out with a new power plan this week that would make electricity more expensive, due to the need for carbon sequestration.

Mr. ROSENDALE. Thank you so much.

Mr. Chair, I see I have expired my time. I would yield back.

Dr. GOSAR. I thank the gentleman. The gentleman from California, Mr. Huffman, is recognized for 5 minutes.

Mr. HUFFMAN. Thank you, Mr. Chairman. And it feels like Groundhog Day here. We are once again discussing the apparently dire need to attack and gut NEPA so that we can speed up fossil fuel development and mining. This is a broken record that we hear played all the time by our friends across the aisle.

And it is like a bad ventriloquist act when their lips are moving, or when the fossil fuel industry's lips are moving, Members of Congress are talking, and you can see it pretty clearly. They don't like NEPA. They don't want to have to do alternative analysis. They don't want to have to limit environmental impacts. They don't want to have to actually provide notice and get public input from disadvantaged communities that are going to be hurt by their projects. So, we have to take this industry speak with a grain of salt.

But when it comes to this latest narrative that has been suggested a couple of times today, that clean energy and electric transmission improvements require gutting NEPA, this is especially false and reckless, and we need to start calling it out. I am leading a task force for the Progressive Caucus that is zeroing in on the real causes for delay.

So, Professor Pleune, I was especially interested in your testimony because you have been doing the same thing. What we are finding is that many of the biggest problems that have delayed the kind of infrastructure and investments that we want to see moving quickly, including transmission, have been resolved by legislation already in the last couple of years, and by new funding, and by executive orders on agency coordination. And there are other major improvements that actually are in the works under existing authorities, including at FERC. And that is why, as you said, Professor, projects actually are beginning to move faster. NEPA is becoming a less compelling boogeyman all the time.

And I would like to just follow up, Professor Pleune, because you talked about this very small percentage of permitting projects that are actually subject to a full NEPA analysis. And let me get this right. It is less than 1 percent, is that right, a full EIS?

Ms. PLEUNE. The GAO estimated, government-wide, it is less than 1 percent. When we looked at the Forest Service, they do about 2 percent.

Mr. HUFFMAN. You also alluded to the possibility that the NEPA process, if it is done right, can actually move complex projects faster by early engagement of communities that may be impacted by that. Could you expand on that a little bit?

Ms. PLEUNE. I can. I actually have two examples of evidence that shows that the NEPA process does speed up decision making.

The first is a natural example that was set up by a circuit split, where in some areas identifying designated critical habitat required a NEPA analysis and in the other circuits it did not. And for those critical habitat designations that went through the NEPA process, they were finished, on average, 3 months faster than the critical habitat designations that did not go through the NEPA process.

There is another very interesting study done by the Office of Investigation on APD permits. They found that the average time to process the APDs was about 228 days. Pinedale was doing it in 49 days. And the reason is that Pinedale had undergone a very thorough programmatic EIS. It thoroughly understood the well fields that it was developing, and so the implementation piece was much faster.

Mr. HUFFMAN. Would you agree that the biggest barrier to renewable energy and electric transmission capacity is not NEPA, but the interconnection, the dysfunctional and broken interconnection queue at FERC?

Ms. PLEUNE. Yes, and it is often conflated with the NEPA process because it is a part of the permitting process, but it has nothing to do with environmental standards or NEPA.

Mr. HUFFMAN. And FERC has proposed some major reforms to fix this problem with the interconnection queue using existing authority. My understanding is that, if the Senate would simply confirm a fifth FERC commissioner, these reforms could move forward. Is that your understanding?

Ms. PLEUNE. That is beyond my expertise.

Mr. HUFFMAN. All right. I also understand——

Ms. PLEUNE. It sounds right.

Mr. HUFFMAN. I also understand a certain Senator that is targeting NEPA is the one holding up the confirmation of that fifth FERC commissioner. So, these things get very interesting.

But perhaps the biggest delay I have heard with some of these projects is market risk and financing, things that we addressed last year in the Inflation Reduction Act. Would you agree that these are game-changers, these new laws that we have put on the books in this space?

Ms. PLEUNE. Absolutely. The new laws are game-changers.

And I will also note that, with the projects that have gone through FAST-41, the things that they have found is, with the structures that are in place, there is more predictability, and that addresses that concern of not knowing when they will get a permit approved.

Mr. HUFFMAN. Thank you very much. I yield back.

Dr. GOSAR. I thank the gentleman from California. The gentleman from Texas is recognized for 5 minutes.

Mr. HUNT. Thank you, Mr. Chairman. And I want to thank the witnesses, as well. Thank you so much for your time. Thank you for being here. We really appreciate it.

In 2019, then-candidate for President Joe Biden said, and I quote, "We are going to ban fossil fuels. I guarantee it." Well, when tyrants speak, believe them. And one could argue this will be the only campaign promise that Biden plans on following in the future. With the radical climate lobby that helped to put Joe Biden in the White House in the first place, no matter how much American families suffer from the increased energy prices directly resulting from Biden's policies, the White House will continue to pursue this flawed and failed agenda.

Energy bills for the people that I serve in Tomball, Texas, which is actually the northern part of my district in a Houston suburb, have doubled under Joe Biden. The Houston Chronicle, a local publication, says that the natural gas bills in Houston are soaring, and yet Biden and the climate cartel have spent the last 2 years propagandizing an energy transition. And we are transitioning all right, right into poverty.

The truth is, if you plan on continuing to live in a free and prosperous society, then fossil fuels aren't going anywhere for the foreseeable future. And here is where fantasy meets reality. You see, on one hand, this Administration killed the Keystone XL pipeline and placed a moratorium on all oil and gas leases on Federal land. But then on the other hand, the Biden administration wrote a letter to several oil companies demanding they increase oil production and refining capacity.

Now, why would an energy company do that, if they know that their industry is going to be killed in the not-so-distant future? That is not capitalism and that is not America. That is ridiculous.

On one hand, Biden wants to mandate that Americans drive EVs. But then, on the other hand, he prohibits the domestic mining of necessary materials to build these vehicles and their batteries. On the one hand, Biden says he wants to reduce carbon emissions, and then turns around and he taxed the energy source that is responsible for lowering U.S. emissions. And that, ladies and gentleman, is natural gas.

On the one hand, Biden says that he wants to transition our entire energy future, and he even said that he wants to make sure that all government vehicles are powered by a renewable source by 2035. And I am an Apache helicopter pilot. Good luck with that. But then on the other hand, he has set up a permitting and regulatory system that, as you mentioned in your statement, Mr. Harrell, is procedurally impossible, and thank you for being honest on that.

I am from Houston, Texas. Houston, Texas is the energy capital of the world. I deem myself to be the energy Congressman of the world. And I want to tell my friends on the opposite side of the aisle something that is very important. Every single oil and gas executive that I have spoken to in my district in the energy corridor has never said anything disparaging about renewable energy. Never. Not one. In fact, we need more electrons. And we in Congress have to stop talking about an energy transition, because that is a lie. This is about energy addition.

In this country, we are sitting on the Marcellus Shale, which I was told is the Saudi Arabia of natural gas. Now, for 2 years, when I was in the military, I was stationed in Saudi Arabia. You want to talk about vast resources? The insinuation that our country is sitting on a resource remotely close to the oil in Saudi Arabia is unbelievable, and it is God-given. But at the same time, New York has outlawed fracking, and the Marcellus Shale is sitting right at their feet. Yet, the Northeast would rather import heating oil from foreign countries than tap the resource that is sitting right beneath them.

This lack of coherent leadership is dangerous for our country and for our children's future. We can only ride the coattails of the good policies of the Trump administration for so long. Joe Biden came for your gas stoves. He came for your cars. Apparently now it is your dishwasher.

And as I said in the very beginning, Joe Biden said, and I quote, "We are going to ban fossil fuels." Promises made, promises kept.

Thank you all so much for being here. And with that I yield back.

Dr. GOSAR. I thank the gentleman from Texas. The gentlewoman from New Mexico, the Ranking Member, is recognized.

Ms. STANSBURY. Thank you, Mr. Chairman. I appreciate the opportunity to have this discussion. It is important that we talk about energy and the energy transition, but it is also important that we remain clear about what we are talking about.

I think that this has been a very muddled conversation this afternoon. I have heard electricity conflated with heating fuel, conflated with oil and gas, conflated with the vehicle sector. So, there is a lot of confusion, I think, in this conversation about what we are talking about in terms of energy security here in the United States. So, I want to just re-clarify.

Again, this is the second highest year of energy production in American history. There has not been a ban on energy production in the United States. In fact, the Administration has approved recently several new lease plans in New Mexico, which is my home state. Not only do we see the highest production of oil and gas ever in our state's history, we are seeing the highest profitability and first quarter profits of oil and gas companies ever in the history of our state. Those are the facts.

And as we have heard today, and I really want to thank Ms. Pleune for being here today, when we look at the science, not just anecdote, and I was really struck by your comments that you analyzed over 41,000 NEPA decisions, 41,000 decisions, the science shows that it is not the regulatory structure of NEPA that causes delays.

So, I want to reiterate the three factors that were shared here today, and then ask a couple of questions: the first is staffing and capacity; the second is operator delays, meaning the individuals who are seeking to have a decision approved may not provide the appropriate paperwork that is required or, in other cases, they may choose to not move forward with a project because they don't have financing, the market conditions are not right, and in fact, when we talk about many of these energy mixes, the reason why they are not moving forward with these projects is they are no longer profitable because technologies have changed or there are other energy mixes coming online; and the other is other laws.

So, I want to just take one moment, because the reason why NEPA passed in 1969 and was signed into law by President Nixon, a Republican administration, why the Council on Environmental Quality was established in NEPA was because there were rivers on fire, because we were strip mining our forests and precious natural resources, and we were destroying sources of water across the United States and significantly impacting what we now call front- line communities, and especially those communities that did not have a voice in these decision-making processes.

And particularly significant for my state, in New Mexico, are tribal voices and tribal consultation. And I do want to take umbrage with something that was said just a few moments ago, because there was a comment made about Resolution Copper being delayed because of NEPA. That is factually untrue. Myself and the Chairman of this Committee, excuse me, Ranking Member of this Committee, Grijalva, had been working along with Congresswoman Leger Ferna´ndez to ensure that there is appropriate tribal consultation to make sure that an international mining company does not strip mine a tribal sacred site that is protected by treaty and over 150 years of Federal law. And that is why that project has been delayed, not because of NEPA.

But let's talk about what can be done. Ms. Pleune, you said in your testimony that good decisions are more important than rush decisions. And I want to ask you, first, we are hearing a lot of comments here, and we continue to hear these arguments that we should just weaken NEPA or take it away, and somehow that will solve the problem. What are the risks to our communities and the environment if we were to do that?

Ms. PLEUNE. The risk of weakening NEPA is that NEPA is a look-before-you-leap statute. It simply says look at what you are going to do, and disclose the impacts. It provides an opportunity to think before you do it, is there a way that I can avoid, reduce, or mitigate risks or harms to communities?

It also provides an opportunity for communities to speak and effect change on things that affect them. Weakening NEPA weakens all of these opportunities.

Furthermore, a lot of delays that projects encounter are due to different laws, there are overlapping laws. There are different jurisdictions, and communities can use those differences to stop a project if it is going to be harming them. That is the whole point of laws; they protect people.

And using NEPA ~~as~~ a structure to bring people to the table to understand the effects of a project, to synthesize the way it is going to be analyzed and come to an agreement about whether or not it goes forward is a productive, transparent process that is appropriate to the democracy we live in.

Ms. STANSBURY. Thank you, Ms. Pleune. I don't think I could state it any more clear, so we appreciate you being here and bringing good science and facts to this conversation.

Ms. PLEUNE. Thank you.

Dr. GOSAR. I thank the gentlelady. The gentlelady from Florida, Mrs. Luna, is recognized. Sorry. You should have been——

Mrs. LUNA. It is OK. Thank you, Chairman. Actually, since we are clarifying facts, I just wanted to add some e-mails to the record showing that CEQ did receive sufficient notice to testify at today's hearing, if that is OK with you.

Dr. GOSAR. So ordered.

[The information follows:]

Varnasidis, Sophia

From:	Varnasidis, Sophia
Sent:	Thursday, April 27, 2023 8:52 AM
To:	'sara.l.jordan@ceq.eop.gov'
Cc:	Lane, Michelle; Knecht, Thomas
Subject:	Invitation to testify before the Sub on Oversight and Investigations 05.11.23
Attachments:	118th Testimony & Disclosure Requirements.pdf; Invite Mallory.pdf

Good morning Sara,
Attached is an invitation for CEQ Chair Mallory from Subcommittee Chairman Gosar, to testify before the House Committee on Natural Resources, Subcommittee on Oversight and Investigations, at an oversight hearing on Thursday, May 11, 2023 at 2:15 p.m. in the 1334 hearing room of the Longworth House Office Building, at a hearing titled *"The Biden Administration's Executive Overreach and its Impact on American Energy Independence."*

Testimony is due to the Committee by Tuesday afternoon, May 9, per the Testimony and Disclosure Requirements pdf, also attached. As a governmental witness, the Chair does not need to complete a disclosure form.

Please contact us with any questions.
Sophia

Sophia A. Varnasidis
Director of Legislative Operations
Committee on Natural Resources
Chairman Bruce Westerman

1324 Longworth HOB
Washington, DC 20515
(202) 225-2761

BRUCE WESTERMAN OF ARKANSAS
CHAIRMAN

VIVIAN MOEGLEIN
STAFF DIRECTOR

RAÚL M. GRIJALVA OF ARIZONA
RANKING DEMOCRAT

LORA SNYDER
DEMOCRAT STAFF DIRECTOR

U.S. House of Representatives
Committee on Natural Resources
Washington, DC 20515

April 27, 2023

The Hon. Brenda Mallory
Chair
Council on Environmental Quality
730 Jackson Place, NW
Washington, DC 20503

Dear Chair Mallory:

The Subcommittee on Oversight and Investigations will hold an oversight hearing titled *"The Biden Administration's Executive Overreach and its Impact on American Energy Independence"* on Thursday, May 11, 2023 at 2:15 p.m. in room 1334 Longworth House Office Building. I cordially invite you to testify at this hearing.

Enclosed with this letter are the parameters regarding written and oral testimony. Should you have any questions or need additional information, please contact Sophia Varnasidis, Director of Legislative Operations, Committee on Natural Resources at (202) 225-2761.

Sincerely,

Paul Gosar, D.D.S.
Chairman
Subcommittee on Oversight and Investigations

Enclosure

http://naturalresources.house.gov

Mrs. LUNA. Thank you very much. I wanted to read you guys some facts. And I had actually heard what Ms. Diana Roth had stated earlier.

When I sit up here and I hear legislators, on both sides sometimes, talking about how we need to fully go green without taking into account the impacts that it has, especially on those that are not as fortunate in this country, I look at that as rather tone deaf. Some of these stats that I was able to look up during this hearing, in 2021 the average salary of Black Americans is $46,400 a year. In 2020, for Hispanics it was about $55,000. And the poverty rate for Indigenous Americans is about a quarter. So, when I see that the average cost for an electric vehicle is $66,000, and then I hear that we need to be pushing for more green initiatives, I think that that is done by people who have maybe never even experienced the struggle.

I just got back recently from Indonesia on a congressional delegation where we met with the President, where we met with their parliament, and where we heard that American companies were going there to mine things like cobalt and lithium, things that we use here in the States to produce our own electric vehicles because we know that this Administration, through the Inflation Reduction Act and their war on fossil fuels, has been pushing this. But when we asked the Indonesians what they were doing for their own people, they said, "Well, we use fossil fuels."

When we were in South Korea, even though they are moving toward nuclear energy, they are still also using fossil fuels. Yes, they have embraced going green. However, they are not stupid enough to completely cut off their arm in the process and hurt their own people in order to do so.

And, of course, then you have China, who doesn't really care about the environment and is busy destroying it. So, when I hear about these climate disasters, I take those things into account.

My question is for you, Ms. Roth. How has the Biden administration enabled our adversaries through these green initiatives like China and Russia?

Ms. FURCHTGOTT-ROTH. China makes most of the components of electric vehicles. They make the batteries, they make the components, they make about 70 or 80 percent of wind turbines and solar panels. So, the big question we have before us is this: Is displacing American energy-intensive manufacturing and sending it to China, does that help the environment at all? Does that help the global climate? Does it help climate change?

And I would argue that the answer is no, because what we are doing is giving up American jobs in energy-intensive manufacturing sectors, giving up auto jobs. And these individuals are hurt, and the Chinese are gaining. The Chinese subsidize labor. They have forced labor from Xinjiang. They subsidize energy. They have these coal-fired power plants that we are not allowed to have here. They subsidize capital with interest, very low interest loans to favored companies. And they are gaining, and we are losing, and it is not helping the environment.

That is the fundamental question before us: Is this harm we are doing to America helping reduce emissions? I would answer that the answer is no.

Mrs. LUNA. Do you believe that politicians that lecture the American people on climate justice should adhere to their own ideologies? For example, should they ride bicycles to Washington, DC? Do you think they should stop eating steak? Because we have heard that cows contribute to and their flatulence contributes to climate change and global warming. I mean, don't you think that those people should embrace their own ideologies?

Ms. FURCHTGOTT-ROTH. Food security is very important and fossil fuels are needed for fertilizers. And we need to keep the price of food low.

Again, it is low-income individuals that are hurt the most because food is a high proportion of their income. And it is a little strange to see people lecturing about the evils of carbon emissions on planes, and then flying their own jets to Davos to take part in conferences.

Mrs. LUNA. Oh, that is my favorite. You watch them going in on their G6 and you can see that they clearly have no clue on how their legislation and their crazy ideologies are impacting the rest of the world.

But thank you for your time. Chairman, I yield back.

Dr. GOSAR. I thank the gentlelady. I will recognize myself.

So, Mr. Harrell, a comment was made that a tribal site was being condemned at Resolution Copper. Is that true, particularly to the historian from the San Carlos Tribe?

Mr. HARRELL. It is not true. And just to clarify my statement in my testimony, I didn't use NEPA as the barrier for Resolution Copper. I used it as the example of where we need to have faith in decisions by the administration.

Over the last 9 years, since Congress enacted the land exchange to move that project forward, a significant NEPA process was done to move it forward. There was a record of decision that was made almost 9 years later, and then 2 months into this Administration they reversed that. There were significant requirements for tribal consultation that were followed by Interior over the course of that EIS, and we are simply just disadvantaging ourselves in producing a robust copper supply in this country.

Dr. GOSAR. Well, I just wanted to make sure that the consultation, based on historical sites, was not part of the deal. So, just to set that straight.

My good friend from California made the comment about financing. Can you tell me a little bit about ESG and some of the threats in regards to getting financing for actually applications for energy?

Mr. HARRELL. Yes, absolutely, happy to talk about financing for clean energy as a whole.

We need regulatory predictability. In the end, the biggest thing that drives up private sector capital moving into projects, whether it is wind, solar, transmission, nuclear power, carbon capture, is regulatory uncertainty. In the end, there needs to be confidence there is going to be a return on investment, and folks simply need predictability.

So, I would argue that we have an environmental imperative to more efficiently move energy and clean energy projects forward.

Dr. GOSAR. So, streamlining the process, I mean, there is always improvement. I mean, as a businessman I have always looked at that. As a dentist I have always looked at that. Is there a way that we might have, like, a point person that actually calibrates this? Instead of doing it linear, you are doing it all at once. Could that be a benefit?

Mr. HARRELL. Absolutely. And NEPA is just one piece of the puzzle, right? There are a wide variety of the permits that need to be done across agencies that have jurisdiction. We need a lead person driving these type of things forward, and kind of being that ombudsman, effectively, and ensuring the agencies are working in tandem to approve things in an expeditious manner.

Dr. GOSAR. So, Ms. Roth, going back to this ESG phenomenon, we had this in Oversight just yesterday. So, if you don't pertain or hold to the particular nomenclature or let's say the Administration's belief, you can't get funding. Does that cause a problem?

Ms. FURCHTGOTT-ROTH. It causes a major problem for companies who are downgraded by the Securities and Exchange Commission because their projects might have climate risks and the Office of the Comptroller of the Currency for banks, because those projects might have climate risks. And it is very difficult to measure these things. That is called Scope 1, which is the immediate greenhouse gas consequences. Then there is Scope 2, which is do your suppliers have any greenhouse gas consequences for their suppliers? Then there are the suppliers to the suppliers.

It is difficult to measure, and it gives our foreign competitors an advantage because they do not have the equivalent of the OCC or the SEC breathing down their necks, so it makes their imports to us, their exports to us less expensive, and it gives them an advantage. It drives down American manufacturing, drives up American prices, hurts the poor. And that is not any kind of justice.

Dr. GOSAR. And I found it fascinating that investors were actually talking that they could actually make up money based upon ESG rulings, because they don't deliver the same kind of punch for the dollar in investment purposes. I find that very disingenuous because maybe we ought to go back to the way the Fisher people look at things, and that is we do better when our clients do better. We ought to be forthcoming about that. So, this is an intricate web that is here.

I am known for my last question: Each one of you, what was the question you most wanted to have asked and what is the answer?

I am going to start with you, Ms. Roth.

Ms. FURCHTGOTT-ROTH. I think you could have asked about the feasibility of net zero. And I just want to say that fossil fuels are going to be with us as far as the eye can see.

First of all, if we imagine that the United States is covered with solar panels and wind turbines, we need backup natural gas plants to operate the source of electricity when the wind doesn't blow and the sun doesn't shine.

Second, we need fossil fuels to actually make the wind turbines and make the solar panels. So, the idea that we can get rid of fossil fuels together is just a myth. It is just a fantasy.

Third, I want to make it clear that the air and water in the United States is getting cleaner. You look at the EPA website, six criteria pollutants. They are all going down. So, shifting manufacturing and energy production offshore where there are more lenient rules for production is not helping. Telling our companies they cannot drill for oil, going to Saudi Arabia and Venezuela, asking them to produce oil rather than our Canadian friends in the north, that does not help or reduce global emissions. It increases global emissions.

We are doing a great job here of being energy independent. We need to stay that way. We don't want to cede our independence to China.

Dr. GOSAR. Thank you.

Mr. Harrell?

Mr. HARRELL. Thank you, Mr. Chairman, for the opportunity on that, a question about the feasibility of meeting our long-term energy needs, both from an affordability and a climate perspective. When push comes to shove, we need to be focused on reducing emissions, not on different fuel types as a whole. Natural gas can play a huge role, both globally in reducing emissions and meeting our energy needs here. We need more transmission, we need more wind and solar, we need more nuclear power. We need all of those technologies. We need to potentially double the capacity of our grid to meet some of these energy and climate objectives.

And one common thing is true: we are simply not building technologies fast enough to meet those goals. So, we need to find ways to cut down some of these timelines from 21⁄2, 41⁄2 years to 18 months, things like that. And that does involve proactive engagement with communities. It is trying to make sure that we are doing a lot of the work up front.

But the one clear fact here is, if we are going to meet those objectives, we simply cannot operate under the status quo. We cannot deploy technologies at the pace and scale needed under the status quo.

Dr. GOSAR. Thank you. Ms. Pleune?

Ms. PLEUNE. Thank you for this opportunity. I think I would have liked to have been asked a little bit more about what are the procedures in FAST-41 that do promote more efficiency, and why is that?

And the main thing I would like to talk about is the way that it has been able to promote coordination between permitting authorities as a way to streamline the permitting process. The procedures have been effective in achieving predictability, transparency, and improved timelines, all without compromising environmental standards or transparency.

Additional promising practices include encouraging the development of MOUs between permitting authorities, especially state, local, and tribal authorities. To do that, the agencies need the funds, the time, the people, and the people with the expertise to get that done.

Second, encouraging details between agencies to facilitate interagency coordination. That way they will understand how each other work, and be able to coordinate better.

Third, developing protocols for data sharing, mapping, and permit sequencing. This way we don't have to re-analyze decisions every time they come up.

Fourth, propagating best practices through training and interagency collaboration, one thing the Permitting Council has already started. And just through that it is already showing improvements, not only in the complex permitting processes but in the more simple ones.

I think these are very promising practices. I think they offer the opportunity to speed the pace of implementation and permitting without compromising environmental standards. Thank you.

Dr. GOSAR. Thank you. You do know I am from Wyoming, from Pinedale, Wyoming, do you not?

Yes, just a reason I understand the Pinedale model. Mr. Stein?

Mr. STEIN. Yes, I would have liked there to have been a little more discussion of the role of litigation in all these delays that we have discussed. A lot of the focus of the NEPA discussion was on the procedural within the administrative state. And while that is slower than it should be, that is not the true delay. It is the process through the courts, it is the uncertainty in the courts, too. Unfortunately, Congress has left far too much of NEPA implementation to court interpretation.

So, a lot of the uncertainty that slows down some of the administrative decisions and slows down financing decisions is that uncertainty of litigation, sometimes litigation that is funded by the Federal Government itself, and lack of clarity about what is required to actually perform a NEPA analysis correctly.

Dr. GOSAR. Thank you. I think we have had a wonderful discussion across the board, and I think there is more that needs to happen in regards to clarity, transparency, and accountability. And I think some of it belongs to us. I think Congress has to recoup its power of the purse and balance this baby off.

I want to thank all the witnesses for their valuable testimony and the Members for their questions.

The members of this Committee may have some additional questions for you. We will ask them to respond with those in writing. Under Committee Rule 3, members of the Committee must submit questions to the Subcommittee Clerk by 5 p.m. on May 16. The hearing record will be held open for 10 business days for those responses.

If there is no further business, without objection, the Committee stands adjourned.

[Whereupon, at 3:44 p.m., the Subcommittee was adjourned.]

Chapter 3

Energy Independence and Security*

Office of Energy Efficiency & Renewable Energy

Energy independence is the state in which a nation does not need to import energy resources to meet its energy demand. Energy security means having enough energy to meet demand and having a power system and infrastructure that are protected against physical and cyber threats. Together, energy independence and energy security enhance national security, American competitiveness, and economic standing.

The Biden-Harris administration's climate laws support U.S. energy independence and security. The Bipartisan Infrastructure Law provides $27 billion to upgrade and modernize the electrical grid to make it more resilient to extreme weather and resistant to cyberattacks. The Inflation Reduction Act increases American energy security through policies to support energy reliability and cleaner production, coupled with historic investments in American clean energy manufacturing.

The U.S. Department of Energy (DOE) Office of Energy Efficiency and Renewable Energy (EERE) Cybersecurity Multiyear Program Plan guides cybersecurity research and development for EERE technologies.

Increasing Energy Independence and Security

EERE is working to achieve U.S. energy independence and increase energy security by supporting and enabling the clean energy transition. The United

---

* This is an edited, reformatted and augmented version of Office of Energy Efficiency & Renewable Energy Publication.

In: The Struggle for Energy Independence in America
Editor: Victoria Whitehead
ISBN: 979-8-89530-558-4

States can achieve energy independence and security by using renewable power; improving the energy efficiency of buildings, vehicles, appliances, and electronics; increasing energy storage capacity; and modernizing the electric grid.

Renewable power supports energy security by increasing:

- Diversity of electricity sources
- Backup energy on the grid and battery storage
- Local electricity generation
- Resistance to threats.

Clean energy will reduce reliance on other countries for energy, technologies, and materials to build clean energy technologies. EERE supports reducing reliance on other countries by:

- Increasing domestic manufacturing to produce American-made clean energy technologies
- Securing the manufacturing supply chain
- Increasing material recycling, reuse, and recovery.

Benefits of Energy Independence and Security

When the United States is self-reliant on energy, it will create more American jobs in the clean power sector and related industries, such as manufacturing. Energy independence will trickle down to the clean energy supply chain, increasing national security, when materials and components are sourced domestically. In turn, this will support U.S. economic growth and prosperity.

Index

D

E

P

R

S

T

U

V

W

Y